MELANIE MITCHELL

Complexity
A Guided Tour

OXFORD
UNIVERSITY PRESS

Oxford University Press, Inc., publishes works that further
Oxford University's objective of excellence
in research, scholarship, and education.

Oxford New York
Auckland Cape Town Dar es Salaam Hong Kong Karachi
Kuala Lumpur Madrid Melbourne Mexico City Nairobi
New Delhi Shanghai Taipei Toronto

With offices in
Argentina Austria Brazil Chile Czech Republic France Greece
Guatemala Hungary Italy Japan Poland Portugal Singapore
South Korea Switzerland Thailand Turkey Ukraine Vietnam

Copyright © 2009 by Melanie Mitchell

The author is grateful to the following publishers for permission to reprint excerpts from the following works that appear as epigraphs in the book. *Gödel, Escher, Bach: an Eternal Braid* by Douglas R. Hofstadter, copyright © 1979 by Basic Books and reprinted by permission of the publisher. *The Dreams of Reason* by Heinz Pagels, copyright © by Heinz Pagels and reprinted by permission of Simon & Schuster Adult Publishing Group. *Arcadia* by Tom Stoppard, copyright © 1993 by Tom Stoppard and reprinted with permission of Faber and Faber, Inc., an affiliate of Farrar, Strauss & Giroux, LLC. "Trading Cities" from *Invisible Cities* by Italo Calvino, published by Secker and Warburg and reprinted by permission of The Random House Group Ltd. *The Ages of Gaia: A Biography of Our Living Earth* by James Lovelock, copyright © 1988 by The Commonwealth Fund Book Program of Memorial Sloan-Kettering Cancer Center and used by permission of W. W. Norton & Company, Inc. *Mine the Harvest: A Collection of New Poems* by Edna St. Vincent Millay, copyright © 1954, 1982 by Norma Millay Ellis and reprinted by permission of Elizabeth Barnett, Literary Executor, the Millay Society. "Trading Cities" from *Invisible Cities* by Italo Calvino, copyright © 1972 by Giulio Einaudi editore s.p.a., English translation by William Weaver, copyright © 1974 by Houghton Mifflin Harcourt Publishing Company, and reprinted by permission of the publisher. *Complexity: Life at the Edge of Chaos* by Roger Lewin, copyright © 1992, 1999 by Roger Lewin and reprinted by permission of the author.

Published by Oxford University Press, Inc.
198 Madison Avenue, New York, New York 10016
www.oup.com

First issued as an Oxford University Press paperback, 2011

Oxford is a registered trademark of Oxford University Press

All rights reserved. No part of this publication may be reproduced,
stored in a retrieval system, or transmitted, in any form or by any means,
electronic, mechanical, photocopying, recording, or otherwise,
without the prior permission of Oxford University Press.

Library of Congress Cataloging-in-Publication Data
Mitchell, Melanie.
Complexity: a guided tour/Melanie Mitchell.
  p. cm.
Includes bibliographical references and index.
ISBN 978-0-19-512441-5 (hardcover); 978-0-19-979810-0 (paperback)
1. Complexity (Philosophy) I. Title.
Q175.32.C65M58 2009
501—dc22   2008023794

To Douglas Hofstadter and John Holland

CONTENTS

Preface ix
Acknowledgments xv

PART ONE **Background and History**
CHAPTER ONE What Is Complexity? 3
CHAPTER TWO Dynamics, Chaos, and Prediction 15
CHAPTER THREE Information 40
CHAPTER FOUR Computation 56
CHAPTER FIVE Evolution 71
CHAPTER SIX Genetics, Simplified 88
CHAPTER SEVEN Defining and Measuring Complexity 94

PART TWO **Life and Evolution in Computers**
CHAPTER EIGHT Self-Reproducing Computer Programs 115
CHAPTER NINE Genetic Algorithms 127

PART THREE **Computation Writ Large**
CHAPTER TEN Cellular Automata, Life, and the Universe 145
CHAPTER ELEVEN Computing with Particles 160
CHAPTER TWELVE Information Processing in Living Systems 169

| CHAPTER THIRTEEN | How to Make Analogies (if You Are a Computer) 186 |
| CHAPTER FOURTEEN | Prospects of Computer Modeling 209 |

PART FOUR **Network Thinking**

CHAPTER FIFTEEN	The Science of Networks 227
CHAPTER SIXTEEN	Applying Network Science to Real-World Networks 247
CHAPTER SEVENTEEN	The Mystery of Scaling 258
CHAPTER EIGHTEEN	Evolution, Complexified 273

PART FIVE **Conclusion**

| CHAPTER NINETEEN | The Past and Future of the Sciences of Complexity 291 |

Notes 304
Bibliography 326
Index 337

PREFACE

REDUCTIONISM is the most natural thing in the world to grasp. It's simply the belief that "a whole can be understood completely if you understand its parts, and the nature of their 'sum.'" No one in her left brain could reject reductionism.
—Douglas Hofstadter, *Gödel, Escher, Bach: an Eternal Golden Braid*

REDUCTIONISM HAS BEEN THE DOMINANT approach to science since the 1600s. René Descartes, one of reductionism's earliest proponents, described his own scientific method thus: "to divide all the difficulties under examination into as many parts as possible, and as many as were required to solve them in the best way" and "to conduct my thoughts in a given order, beginning with the *simplest* and most easily understood objects, and gradually ascending, as it were step by step, to the knowledge of the most *complex*."[1]

Since the time of Descartes, Newton, and other founders of the modern scientific method until the beginning of the twentieth century, a chief goal of science has been a reductionist explanation of all phenomena in terms of fundamental physics. Many late nineteenth-century scientists agreed with the well-known words of physicist Albert Michelson, who proclaimed in 1894 that "it seems probable that most of the grand underlying principles have been firmly established and that further advances are to be sought chiefly in

---

1. Full references for all quotations are given in the notes.

the rigorous application of these principles to all phenomena which come under our notice."

Of course within the next thirty years, physics would be revolutionized by the discoveries of relativity and quantum mechanics. But twentieth-century science was also marked by the demise of the reductionist dream. In spite of its great successes explaining the very large and very small, fundamental physics, and more generally, scientific reductionism, have been notably mute in explaining the complex phenomena closest to our human-scale concerns.

Many phenomena have stymied the reductionist program: the seemingly irreducible unpredictability of weather and climate; the intricacies and adaptive nature of living organisms and the diseases that threaten them; the economic, political, and cultural behavior of societies; the growth and effects of modern technology and communications networks; and the nature of intelligence and the prospect for creating it in computers. The antireductionist catch-phrase, "the whole is more than the sum of its parts," takes on increasing significance as new sciences such as chaos, systems biology, evolutionary economics, and network theory move beyond reductionism to explain how complex behavior can arise from large collections of simpler components.

By the mid-twentieth century, many scientists realized that such phenomena cannot be pigeonholed into any single discipline but require an interdisciplinary understanding based on scientific foundations that have not yet been invented. Several attempts at building those foundations include (among others) the fields of cybernetics, synergetics, systems science, and, more recently, the science of complex systems.

In 1984, a diverse interdisciplinary group of twenty-four prominent scientists and mathematicians met in the high desert of Santa Fe, New Mexico, to discuss these "emerging syntheses in science." Their goal was to plot out the founding of a new research institute that would "pursue research on a large number of highly complex and interactive systems which can be properly studied only in an interdisciplinary environment" and "promote a unity of knowledge and a recognition of shared responsibility that will stand in sharp contrast to the present growing polarization of intellectual cultures." Thus the Santa Fe Institute was created as a center for the study of complex systems.

In 1984 I had not yet heard the term *complex systems*, though these kinds of ideas were already in my head. I was a first-year graduate student in Computer Science at the University of Michigan, where I had come to study *artificial intelligence*; that is, how to make computers think like people. One of my motivations was, in fact, to understand how *people* think—how abstract reasoning, emotions, creativity, and even consciousness emerge from trillions of tiny brain cells and their electrical and chemical communications. Having been deeply enamored of physics and reductionist goals, I was going through my own antireductionist epiphany, realizing that not only did current-day physics have little, if anything, to say on the subject of intelligence but that even neuroscience, which actually focused on those brain cells, had very little understanding of how thinking arises from brain activity. It was becoming clear that the reductionist approach to cognition was misguided—we just couldn't understand it at the level of individual neurons, synapses, and the like.

Therefore, although I didn't yet know what to call it, the program of complex systems resonated strongly with me. I also felt that my own field of study, computer science, had something unique to offer. Influenced by the early pioneers of computation, I felt that *computation* as an idea goes much deeper than operating systems, programming languages, databases, and the like; the deep ideas of computation are intimately related to the deep ideas of life and intelligence. At Michigan I was lucky enough to be in a department in which "computation in natural systems" was as much a part of the core curriculum as software engineering or compiler design.

In 1989, at the beginning of my last year of graduate school, my Ph.D. advisor, Douglas Hofstadter, was invited to a conference in Los Alamos, New Mexico, on the subject of "emergent computation." He was too busy to attend, so he sent me instead. I was both thrilled and terrified to present work at such a high-profile meeting. It was at that meeting that I first encountered a large group of people obsessed with the same ideas that I had been pondering. I found that they not only had a name for this collection of ideas—complex systems—but that their institute in nearby Santa Fe was exactly the place I wanted to be. I was determined to find a way to get a job there.

Persistence, and being in the right place at the right time, eventually won me an invitation to visit the Santa Fe Institute for an entire summer. The summer stretched into a year, and that stretched into additional years. I eventually became one of the institute's resident faculty. People from many different countries and academic disciplines were there, all exploring different sides of the same question. How do we move beyond the traditional paradigm of reductionism toward a new understanding of seemingly irreducibly complex systems?

The idea for this book came about when I was invited to give the Ulam Memorial Lectures in Santa Fe—an annual set of lectures on complex systems for a general audience, given in honor of the great mathematician Stanislaw Ulam. The title of my lecture series was "The Past and Future of the Sciences of Complexity." It was very challenging to figure out how to introduce the audience of nonspecialists to the vast territory of complexity, to give them a feel for what is already known and for the daunting amount that remains to be learned. My role was like that of a tour guide in a large, culturally rich foreign country. Our schedule permitted only a short time to hear about the historical background, to visit some important sites, and to get a feel for the landscape and culture of the place, with translations provided from the native language when necessary.

This book is meant to be a much expanded version of those lectures—indeed, a written version of such a tour. It is about the questions that fascinate me and others in the complex systems community, past and present: How is it that those systems in nature we call *complex* and *adaptive*—brains, insect colonies, the immune system, cells, the global economy, biological evolution—produce such complex and adaptive behavior from underlying, simple rules? How can interdependent yet self-interested organisms come together to cooperate on solving problems that affect their survival as a whole? And are there any general principles or laws that apply to such phenomena? Can life, intelligence, and adaptation be seen as mechanistic and computational? If so, could we build truly intelligent and *living* machines? And if we could, would we want to?

I have learned that as the lines between disciplines begin to blur, the content of scientific discourse also gets fuzzier. People in the field of complex systems talk about many vague and imprecise notions such as spontaneous order, self-organization, and emergence (as well as "complexity" itself). A central purpose of this book is to provide a clearer picture of what these people are talking about and to ask whether such interdisciplinary notions and methods are likely to lead to useful science and to new ideas for addressing the most difficult problems faced by humans, such as the spread of disease, the unequal distribution of the world's natural and economic resources, the proliferation of weapons and conflicts, and the effects of our society on the environment and climate.

The chapters that follow give a guided tour, flavored with my own perspectives, of some of the core ideas of the sciences of complexity—where they came from and where they are going. As in any nascent, expanding, and vital area of science, people's opinions will differ (to put it mildly) about what the core ideas are, what their significance is, and what they will lead to. Thus my perspective may differ from that of my colleagues. An important part of this book will be spelling out some of those differences, and I'll do my best to provide glimpses of areas in which we are all in the dark or just beginning to see some light. These are the things that make science of this kind so stimulating, fun, and worthwhile both to practice and to read about. Above all else, I hope to communicate the deep enchantment of the ideas and debates and the incomparable excitement of pursuing them.

This book has five parts. In part I I give some background on the history and content of four subject areas that are fundamental to the study of complex systems: information, computation, dynamics and chaos, and evolution. In parts II–IV I describe how these four areas are being woven together in the science of complexity. I describe how life and evolution can be mimicked in computers, and conversely how the notion of *computation* itself is being imported to explain the behavior of natural systems. I explore the new science of networks and how it is discovering deep commonalities among systems as disparate as social communities, the Internet, epidemics, and metabolic systems in organisms. I describe several examples of how complexity can be measured in nature, how it is changing our view of living systems, and how this new view might inform the design of intelligent machines. I look at prospects of computer modeling of complex systems, as well as the perils of such models. Finally, in the last part I take on the larger question of the search for general principles in the sciences of complexity.

No background in math or science is needed to grasp what follows, though I will guide you gently and carefully through explorations in both. I hope to offer value to scientists and nonscientists alike. Although the discussion is not technical, I have tried in all cases to make it substantial. The notes give references to quotations, additional information on the discussion, and pointers to the scientific literature for those who want even more in-depth reading.

Have you been curious about the sciences of complexity? Would you like to come on such a guided tour? Let's begin.

ACKNOWLEDGMENTS

I AM GRATEFUL TO THE SANTA FE INSTITUTE (SFI) for inviting me to direct the Complex Systems Summer School and to give the Ulam Memorial Lectures, both of which spurred me to write this book. I am also grateful to SFI for providing me with a most stimulating and productive scientific home for many years. The various scientists who are part of the SFI family have been inspiring and generous in sharing their ideas, and I thank them all, too numerous to list here. I also thank the SFI staff for the ever-friendly and essential support they have given me during my association with the institute.

Many thanks to the following people for answering questions, commenting on parts of the manuscript, and helping me think more clearly about the issues in this book: Bob Axelrod, Liz Bradley, Jim Brown, Jim Crutchfield, Doyne Farmer, Stephanie Forrest, Bob French, Douglas Hofstadter, John Holland, Greg Huber, Ralf Juengling, Garrett Kenyon, Tom Kepler, David Krakauer, Will Landecker, Manuel Marques-Pita, Dan McShea, John Miller, Jack Mitchell, Norma Mitchell, Cris Moore, David Moser, Mark Newman, Norman Packard, Lee Segel, Cosma Shalizi, Eric Smith, Kendall Springer, J. Clint Sprott, Mick Thomure, Andreas Wagner, and Chris Wood. Of course any errors in this book are my own responsibility.

Thanks are also due to Kirk Jensen and Peter Prescott, my editors at Oxford, for their constant encouragement and superhuman patience, and to Keith Faivre and Tisse Takagi at Oxford, for all their help. I am also grateful to Google Scholar, Google Books, Amazon.com, and the often maligned but tremendously useful Wikipedia.org for making scholarly research so much easier.

This book is dedicated to Douglas Hofstadter and John Holland, who have done so much to inspire and encourage me in my work and life. I am very lucky to have had the benefit of their guidance and friendship.

Finally, much gratitude to my family: my parents, Jack and Norma Mitchell, my brother, Jonathan Mitchell, and my husband, Kendall Springer, for all their love and support. And I am grateful for Jacob and Nicholas Springer; although their births delayed the writing of this book, they have brought extraordinary joy and delightful complexity into our lives.

PART I | Background and History

Science has explored the microcosmos and the macrocosmos; we have a good sense of the lay of the land. The great unexplored frontier is complexity.

—Heinz Pagels, *The Dreams of Reason*

CHAPTER 1 | What Is Complexity?

Ideas thus made up of several simple ones put together, I call Complex; such as are Beauty, Gratitude, a Man, an Army, the Universe.

—John Locke, *An Essay Concerning Human Understanding*

Brazil: The Amazon rain forest. Half a million army ants are on the march. No one is in charge of this army; it has no commander. Each individual ant is nearly blind and minimally intelligent, but the marching ants together create a coherent fan-shaped mass of movement that swarms over, kills, and efficiently devours all prey in its path. What cannot be devoured right away is carried with the swarm. After a day of raiding and destroying the edible life over a dense forest the size of a football field, the ants build their nighttime shelter—a chain-mail ball a yard across made up of the workers' linked bodies, sheltering the young larvae and mother queen at the center. When dawn arrives, the living ball melts away ant by ant as the colony members once again take their places for the day's march.

Nigel Franks, a biologist specializing in ant behavior, has written, "The solitary army ant is behaviorally one of the least sophisticated animals imaginable," and, "If 100 army ants are placed on a flat surface, they will walk around and around in never decreasing circles until they die of exhaustion." Yet put half a million of them together, and the group as a whole becomes what some have called a "superorganism" with "collective intelligence."

How does this come about? Although many things are known about ant colony behavior, scientists still do not fully understand all the mechanisms underlying a colony's collective intelligence. As Franks comments further, "I have studied *E. burchelli* [a common species of army ant] for many years, and for me the mysteries of its social organization still multiply faster than the rate at which its social structure can be explored."

The mysteries of army ants are a microcosm for the mysteries of many natural and social systems that we think of as "complex." No one knows exactly how any community of social organisms—ants, termites, humans— come together to collectively build the elaborate structures that increase the survival probability of the community as a whole. Similarly mysterious is how the intricate machinery of the immune system fights disease; how a group of cells organizes itself to be an eye or a brain; how independent members of an economy, each working chiefly for its own gain, produce complex but structured global markets; or, most mysteriously, how the phenomena we call "intelligence" and "consciousness" emerge from nonintelligent, nonconscious material substrates.

Such questions are the topics of *complex systems*, an interdisciplinary field of research that seeks to explain how large numbers of relatively simple entities organize themselves, without the benefit of any central controller, into a collective whole that creates patterns, uses information, and, in some cases, evolves and learns. The word *complex* comes from the Latin root *plectere*: to weave, entwine. In complex systems, many simple parts are irreducibly entwined, and the field of complexity is itself an entwining of many different fields.

Complex systems researchers assert that different complex systems in nature, such as insect colonies, immune systems, brains, and economies, have much in common. Let's look more closely.

Insect Colonies

Colonies of social insects provide some of the richest and most mysterious examples of complex systems in nature. An ant colony, for instance, can consist of hundreds to millions of individual ants, each one a rather simple creature that obeys its genetic imperatives to seek out food, respond in simple ways to the chemical signals of other ants in its colony, fight intruders, and so forth. However, as any casual observer of the outdoors can attest, the ants in a colony, each performing its own relatively simple actions, work together to build astoundingly complex structures that are clearly of great importance for the survival of the colony as a whole. Consider, for example, their use of soil, leaves, and twigs to construct huge nests of great strength and stability, with large networks of underground passages and dry, warm, brooding chambers whose temperatures are carefully controlled by decaying nest materials and the ants' own bodies. Consider also the long bridges certain species of ants build with their own bodies to allow emigration from one nest site to another via tree branches separated by great distances (to an ant, that is) (figure 1.1). Although much is now understood about ants and their social structures, scientists still can fully explain neither their individual nor group behavior: exactly how the individual actions of the ants produce large, complex structures, how the ants signal one another, and how the colony as a whole adapts to changing circumstances (e.g., changing weather or attacks on the colony). And how did biological evolution produce creatures with such an enormous contrast between their individual simplicity and their collective sophistication?

The Brain

The cognitive scientist Douglas Hofstadter, in his book *Gödel, Escher, Bach*, makes an extended analogy between ant colonies and brains, both being

FIGURE 1.1. Ants build a bridge with their bodies to allow the colony to take the shortest path across a gap. (Photograph courtesy of Carl Rettenmeyer.)

complex systems in which relatively simple components with only limited communication among themselves collectively give rise to complicated and sophisticated system-wide ("global") behavior. In the brain, the simple components are cells called *neurons*. The brain is made up of many different types of cells in addition to neurons, but most brain scientists believe that the actions of neurons and the patterns of connections among groups of neurons are what cause perception, thought, feelings, consciousness, and the other important large-scale brain activities.

Neurons are pictured in figure 1.2 (top). Neurons consists of three main parts: the cell body (*soma*), the branches that transmit the cell's input from other neurons (*dendrites*), and the single trunk transmitting the cell's output to other neurons (*axon*). Very roughly, a neuron can be either in an active state (*firing*) or an inactive state (*not firing*). A neuron fires when it receives enough signals from other neurons through its dendrites. *Firing* consists of sending an electric pulse through the axon, which is then converted into a chemical signal via chemicals called *neurotransmitters*. This chemical signal in turn activates other neurons through their dendrites. The firing frequency and the resulting chemical output signals of a neuron can vary over time according to both its input and how much it has been firing recently.

These actions recall those of ants in a colony: individuals (neurons or ants) perceive signals from other individuals, and a sufficient summed strength of these signals causes the individuals to act in certain ways that produce additional signals. The overall effects can be very complex. We saw that an explanation of ants and their social structures is still incomplete; similarly, scientists don't yet understand how the actions of individual or dense networks of neurons give rise to the large-scale behavior of the brain (figure 1.2, bottom). They don't understand what the neuronal signals mean, how large numbers of neurons work together to produce global cognitive behavior, or how exactly they cause the brain to think thoughts and learn new things. And again, perhaps most puzzling is how such an elaborate signaling system with such powerful collective abilities ever arose through evolution.

The Immune System

The immune system is another example of a system in which relatively simple components collectively give rise to very complex behavior involving signaling and control, and in which adaptation occurs over time. A photograph illustrating the immune system's complexity is given in figure 1.3.

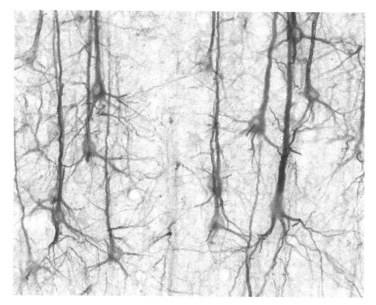

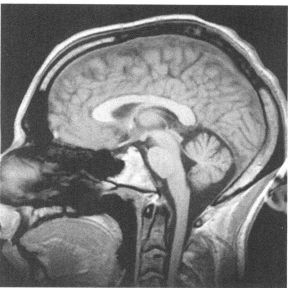

FIGURE 1.2. Top: microscopic view of neurons, visible via staining. Bottom: a human brain. How does the behavior at one level give rise to that of the next level? (Neuron photograph from brainmaps.org [http://brainmaps.org/smi32-pic.jpg], licensed under Creative Commons [http://creativecommons.org/licenses/by/3.0/]. Brain photograph courtesy of Christian R. Linder.)

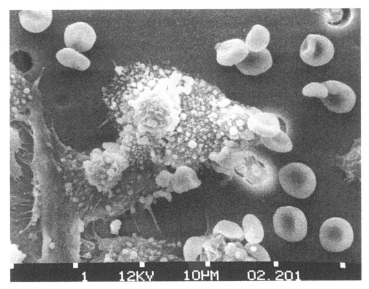

FIGURE 1.3. Immune system cells attacking a cancer cell. (Photograph by Susan Arnold, from National Cancer Institute Visuals Online [http://visualsonline.cancer.gov/details.cfm?imageid=2370].)

The immune system, like the brain, differs in sophistication in different animals, but the overall principles are the same across many species. The immune system consists of many different types of cells distributed over the entire body (in blood, bone marrow, lymph nodes, and other organs). This collection of cells works together in an effective and efficient way without any central control.

The star players of the immune system are white blood cells, otherwise known as *lymphocytes*. Each lymphocyte can recognize, via receptors on its cell body, molecules corresponding to certain possible invaders (e.g., bacteria). Some one trillion of these patrolling sentries circulate in the blood at a given time, each ready to sound the alarm if it is *activated*—that is, if its particular receptors encounter, by chance, a matching invader. When a lymphocyte is activated, it secretes large numbers of molecules—*antibodies*—that can identify similar invaders. These antibodies go out on a seek-and-destroy mission throughout the body. An activated lymphocyte also divides at an increased rate, creating daughter lymphocytes that will help hunt out invaders and secrete antibodies against them. It also creates daughter lymphocytes that will hang around and remember the particular invader that was seen, thus giving the body immunity to pathogens that have been previously encountered.

One class of lymphocytes are called *B cells* (the *B* indicates that they develop in the bone marrow) and have a remarkable property: the better the match between a B cell and an invader, the more antibody-secreting daughter cells the B cell creates. The daughter cells each differ slightly from the mother cell in random ways via mutations, and these daughter cells go on to create their own daughter cells in direct proportion to how well they match the invader. The result is a kind of Darwinian natural selection process, in which the match between B cells and invaders gradually gets better and better, until the antibodies being produced are extremely efficient at seeking and destroying the culprit microorganisms.

Many other types of cells participate in the orchestration of the immune response. *T cells* (which develop in the thymus) play a key role in regulating the response of B cells. *Macrophages* roam around looking for substances that have been tagged by antibodies, and they do the actual work of destroying the invaders. Other types of cells help effect longer-term immunity. Still other parts of the system guard against attacking the cells of one's own body.

Like that of the brain and ant colonies, the immune system's behavior arises from the independent actions of myriad simple players with no one actually in charge. The actions of the simple players—B cells, T cells, macrophages, and the like—can be viewed as a kind of chemical signal-processing network in which the recognition of an invader by one cell triggers a cascade of signals among cells that put into play the elaborate complex response. As yet many crucial aspects of this signal-processing system are not well understood. For example, it is still to be learned what, precisely, are the relevant signals, their specific functions, and how they work together to allow the system as a whole to "learn" what threats are present in the environment and to produce long-term immunity to those threats. We do not yet know precisely how the system avoids attacking the body; or what gives rise to flaws in the system, such as autoimmune diseases, in which the system does attack the body; or the detailed strategies of the human immunodeficiency virus (HIV), which is able to get by the defenses by attacking the immune system itself. Once again, a key question is how such an effective complex system arose in the first place in living creatures through biological evolution.

Economies

Economies are complex systems in which the "simple, microscopic" components consist of people (or companies) buying and selling goods, and the collective behavior is the complex, hard-to-predict behavior of markets as a whole, such as changes in the price of housing in different areas of the country or fluctuations in stock prices (figure 1.4). Economies are thought by some economists to be adaptive on both the microscopic and macroscopic level. At the microscopic level, individuals, companies, and markets try to increase their profitability by learning about the behavior of other individuals and companies. This microscopic self-interest has historically been thought to push markets as a whole—on the macroscopic level—toward an equilibrium state in which the prices of goods are set so there is no way to change production or consumption patterns to make everyone better off. In terms of profitability or consumer satisfaction, if someone is made better off, someone else will be made worse off. The process by which markets obtain this equilibrium is called *market efficiency*. The eighteenth-century economist Adam Smith called this self-organizing behavior of markets the "invisible hand": it arises from the myriad microscopic actions of individual buyers and sellers.

Economists are interested in how markets become efficient, and conversely, what makes efficiency fail, as it does in real-world markets. More recently, economists involved in the field of complex systems have tried to explain market behavior in terms similar to those used previously in the descriptions of other complex systems: dynamic hard-to-predict patterns in global behavior, such as patterns of market bubbles and crashes; processing of signals and information, such as the decision-making processes of individual buyers and sellers, and the resulting "information processing" ability of the market as a whole to "calculate" efficient prices; and adaptation and learning, such as individual sellers adjusting their production to adapt to changes in buyers' needs, and the market as a whole adjusting global prices.

The World Wide Web

The World Wide Web came on the world scene in the early 1990s and has experienced exponential growth ever since. Like the systems described above, the Web can be thought of as a self-organizing social system: individuals, with little or no central oversight, perform simple tasks: posting Web pages and linking to other Web pages. However, complex systems scientists have discovered that the network as a whole has many unexpected large-scale properties involving its overall structure, the way in which it grows, how information propagates over its links, and the coevolutionary relationships between the behavior of search engines and the Web's link structure, all of which lead to what could be called "adaptive" behavior for the system as a whole. The

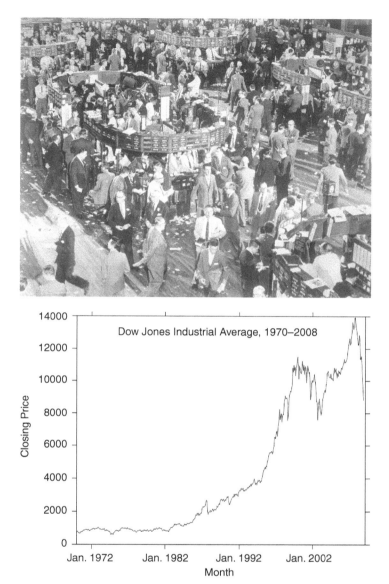

FIGURE 1.4. Individual actions on a trading floor give rise to the hard-to-predict large-scale behavior of financial markets. Top: New York Stock Exchange (photograph from Milstein Division of US History, Local History and Genealogy, The New York Public Library, Astor, Lenox, and Tilden Foundations, used by permission). Bottom: Dow Jones Industrial Average closing price, plotted monthly 1970–2008.

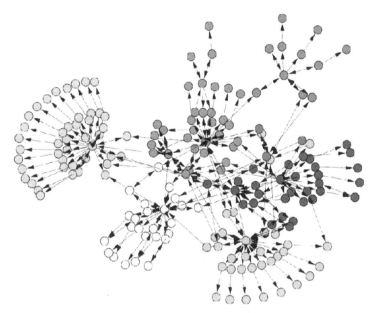

FIGURE 1.5. Network structure of a section of the World Wide Web. (Reprinted with permission from M.E.J. Newman and M. Girvin, *Physical Review Letters E*, 69,026113, 2004. Copyright 2004 by the American Physical Society.)

complex behavior emerging from simple rules in the World Wide Web is currently a hot area of study in complex systems. Figure 1.5 illustrates the structure of one collection of Web pages and their links. It seems that much of the Web looks very similar; the question is, *why?*

Common Properties of Complex Systems

When looked at in detail, these various systems are quite different, but viewed at an abstract level they have some intriguing properties in common:

1. **Complex collective behavior**: All the systems I described above consist of large networks of individual components (ants, B cells, neurons, stock-buyers, Web-site creators), each typically following relatively simple rules with no central control or leader. It is the collective actions of vast numbers of components that give rise to the complex, hard-to-predict, and changing patterns of behavior that fascinate us.

2. **Signaling and information processing**: All these systems produce and use information and signals from both their internal and external environments.
3. **Adaptation**: All these systems adapt—that is, change their behavior to improve their chances of survival or success—through learning or evolutionary processes.

Now I can propose a definition of the term *complex system*: **a system in which large networks of components with no central control and simple rules of operation give rise to complex collective behavior, sophisticated information processing, and adaptation via learning or evolution.** (Sometimes a differentiation is made between *complex adaptive systems*, in which adaptation plays a large role, and nonadaptive complex systems, such as a hurricane or a turbulent rushing river. In this book, as most of the systems I do discuss are adaptive, I do not make this distinction.)

Systems in which organized behavior arises without an internal or external controller or leader are sometimes called *self-organizing*. Since simple rules produce complex behavior in hard-to-predict ways, the macroscopic behavior of such systems is sometimes called *emergent*. Here is an alternative definition of a *complex system*: **a system that exhibits nontrivial emergent and self-organizing behaviors.** The central question of the sciences of complexity is how this emergent self-organized behavior comes about. In this book I try to make sense of these hard-to-pin-down notions in different contexts.

How Can Complexity Be Measured?

In the paragraphs above I have sketched some qualitative common properties of complex systems. But more quantitative questions remain: Just how *complex* is a particular complex system? That is, how do we measure *complexity*? Is there any way to say precisely how much more complex one system is than another?

These are key questions, but they have not yet been answered to anyone's satisfaction and remain the source of many scientific arguments in the field. As I describe in chapter 7, many different measures of complexity have been proposed; however, none has been universally accepted by scientists. Several of these measures and their usefulness are described in various chapters of this book.

But how can there be a science of complexity when there is no agreed-on quantitative definition of complexity?

I have two answers to this question. First, neither a single *science of complexity* nor a single *complexity theory* exists yet, in spite of the many articles and books that have used these terms. Second, as I describe in many parts of this book, an essential feature of forming a new science is a struggle to define its central terms. Examples can be seen in the struggles to define such core concepts as *information*, *computation*, *order*, and *life*. In this book I detail these struggles, both historical and current, and tie them in with our struggles to understand the many facets of complexity. This book is about cutting-edge science, but it is also about the history of core concepts underlying this cutting-edge science. The next four chapters provide this history and background on the concepts that are used throughout the book.

CHAPTER 2 | Dynamics, Chaos, and Prediction

It makes me so happy. To be at the beginning again, knowing almost nothing.... The ordinary-sized stuff which is our lives, the things people write poetry about—clouds—daffodils—waterfalls....these things are full of mystery, as mysterious to us as the heavens were to the Greeks...It's the best possible time to be alive, when almost everything you thought you knew is wrong.
—Tom Stoppard, *Arcadia*

D YNAMICAL SYSTEMS THEORY (or *dynamics*) concerns the description and prediction of systems that exhibit complex *changing* behavior at the macroscopic level, emerging from the collective actions of many interacting components. The word *dynamic* means changing, and dynamical systems are systems that change over time in some way. Some examples of dynamical systems are [handwritten: doesn't everything change with time? Each thing is interacting with every other thing.]

The solar system (the planets change position over time)

The heart of a living creature (it beats in a periodic fashion rather than standing still)

The brain of a living creature (neurons are continually firing, neurotransmitters are propelled from one neuron to another, synapse strengths are changing, and generally the whole system is in a continual state of flux)

The stock market

The world's population

The global climate

Dynamical systems include these and most other systems that you probably can think of. Even rocks change over geological time. Dynamical systems theory describes in general terms the ways in which systems can change, what types of macroscopic behavior are possible, and what kinds of predictions about that behavior can be made.

Dynamical systems theory has recently been in vogue in popular science because of the fascinating results coming from one of its intellectual offspring, the study of chaos. However, it has a long history, starting, as many sciences did, with the Greek philosopher Aristotle.

Early Roots of Dynamical Systems Theory

Aristotle was the author of one of the earliest recorded theories of motion, one that was accepted widely for over 1,500 years. His theory rested on two main principles, both of which turned out to be wrong. First, he believed that motion on Earth differs from motion in the heavens. He asserted that on

Aristotle, 384–322 B.C.
(Ludovisi Collection)

Earth objects move in straight lines and only when something forces them to; when no forces are applied, an object comes to its natural resting state. In the heavens, however, planets and other celestial objects move continuously in perfect circles centered about the Earth. Second, Aristotle believed that earthly objects move in different ways depending on what they are made of. For example, he believed that a rock will fall to Earth because it is mainly composed of the element *earth*, whereas smoke will rise because it is mostly composed of the element *air*. Likewise, heavier objects, presumably containing more earth, will fall faster than lighter objects.

Clearly Aristotle (like many theorists since) was not one to let experimental results get in the way of his theorizing. His scientific method was to let logic and common sense direct theory; the importance of testing the resulting theories by experiments is a more modern notion. The influence of Aristotle's ideas was strong and continued to hold sway over most of Western science until the sixteenth century—the time of Galileo.

[margin note: Common sense can be wrong. Our senses are imperfect.]

Galileo was a pioneer of experimental, empirical science, along with his predecessor Copernicus and his contemporary Kepler. Copernicus established that the motion of the planets is centered not about the Earth but about the sun. (Galileo got into big trouble with the Catholic Church for promoting this view and was eventually forced to publicly renounce it; only in 1992 did the Church officially admit that Galileo had been unfairly persecuted.) In the early 1600s, Kepler discovered that the motion of the planets is not circular but rather elliptical, and he discovered laws describing this elliptical motion.

Whereas Copernicus and Kepler focused their research on celestial motion, Galileo studied motion not only in the heavens but also here on Earth by experimenting with the objects one now finds in elementary physics courses: pendula, balls rolling down inclined planes, falling objects, light reflected by mirrors. Galileo did not have the sophisticated experimental devices we have today: he is said to have timed the swinging of a pendulum by counting his heartbeats and to have measured the effects of gravity by dropping objects off the leaning tower of Pisa. These now-classic experiments revolutionized ideas about motion. In particular, Galileo's studies directly contradicted Aristotle's long-held principles of motion. Against common sense, rest is *not* the natural state of objects; rather it takes *force* to stop a moving object. Heavy and light objects in a vacuum fall at the same rate. And perhaps most revolutionary of all, laws of motion on the Earth could explain some aspects of motions in the heavens. With Galileo, the scientific revolution, with experimental observations at its core, was definitively launched.

The most important person in the history of dynamics was Isaac Newton. Newton, who was born the year after Galileo died, can be said to have

Galileo, 1564–1642 (AIP Emilio Segre Visual Archives, E. Scott Barr Collection)

Isaac Newton, 1643–1727 (Original engraving by unknown artist, courtesy AIP Emilio Segre Visual Archives)

invented, on his own, the science of dynamics. Along the way he also had to invent calculus, the branch of mathematics that describes motion and change.

Physicists call the general study of motion *mechanics*. This is a historical term dating from ancient Greece, reflecting the classical view that all motion could be explained in terms of the combined actions of simple "machines" (e.g., lever, pulley, wheel and axle). Newton's work is known today as *classical mechanics*. Mechanics is divided into two areas: kinematics, which describes how things move, and dynamics, which explains why things obey the laws of kinematics. For example, Kepler's laws are kinematic laws—they describe *how* the planets move (in ellipses with the sun at one focus)—but not *why* they move in this particular way. Newton's laws are the foundations of dynamics: they explain the motion of the planets, and everything else, in terms of the basic notions of force and mass.

Newton's famous three laws are as follows:

1. Constant motion: Any object not subject to a force moves with unchanging speed.
2. Inertial mass: When an object is subject to a force, the resulting change in its motion is inversely proportional to its mass.
3. Equal and opposite forces: If object A exerts a force on object B, then object B must exert an equal and opposite force on object A.

One of Newton's greatest accomplishments was to realize that these laws applied not just to earthly objects but to those in the heavens as well. Galileo was the first to state the constant-motion law, but he believed it applied only to objects on Earth. Newton, however, understood that this law should apply to the planets as well, and realized that elliptical orbits, which exhibit a constantly *changing* direction of motion, require explanation in terms of a force, namely gravity. Newton's other major achievement was to state a universal law of gravity: the force of gravity between two objects is proportional to the product of their masses divided by the square of the distance between them. Newton's insight—now the backbone of modern science—was that this law applies everywhere in the universe, to falling apples as well as to planets. As he wrote: "nature is exceedingly simple and conformable to herself. Whatever reasoning holds for greater motions, should hold for lesser ones as well."

Newtonian mechanics produced a picture of a "clockwork universe," one that is wound up with the three laws and then runs its mechanical course. The mathematician Pierre Simon Laplace saw the implication of this clockwork view for prediction: in 1814 he asserted that, given Newton's laws and the current position and velocity of every particle in the universe, it was possible, in principle, to predict everything for all time. With the invention of electronic computers in the 1940s, the "in principle" might have seemed closer to "in practice."

Revised Views of Prediction

However, two major discoveries of the twentieth century showed that Laplace's dream of complete prediction is not possible, even in principle. One discovery was Werner Heisenberg's 1927 "uncertainty principle" in quantum mechanics, which states that one cannot measure the exact values of the position and the momentum (mass times velocity) of a particle at the same time. The more certain one is about where a particle is located at a given time, the less one can know about its momentum, and vice versa. However, effects of Heisenberg's principle exist only in the quantum world of tiny particles, and most people viewed it as an interesting curiosity, but not one that would have much implication for prediction at a larger scale—predicting the weather, say.

It was the understanding of *chaos* that eventually laid to rest the hope of perfect prediction of all complex systems, quantum or otherwise. The defining idea of chaos is that there are some systems—*chaotic* systems—in which even minuscule uncertainties in measurements of initial position and momentum can result in huge errors in long-term predictions of these quantities. This is known as "sensitive dependence on initial conditions."

In parts of the natural world such small uncertainties will not matter. If your initial measurements are fairly but not perfectly precise, your predictions will likewise be close to right if not exactly on target. For example, astronomers can predict eclipses almost perfectly in spite of even relatively large uncertainties in measuring the positions of planets. But sensitive dependence on initial conditions says that in chaotic systems, even the tiniest errors in your initial measurements will eventually produce huge errors in your prediction of the future motion of an object. In such systems (and hurricanes may well be an example) *any* error, no matter how small, will make long-term predictions vastly inaccurate.

This kind of behavior is counterintuitive; in fact, for a long time many scientists denied it was possible. However, chaos in this sense has been observed in cardiac disorders, turbulence in fluids, electronic circuits, dripping faucets, and many other seemingly unrelated phenomena. These days, the existence of chaotic systems is an accepted fact of science.

It is hard to pin down who first realized that such systems might exist. The possibility of sensitive dependence on initial conditions was proposed by a number of people long before quantum mechanics was invented. For example, the physicist James Clerk Maxwell hypothesized in 1873 that there are classes of phenomena affected by "influences whose physical magnitude is too small to be taken account of by a finite being, [but which] may produce results of the highest importance."

Possibly the first clear example of a chaotic system was given in the late nineteenth century by the French mathematician Henri Poincaré. Poincaré was the founder of and probably the most influential contributor to the modern field of dynamical systems theory, which is a major outgrowth of Newton's science of dynamics. Poincaré discovered sensitive dependence on initial conditions when attempting to solve a much simpler problem than predicting the motion of a hurricane. He more modestly tried to tackle the so-called three-body problem: to determine, using Newton's laws, the long-term motions of three masses exerting gravitational forces on one another. Newton solved the *two*-body problem, but the three-body problem turned out to be much harder. Poincaré tackled it in 1887 as part of a mathematics contest held in honor of the king of Sweden. The contest offered a prize of 2,500 Swedish crowns for a solution to the "many body" problem: predicting the future positions of arbitrarily many masses attracting one another under Newton's laws. This problem was inspired by the question of whether or not the solar system is stable: will the planets remain in their current orbits, or will they wander from them? Poincaré started off by seeing whether he could solve it for merely three bodies.

He did not completely succeed—the problem was too hard. But his attempt was so impressive that he was awarded the prize anyway. Like Newton with calculus, Poincaré had to invent a new branch of mathematics, *algebraic topology*, to even tackle the problem. Topology is an extended form of geometry, and it was in looking at the geometric consequences of the three-body problem that he discovered the possibility of sensitive dependence on initial conditions. He summed up his discovery as follows:

> If we knew exactly the laws of nature and the situation of the universe at the initial moment, we could predict exactly the situation of that same universe at a succeeding moment. But even if it were the case that the natural laws had no longer any secret for us, we could still only know the initial situation approximately. If that enabled us to predict the succeeding situation with the same approximation, that is all we require, and we should say that the phenomenon has been predicted, that it is governed by laws. But it is not always so; it may happen that small differences in the initial conditions produce very great ones in the final phenomenon. A small error in the former will produce an enormous error in the latter. Prediction becomes impossible....

In other words, even if we know the laws of motion perfectly, two different sets of initial conditions (here, initial positions, masses, and velocities for

Henri Poincaré, 1854–1912 (AIP Emilio Segre Visual Archives)

objects), even if they differ in a minuscule way, can sometimes produce greatly different results in the subsequent motion of the system. Poincaré found an example of this in the three-body problem.

It was not until the invention of the electronic computer that the scientific world began to see this phenomenon as significant. Poincaré, way ahead of his time, had guessed that sensitive dependence on initial conditions would stymie attempts at long-term weather prediction. His early hunch gained some evidence when, in 1963, the meteorologist Edward Lorenz found that even simple computer models of weather phenomena were subject to sensitive dependence on initial conditions. Even with today's modern, highly complex meteorological computer models, weather predictions are at best reasonably accurate only to about one week in the future. It is not yet known whether this limit is due to fundamental chaos in the weather, or how much this limit can be extended by collecting more data and building even better models.

Linear versus Nonlinear Rabbits

Let's now look more closely at sensitive dependence on initial conditions. How, precisely, does the huge magnification of initial uncertainties come about in chaotic systems? The key property is *nonlinearity*. A linear system is one you can understand by understanding its parts individually and then putting them together. When my two sons and I cook together, they like to take turns adding ingredients. Jake puts in two cups of flour. Then Nicky puts in a cup of sugar. The result? Three cups of flour/sugar mix. The whole is equal to the sum of the parts.

A nonlinear system is one in which the whole is different from the sum of the parts. Jake puts in two cups of baking soda. Nicky puts in a cup of vinegar. The whole thing explodes. (You can try this at home.) The result? *More* than three cups of vinegar-and-baking-soda-and-carbon-dioxide fizz.

The difference between the two examples is that in the first, the flour and sugar don't really interact to create something new, whereas in the second, the vinegar and baking soda interact (rather violently) to create a lot of carbon dioxide.

Linearity is a reductionist's dream, and nonlinearity can sometimes be a reductionist's nightmare. Understanding the distinction between linearity and nonlinearity is very important and worthwhile. To get a better handle on this distinction, as well as on the phenomenon of chaos, let's do a bit of very simple mathematical exploration, using a classic illustration of linear and nonlinear systems from the field of biological population dynamics.

Suppose you have a population of breeding rabbits in which every year all the rabbits pair up to mate, and each pair of rabbit parents has exactly four offspring and then dies. The population growth, starting from two rabbits, is illustrated in figure 2.1.

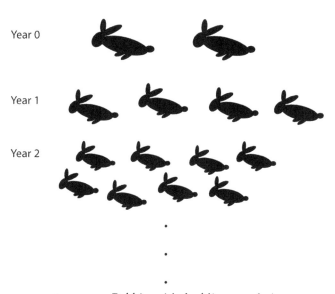

FIGURE 2.1. Rabbits with doubling population.

DYNAMICS, CHAOS, AND PREDICTION | 23

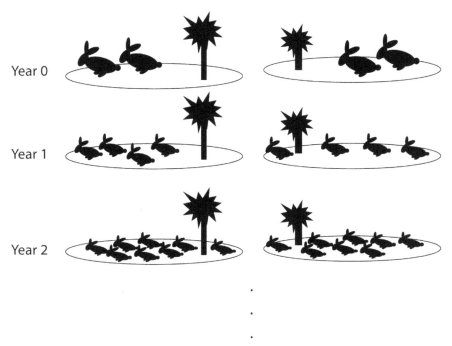

FIGURE 2.2. Rabbits with doubling population, split on two islands.

It is easy to see that the population doubles every year without limit (which means the rabbits would quickly take over the planet, solar system, and universe, but we won't worry about that for now).

This is a linear system: the whole is equal to the sum of the parts. What do I mean by this? Let's take a population of four rabbits and split them between two separate islands, two rabbits on each island. Then let the rabbits proceed with their reproduction. The population growth over two years is illustrated in figure 2.2.

Each of the two populations doubles each year. At each year, if you add the populations on the two islands together, you'll get the same number of rabbits that you would have gotten had there been no separation—that is, had they all lived on one island.

If you make a plot with the current year's population size on the horizontal axis and the next-year's population size on the vertical axis, you get a straight line (figure 2.3). This is where the term *linear system* comes from.

But what happens when, more realistically, we consider limits to population growth? This requires us to make the growth rule nonlinear. Suppose that, as before, each year every pair of rabbits has four offspring and then dies. But now suppose that some of the offspring die before they reproduce

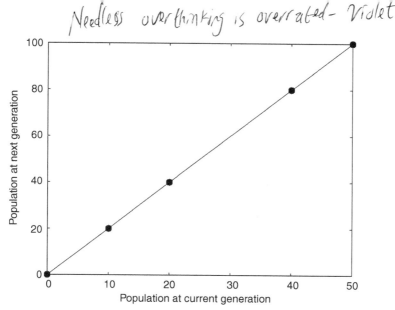

FIGURE 2.3. A plot of how the population size next year depends on the population size this year for the linear model.

because of overcrowding. Population biologists sometimes use an equation called the *logistic model* as a description of population growth in the presence of overcrowding. This sense of the word *model* means a mathematical formula that describes population growth in a simplified way.

In order to use the logistic model to calculate the size of the next generation's population, you need to input to the logistic model the current generation's population size, the *birth rate*, the *death rate* (the probability of an individual will die due to overcrowding), and the maximum *carrying capacity* (the strict upper limit of the population that the habitat will support.)

I won't give the actual equation for the logistic model here (it is given in the notes), but you can see its behavior in figure 2.4.

As a simple example, let's set *birth rate* = 2 and *death rate* = 0.4, assume the carrying capacity is thirty-two, and start with a population of twenty rabbits in the first generation. Using the logistic model, I calculate that the number of surviving offspring in the second generation is twelve. I then plug this new population size into the model, and find that there are still exactly twelve surviving rabbits in the third generation. The population will stay at twelve for all subsequent years.

If I reduce the death rate to 0.1 (keeping everything else the same), things get a little more interesting. From the model I calculate that the second generation has 14.25 rabbits and the third generation has 15.01816.

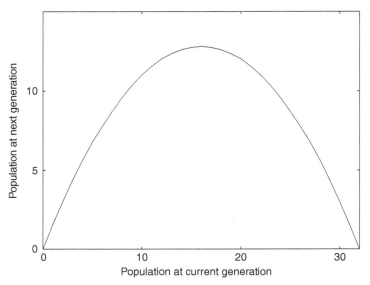

FIGURE 2.4. A plot of how the population size next year depends on the population size this year under the logistic model, with birth rate equal to 2, death rate equal to 0.4, and carrying capacity equal to 32. The plot will also be a parabola for other values of these parameters.

Wait a minute! How can we have 0.25 of a rabbit, much less 0.01816 of a rabbit? Obviously in real life we cannot, but this is a mathematical model, and it allows for fractional rabbits. This makes it easier to do the math, and can still give reasonable predictions of the actual rabbit population. So let's not worry about that for now.

This process of calculating the size of the next population again and again, starting each time with the immediately previous population, is called "iterating the model."

What happens if the death rate is set back to 0.4 and carrying capacity is doubled to sixty-four? The model tells me that, starting with twenty rabbits, by year nine the population reaches a value close to twenty-four and stays there.

You probably noticed from these examples that the behavior is more complicated than when we simply doubled the population each year. That's because the logistic model is nonlinear, due to its inclusion of death by overcrowding. Its plot is a parabola instead of a line (figure 2.4). The logistic population growth is not simply equal to the sum of its parts. To show this, let's see what happens if we take a population of twenty rabbits and segregate it

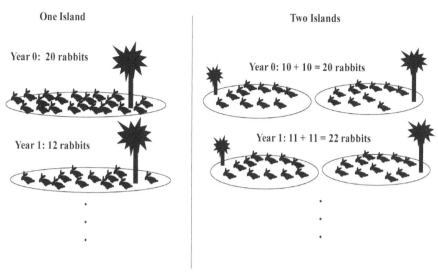

FIGURE 2.5. Rabbit population split on two islands, following the logistic model.

into populations of ten rabbits each, and iterate the model for each population (with *birth rate* = 2 and *death rate* = .4, and carrying capacity of each island equal to 32, as in the first example above). The result is illustrated in figure 2.5.

At year one, the original twenty-rabbit population has been cut down to twelve rabbits, but each of the original ten-rabbit populations now has eleven rabbits, for a total of twenty-two rabbits. The behavior of the whole is clearly not equal to the sum of the behavior of the parts.

The Logistic Map

Many scientists and mathematicians who study this sort of thing have used a simpler form of the logistic model called the *logistic map*, which is perhaps the most famous equation in the science of dynamical systems and chaos. The logistic model is simplified by combining the effects of birth rate and death rate into one number, called R. *Population size* is replaced by a related concept called "fraction of carrying capacity," called x. Given this simplified model, scientists and mathematicians promptly forget all about population growth, carrying capacity, and anything else connected to the real world, and simply get lost in the astounding behavior of the equation itself. We will do the same.

Here is the equation, where x_t is the current value of x and x_{t+1} is its value at the next time step:[1]

$$x_{t+1} = R x_t (1 - x_t).$$

I give the equation for the logistic map to show you how simple it is. In fact, it is one of the simplest systems to capture the essence of chaos: sensitive dependence on initial conditions. The logistic map was brought to the attention of population biologists in a 1971 article by the mathematical biologist Robert May in the prestigious journal *Nature*. It had been previously analyzed in detail by several mathematicians, including Stanislaw Ulam, John von Neumann, Nicholas Metropolis, Paul Stein, and Myron Stein. But it really achieved fame in the 1980s when the physicist Mitchell Feigenbaum used it to demonstrate *universal* properties common to a very large class of chaotic systems. Because of its apparent simplicity and rich history, it is a perfect vehicle to introduce some of the major concepts of dynamical systems theory and chaos.

The logistic map gets very interesting as we vary the value of R. Let's start with $R = 2$. We also need to start out with some value between 0 and 1 for x_0, say 0.5. If you plug those numbers into the logistic map, the answer for x_1 is 0.5. Likewise, $x_2 = 0.5$, and so on. Thus, if $R = 2$ and the population starts out at half the maximum size, it will stay there forever.

Now let's try $x_0 = 0.2$. You can use your calculator to compute this one. (I'm using one that reads off at most seven decimal places.) The results are more interesting:

$$x_0 = 0.2$$
$$x_1 = 0.32$$
$$x_2 = 0.4352$$
$$x_3 = 0.4916019$$
$$x_4 = 0.4998589$$
$$x_5 = 0.5$$
$$x_6 = 0.5$$
$$\vdots$$

---

1. Authors of popular-audience science books are always warned of the following rule: every equation in your book will cut the readership by one-half. I'm no exception—my editor told me this fact very clearly. I'm going to give the logistic map equation here anyway, so the half of you who would throw the book out the window if you ever encountered an equation, please skip over the next line.

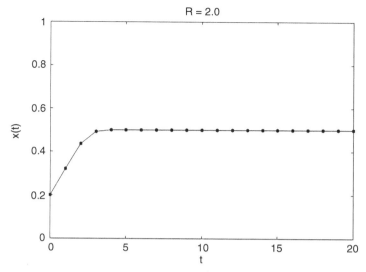

FIGURE 2.6. Behavior of the logistic map for $R = 2$ and $x_0 = 0.2$.

The same eventual result ($x_t = 0.5$ forever) occurs but here it takes five iterations to get there.

It helps to see these results visually. A plot of the value of x_t at each time t for 20 time steps is shown in figure 2.6. I've connected the points by lines to better show how as time increases, x quickly converges to 0.5.

What happens if x_0 is large, say, 0.99? Figure 2.7 shows a plot of the results.

Again the same ultimate result occurs, but with a longer and more dramatic path to get there.

You may have guessed it already: if $R = 2$ then x_t eventually always gets to 0.5 and stays there. The value 0.5 is called a *fixed point*: how long it takes to get there depends on where you start, but once you are there, you are fixed.

If you like, you can do a similar set of calculations for $R = 2.5$, and you will find that the system also always goes to a fixed point, but this time the fixed point is 0.6.

For even more fun, let $R = 3.1$. The behavior of the logistic map now gets more complicated. Let $x_0 = 0.2$. The plot is shown in figure 2.8.

In this case x never settles down to a fixed point; instead it eventually settles into an oscillation between two values, which happen to be 0.5580141 and 0.7645665. If the former is plugged into the formula the latter is produced, and vice versa, so this oscillation will continue forever. This oscillation will be reached eventually no matter what value is given for x_0. This kind of regular

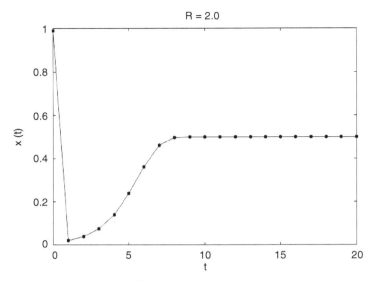

FIGURE 2.7. Behavior of the logistic map for $R = 2$ and $x_0 = 0.99$.

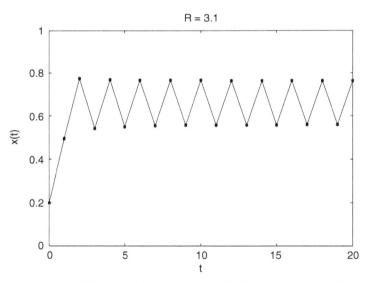

FIGURE 2.8. Behavior of the logistic map for $R = 3.1$ and $x_0 = 0.2$.

final behavior (either fixed point or oscillation) is called an "attractor," since, loosely speaking, any initial condition will eventually be "attracted to it."

For values of R from 3.0 up to around 3.4 the logistic map will have similar behavior: after a certain number of iterations, the system will oscillate between two different values. (The final pair of values will be different for each value of

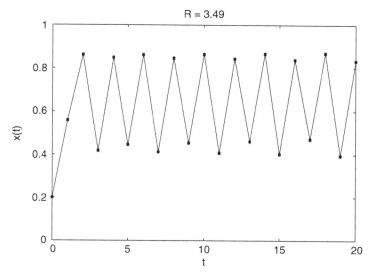

FIGURE 2.9. Behavior of the logistic map for $R = 3.49$ and $x_0 = 0.2$.

R.) Because it oscillates between two values, the system is said to have *period* equal to 2.

But at a value between $R = 3.4$ and $R = 3.5$ an abrupt change occurs. Given any value of x_0, the system will eventually reach an oscillation among *four* distinct values instead of two. For example, if we set $R = 3.49, x_0 = 0.2$, we see the results in figure 2.9.

Indeed, the values of x fairly quickly reach an oscillation among four different values (which happen to be approximately 0.872, 0.389, 0.829, and 0.494, if you're interested). That is, at some R between 3.4 and 3.5, the period of the final oscillation has abruptly doubled from 2 to 4.

Somewhere between $R = 3.54$ and $R = 3.55$ the period abruptly doubles again, jumping to 8. Somewhere between 3.564 and 3.565 the period jumps to 16. Somewhere between 3.5687 and 3.5688 the period jumps to 32. The period doubles again and again after smaller and smaller increases in R until, in short order, the period becomes effectively infinite, at an R value of approximately 3.569946. Before this point, the behavior of the logistic map was roughly predictable. If you gave me the value for R, I could tell you the ultimate long-term behavior from any starting point x_0: fixed points are reached when R is less than about 3.1, period-two oscillations are reached when R is between 3.1 and 3.4, and so on.

When R is approximately 3.569946, the values of x no longer settle into an oscillation; rather, they become chaotic. Here's what this means. Let's call the series of values x_0, x_1, x_2, and so on the *trajectory* of x. At values of

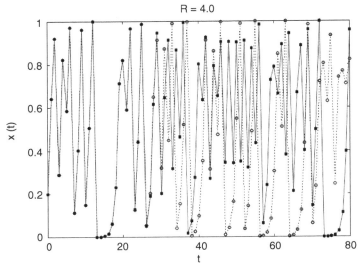

FIGURE 2.10. Two trajectories of the logistic map for $R = 4.0$: $x_0 = 0.2$ and $x_0 = 0.2000000001$.

R that yield chaos, two trajectories starting from very similar values of x_0, rather than converging to the same fixed point or oscillation, will instead progressively diverge from each other. At $R = 3.569946$ this divergence occurs very slowly, but we can see a more dramatic sensitive dependence on x_0 if we set $R = 4.0$. First I set $x_0 = 0.2$ and iterate the logistic map to obtain a trajectory. Then I restarted with a new x_0, increased slightly by putting a 1 in the tenth decimal place, $x_0 = 0.2000000001$, and iterated the map again to obtain a second trajectory. In figure 2.10 the first trajectory is the dark curve with black circles, and the second trajectory is the light line with open circles.

The two trajectories start off very close to one another (so close that the first, solid-line trajectory blocks our view of the second, dashed-line trajectory), but after 30 or so iterations they start to diverge significantly, and soon after there is no correlation between them. This is what is meant by "sensitive dependence on initial conditions."

So far we have seen three different classes of final behavior (attractors): fixed-point, periodic, and chaotic. (Chaotic attractors are also sometimes called "strange attractors.") *Type of attractor* is one way in which dynamical systems theory characterizes the behavior of a system.

Let's pause a minute to consider how remarkable the chaotic behavior really is. The logistic map is an extremely simple equation and is completely deterministic: every x_t maps onto one and only one value of x_{t+1}. And yet the chaotic trajectories obtained from this map, at certain values of R, look very random—enough so that the logistic map has been used as a basis for generating pseudo-random numbers on a computer. Thus apparent randomness can arise from very simple deterministic systems.

Moreover, for the values of R that produce chaos, if there is any uncertainty in the initial condition x_0, there exists a time beyond which the future value cannot be predicted. This was demonstrated above with $R = 4$. If we don't know the value of the tenth and higher decimal places of x_0—a quite likely limitation for many experimental observations—then by $t = 30$ or so the value of x_t is unpredictable. For any value of R that yields chaos, uncertainty in any decimal place of x_0, however far out in the decimal expansion, will result in unpredictability at some value of t.

Robert May, the mathematical biologist, summed up these rather surprising properties, echoing Poincaré:

> The fact that the simple and deterministic equation (1) [i.e., the logistic map] can possess dynamical trajectories which look like some sort of random noise has disturbing practical implications. It means, for example, that apparently erratic fluctuations in the census data for an animal population need not necessarily betoken either the vagaries of an unpredictable environment or sampling errors: they may simply derive from a rigidly deterministic population growth relationship such as equation (1).... Alternatively, it may be observed that in the chaotic regime arbitrarily close initial conditions can lead to trajectories which, after a sufficiently long time, diverge widely. This means that, even if we have a simple model in which all the parameters are determined exactly, long-term prediction is nevertheless impossible.

In short, the presence of chaos in a system implies that perfect prediction *à la* Laplace is impossible not only in practice but also *in principle*, since we can never know x_0 to infinitely many decimal places. This is a profound negative result that, along with quantum mechanics, helped wipe out the optimistic nineteenth-century view of a clockwork Newtonian universe that ticked along its predictable path.

But is there a more positive lesson to be learned from studies of the logistic map? Can it help the goal of dynamical systems theory, which attempts to discover general principles concerning systems that change over time? In fact, deeper studies of the logistic map and related maps have resulted in an equally surprising and profound positive result—the discovery of universal characteristics of chaotic systems.

Universals in Chaos

The term *chaos*, as used to describe dynamical systems with sensitive dependence on initial conditions, was first coined by physicists T. Y. Li and James Yorke. The term seems apt: the colloquial sense of the word "chaos" implies randomness and unpredictability, qualities we have seen in the chaotic version of logistic map. However, unlike colloquial chaos, there turns out to be substantial order in mathematical chaos in the form of so-called *universal* features that are common to a wide range of chaotic systems.

THE FIRST UNIVERSAL FEATURE: THE PERIOD-DOUBLING ROUTE TO CHAOS

In the mathematical explorations we performed above, we saw that as R was increased from 2.0 to 4.0, iterating the logistic map for a given value of R first yielded a fixed point, then a period-two oscillation, then period four, then eight, and so on, until chaos was reached. In dynamical systems theory, each of these abrupt period doublings is called a *bifurcation*. This succession of bifurcations culminating in chaos has been called the "period doubling route to chaos."

These bifurcations are often summarized in a so-called bifurcation diagram that plots the attractor the system ends up in as a function of the value of a "control parameter" such as R. Figure 2.11 gives such a bifurcation diagram

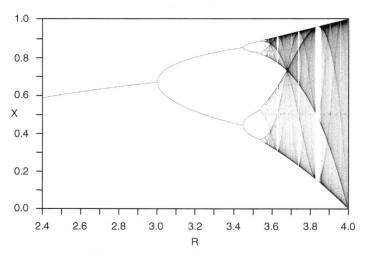

FIGURE 2.11. Bifurcation diagram for the logistic map, with attractor plotted as a function of R.

for the logistic map. The horizontal axis gives R. For each value of R, the final (attractor) values of x are plotted. For example, for $R = 2.9$, x reaches a fixed-point attractor of $x = 0.655$. At $R = 3.0$, x reaches a period-two attractor. This can be seen as the first branch point in the diagram, when the fixed-point attractors give way to the period-two attractors. For R somewhere between 3.4 and 3.5, the diagram shows a bifurcation to a period-four attractor, and so on, with further period doublings, until the onset of chaos at R approximately equal to 3.569946.

The period-doubling route to chaos has a rich history. Period doubling bifurcations had been observed in mathematical equations as early as the 1920s, and a similar cascade of bifurcations was described by P. J. Myrberg, a Finnish mathematician, in the 1950s. Nicholas Metropolis, Myron Stein, and Paul Stein, working at Los Alamos National Laboratory, showed that not just the logistic map but *any* map whose graph is parabola-shaped will follow a similar period-doubling route. Here, "parabola-shaped" means that plot of the map has just one hump—in mathematical terms, it is "unimodal."

THE SECOND UNIVERSAL FEATURE: FEIGENBAUM'S CONSTANT

The discovery that gave the period-doubling route its renowned place among mathematical universals was made in the 1970s by the physicist Mitchell Feigenbaum. Feigenbaum, using only a programmable calculator, made a list of the R values at which the period-doubling bifurcations occur (where $\approx$ means "approximately equal to"):

$R_1 \approx 3.0$
$R_2 \approx 3.44949$
$R_3 \approx 3.54409$
$R_4 \approx 3.564407$
$R_5 \approx 3.568759$
$R_6 \approx 3.569692$
$R_7 \approx 3.569891$
$R_8 \approx 3.569934$
$\vdots$
$R_\infty \approx 3.569946$

Here, R_1 corresponds to period $2^1 (= 2)$, R_2 corresponds to period $2^2 (= 4)$, and in general, R_n corresponds to period 2^n. The symbol ∞

("infinity") is used to denote the onset of chaos—a trajectory with an infinite period.

Feigenbaum noticed that as the period increases, the R values get closer and closer together. This means that for each bifurcation, R has to be increased less than it had before to get to the next bifurcation. You can see this in the bifurcation diagram of Figure 2.11: as R increases, the bifurcations get closer and closer together. Using these numbers, Feigenbaum measured the *rate* at which the bifurcations get closer and closer; that is, the rate at which the R values *converge*. He discovered that the rate is (approximately) the constant value 4.6692016. What this means is that as R increases, each new period doubling occurs about 4.6692016 times faster than the previous one.

This fact was interesting but not earth-shaking. Things started to get a lot more interesting when Feigenbaum looked at some other maps—the logistic map is just one of many that have been studied. As I mentioned above, a few years before Feigenbaum made these calculations, his colleagues at Los Alamos, Metropolis, Stein, and Stein, had shown that any unimodal map will follow a similar period-doubling cascade. Feigenbaum's next step was to calculate the rate of convergence for some other unimodal maps. He started with the so-called sine map, an equation similar to the logistic map but which uses the trigonometric sine function.

Feigenbaum repeated the steps I sketched above: he calculated the values of R at the period-doubling bifurcations in the sine map, and then calculated the rate at which these values converged. He found that the rate of convergence was 4.6692016.

Feigenbaum was amazed. The rate was the same. He tried it for other unimodal maps. It was still the same. No one, including Feigenbaum, had expected this at all. But once the discovery had been made, Feigenbaum went on to develop a mathematical theory that explained why the common value of 4.6692016, now called *Feigenbaum's constant*, is universal—which here means the same for all unimodal maps. The theory used a sophisticated mathematical technique called *renormalization* that had been developed originally in the area of quantum field theory and later imported to another field of physics: the study of phase transitions and other "critical phenomena." Feigenbaum adapted it for dynamical systems theory, and it has become a cornerstone in the understanding of chaos.

It turned out that this is not just a mathematical curiosity. In the years since Feigenbaum's discovery, his theory has been verified in several laboratory experiments on physical dynamical systems, including fluid flow, electronic circuits, lasers, and chemical reactions. Period-doubling cascades

Mitchell Feigenbaum (AIP Emilio Segre Visual Archives, *Physics Today* Collection)

have been observed in these systems, and values of Feigenbaum's constant have been calculated in steps similar to those we saw above. It is often quite difficult to get accurate measurements of, say, what corresponds to R values in such experiments, but even so, the values of Feigenbaum's constant found by the experimenters agree well within the margin of error to Feigenbaum's value of approximately 4.6692016. This is impressive, since Feigenbaum's theory, which yields this number, involves only abstract math, no physics. As Feigenbaum's colleague Leo Kadanoff said, this is "the best thing that can happen to a scientist, realizing that something that's happened in his or her mind exactly corresponds to something that happens in nature."

Large-scale systems such as the weather are, as yet, too hard to experiment with directly, so no one has *directly* observed period doubling or chaos in their behavior. However, certain computer models of weather have displayed the period-doubling route to chaos, as have computer models of electrical power systems, the heart, solar variability, and many other systems.

There is one more remarkable fact to mention about this story. Similar to many important scientific discoveries, Feigenbaum's discoveries were also made, independently and at almost the same time, by another research team.

This team consisted of the French scientists Pierre Coullet and Charles Tresser, who also used the technique of renormalization to study the period-doubling cascade and discovered the universality of 4.6692016 for unimodal maps. Feigenbaum may actually have been the first to make the discovery and was also able to more widely and clearly disseminate the result among the international scientific community, which is why he has received most of the credit for this work. However, in many technical papers, the theory is referred to as the "Feigenbaum-Coullet-Tresser theory" and Feigenbaum's constant as the "Feigenbaum-Coullet-Tresser constant." In the course of this book I point out several other examples of independent, simultaneous discoveries using ideas that are "in the air" at a given time.

Revolutionary Ideas from Chaos

The discovery and understanding of chaos, as illustrated in this chapter, has produced a rethinking of many core tenets of science. Here I summarize some of these new ideas, which few nineteenth-century scientists would have believed.

- Seemingly random behavior can emerge from deterministic systems, with no external source of randomness.
- The behavior of some simple, deterministic systems can be impossible, *even in principle*, to predict in the long term, due to sensitive dependence on initial conditions.
- Although the detailed behavior of a chaotic system cannot be predicted, there is some "order in chaos" seen in universal properties common to large sets of chaotic systems, such as the period-doubling route to chaos and Feigenbaum's constant. Thus even though "prediction becomes impossible" at the detailed level, there are some higher-level aspects of chaotic systems that are indeed predictable.

In summary, changing, hard-to-predict macroscopic behavior is a hallmark of complex systems. Dynamical systems theory provides a mathematical vocabulary for characterizing such behavior in terms of bifurcations, attractors, and universal properties of the ways systems can change. This vocabulary is used extensively by complex systems researchers.

The logistic map is a simplified model of population growth, but the detailed study of it and similar model systems resulted in a major revamping of the scientific understanding of order, randomness, and predictability. This illustrates the power of *idea models*—models that are simple enough to study via mathematics or computers but that nonetheless capture fundamental properties of natural complex systems. Idea models play a central role in this book, as they do in the sciences of complex systems.

Characterizing the dynamics of a complex system is only one step in understanding it. We also need to understand how these dynamics are used in living systems to process information and adapt to changing environments. The next three chapters give some background on these subjects, and later in the book we see how ideas from dynamics are being combined with ideas from information theory, computation, and evolution.

CHAPTER 3 | Information

The law that entropy increases—the Second Law of Thermodynamics—holds, I think, the supreme position among the laws of Nature...[I]f your theory is found to be against the Second Law of Thermodynamics I can give you no hope; there is nothing for it but to collapse in deepest humiliation.
—Sir Arthur Eddington, *The Nature of the Physical World*

COMPLEX SYSTEMS ARE OFTEN said to be "self-organizing": consider, for example, the strong, structured bridges made by army ants; the synchronous flashing of fireflies; the mutually sustaining markets of an economy; and the development of specialized organs by stem cells—all are examples of self-organization. Order is created out of disorder, upending the usual turn of events in which order decays and disorder (or *entropy*) wins out.

A complete account of how such entropy-defying self-organization takes place is the holy grail of complex systems science. But before this can be tackled, we need to understand what is meant by "order" and "disorder" and how people have thought about measuring such abstract qualities.

Many complex systems scientists use the concept of *information* to characterize and measure order and disorder, complexity and simplicity. The immunologist Irun Cohen states that "complex systems sense, store, and deploy more information than do simple systems." The economist Eric Beinhocker writes that "evolution can perform its tricks not just in the 'substrate' of DNA but in any system that has the right information processing and information storage characteristics." The physicist Murray Gell-Mann said of complex adaptive systems that "Although they differ widely in their physical attributes, they resemble one another in the way they handle information. That common feature is perhaps the best starting point for exploring how they operate."

But just what is meant by "information"?

What Is Information?

You see the word "information" all over the place these days: the "information revolution," the "information age," "information technology" (often simply "IT"), the "information superhighway," and so forth. "Information" is used colloquially to refer to any medium that presents knowledge or facts: newspapers, books, my mother on the phone gossiping about relatives, and, most prominently these days, the Internet. More technically, it is used to describe a vast array of phenomena ranging from the fiber-optic transmissions that constitute signals from one computer to another on the Internet to the tiny molecules that neurons use to communicate with one another in the brain.

The different examples of complex systems I described in chapter 1 are all centrally concerned with the communication and processing of information in various forms. Since the beginning of the computer age, computer scientists have thought of information transmission and computation as something that takes place not only in electronic circuits but also in living systems.

In order to understand the information and computation in these systems, the first step, of course, is to have a precise definition of what is meant by the terms *information* and *computation*. These terms have been mathematically defined only in the twentieth century. Unexpectedly, it all began with a late nineteenth-century puzzle in physics involving a very smart "demon" who seemed to get a lot done without expending any energy. This little puzzle got many physicists quite worried that one of their fundamental laws might be wrong. How did the concept of *information* save the day? Before getting there, we need a little bit of background on the physics notions of *energy*, *work*, and *entropy*.

Energy, Work, and Entropy

The scientific study of information really begins with the science of thermodynamics, which describes energy and its interactions with matter. Physicists of the nineteenth century considered the universe to consist of two different types of entities: *matter* (e.g., solids, liquids, and vapors) and *energy* (e.g., heat, light, and sound).

Energy is roughly defined as a system's potential to "do work," which correlates well with our intuitive notion of energy, especially in this age of high-energy workaholics. The origin of the term is the Greek word, *energeia*, which roughly means "work." However, physicists have a specific meaning of "work" done by an object: the amount of force applied to the object multiplied by the distance traveled by the object in the direction that force was applied.

For example, suppose your car breaks down on a flat road and you have to push it for a quarter of a mile to the nearest gas station. In physics terms, the amount of work that you expend is the amount of force with which you push the car multiplied by the distance to the gas station. In pushing the car, you transform energy stored in your body into the kinetic energy (i.e., movement) of the car, and the amount of energy that is transformed is equal to the amount of work that is done plus whatever energy is converted to heat, say, by the friction of the wheels on the road, or by your own body warming up. This so-called heat loss is measured by a quantity called *entropy*. Entropy is a measure of the energy that cannot be converted into additional work. The term "entropy" comes from another Greek word—"trope"—meaning "turning into" or "transformation."

By the end of the nineteenth century two fundamental laws concerning energy had been discovered, the so-called *laws of thermodynamics*. These laws apply to "isolated systems"—ones that do not exchange energy with any outside entity.

> First law: *Energy is conserved.* The total amount of energy in the universe is constant. Energy can be transformed from one form to another, such as the transformation of stored body energy to kinetic energy of a pushed car plus the heat generated by this action. However, energy can never be created or destroyed. Thus it is said to be "conserved."
>
> Second law: *Entropy always increases until it reaches a maximum value.* The total entropy of a system will always increase until it reaches its maximum possible value; it will never decrease on its own unless an outside agent works to decrease it.

As you've probably noticed, a room does not clean itself up, and Cheerios spilled on the floor, left to their own devices, will never find their way back into the cereal box. Someone or something has to do work to turn disorder into order.

Furthermore, transformations of energy, such as the car-pushing example above, will always produce some heat that cannot be put to work. This is why, for example, no one has found a way to take the heat generated by the back of your refrigerator and use it to produce new power for cooling the inside of the refrigerator so that it will be able to power itself. This explains why the proverbial "perpetual motion machine" is a myth.

The second law of thermodynamics is said to define the "arrow of time," in that it proves there are processes that cannot be reversed in time (e.g., heat spontaneously returning to your refrigerator and converting to electrical energy to cool the inside). The "future" is defined as the direction of time in which entropy increases. Interestingly, the second law is the only fundamental law of physics that distinguishes between past and future. All other laws are reversible in time. For example, consider filming an interaction between elementary particles such as electrons, and then showing this movie to a physicist. Now run the movie backward, and ask the physicist which version was the "real" version. The physicist won't be able to guess, since the forward and backward interactions both obey the laws of physics. This is what *reversible* means. In contrast, if you make an infrared film of heat being produced by your refrigerator, and show it forward and backward, any physicist will identify the forward direction as "correct" since it obeys the second law, whereas the backward version does not. This is what *irreversible* means. Why is the second law different from all other physical laws? This is a profound question. As the physicist Tony Rothman points out, "Why the second law should distinguish between past and future while all the other laws of nature do not is perhaps the greatest mystery in physics."

Maxwell's Demon

The British physicist James Clerk Maxwell is most famous for his discovery of what are now called Maxwell's Equations: compact expressions of Maxwell's theory that unified electricity and magnetism. During his lifetime, he was one of the world's most highly regarded scientists, and today would be on any top fifty list of all-time greats of science.

In his 1871 book, *Theory of Heat*, Maxwell posed a puzzle under the heading "Limitation of the Second Law of Thermodynamics." Maxwell proposed a box that is divided into two halves by a wall with a hinged door, as illustrated in figure 3.1. The door is controlled by a "demon," a very small being who

FIGURE 3.1. Top: James Clerk Maxwell, 1831–1879 (AIP Emilio Segre Visual Archives) Bottom: Maxwell's Demon, who opens the door for fast (white) particles moving to the left and for slow (black) particles moving to the right.

measures the velocity of air molecules as they whiz past him. He opens the door to let the fast ones go from the right side to the left side, and closes it when slow ones approach it from the right. Likewise, he opens the door for slow molecules moving from left to right and closes it when fast molecules approach it from the left. After some time, the box will be well organized, with all the fast molecules on the left and all the slow ones on the right. Thus entropy will have been decreased.

According to the second law, work has to be done to decrease entropy. What work has been done by the demon? To be sure, he has opened and closed the door many times. However, Maxwell assumed that a massless and frictionless "slide" could be used as a door by the demon, so that opening and closing it would require negligible work, which we can ignore. (Feasible designs for such a door have been proposed.) Has any other work been done by the demon?

Maxwell's answer was no: "the hot system [the left side] has gotten hotter and the cold [right side] colder and yet no work has been done, only the intelligence of a very observant and neat-fingered being has been employed."

How did entropy decrease with little or no work being done? Doesn't this directly violate the second law of thermodynamics? Maxwell's demon puzzled many of the great minds of the late nineteenth and early twentieth centuries. Maxwell's own answer to his puzzle was that the second law (the increase of entropy over time) is not really a law at all, but rather a statistical effect that holds for large collections of molecules, like the objects we encounter in day-to-day life, but does not necessarily hold at the scale of individual molecules.

However, many physicists of his day and long after vehemently disagreed. They believed that the second law has to remain inviolate; instead there must be something fishy about the demon. For entropy to decrease, work must actually have been done in some subtle, nonapparent way.

Many people tried to resolve the paradox, but no one was able to offer a satisfactory solution for nearly sixty years. In 1929, a breakthrough came: the great Hungarian physicist Leo Szilard (pronounced "siLARD") proposed that it is the "intelligence" of the demon, or more precisely, the act of obtaining information through measurement, that constitutes the missing work.

Szilard was the first to make a link between entropy and *information*, a link that later became the foundation of information theory and a key idea in complex systems. In a famous paper entitled "On the Decrease of Entropy in a Thermodynamic System by the Intervention of Intelligent Beings," Szilard argued that the measurement process, in which the demon acquires a single "bit" of information (i.e., the information as to whether an approaching molecule is a slow one or a fast one) requires energy and must produce at least as much entropy as is decreased by the sorting of that molecule into the left or right side of the box. Thus the entire system, comprising the box, the molecules, and the demon, obeys the second law of thermodynamics.

In coming up with his solution, Szilard was perhaps the first to define the notion of a *bit of information*—the information obtained from the answer to a yes/no (or, in the demon's case, "fast/slow") question.

Leo Szilard, 1898–1964 (AIP Emilio Segre Visual Archives)

From our twenty-first-century vantage, it may seem obvious (or at least unsurprising) that the acquisition of information requires expenditure of work. But at the time of Maxwell, and even sixty years later when Szilard wrote his famous paper, there was still a strong tendency in people's minds to view physical and mental processes as completely separate. This highly ingrained intuition may be why Maxwell, as astute as he was, did not see the "intelligence" or "observing powers" of the demon as relating to the thermodynamics of the box-molecules-demon system. Such relationships between information and physics became clear only in the twentieth century, beginning with the discovery that the "observer" plays a key role in quantum mechanics.

Szilard's theory was later extended and generalized by the French physicists Leon Brillouin and Denis Gabor. Many scientists of the 1950s and later believed that Brillouin's theory in particular had definitively finished off the demon by demonstrating in detail how making a measurement entails an increase of entropy.

However, it wasn't over yet. Fifty years after Szilard's paper, it was discovered that there were some holes in Szilard's and Brillouin's solutions as well. In the 1980s, the mathematician Charles Bennett showed that there are very clever ways to observe and remember information—in the demon's case, whether an air molecule is fast or slow—*without* increasing entropy. Bennett's remarkable demonstration of this formed the basis for *reversible computing*, which says that, in theory, any computation can be done without expending energy. Bennett's discoveries might seem to imply that we are back at square one with the demon, since measurement can, in fact, be done without increasing entropy. However, Bennett noted that the second law of thermodynamics was saved again by an earlier discovery made in the 1960s by physicist Rolf Landauer: it is not the act of measurement, but rather the act of *erasing* memory that necessarily increases entropy. Erasing memory is not reversible; if there is true erasure, then once the information is gone, it cannot be restored without additional measurement. Bennett showed that for the demon to work, its memory must be erased at some point, and when it is, the physical act of this erasure will produce heat, thus increasing entropy by an amount exactly equal to the amount entropy was decreased by the demon's sorting actions.

Landauer and Bennett's solution to the paradox of Maxwell's demon fixed holes in Szilard's solution, but it was in the same spirit: the demon's act of measurement and decision making, which requires erasure, will inevitably increase entropy, and the second law is saved. (I should say here that there are still some physicists who don't buy the Landauer and Bennett solution; the demon remains controversial to this day.)

Maxwell invented his demon as a simple thought experiment to demonstrate his view that the second law of thermodynamics was not a law but a statistical effect. However, like many of the best thought-experiments in science, the demon's influence was much broader: resolutions to the demon paradox became the foundations of two new fields: information theory and the physics of information.

Statistical Mechanics in a Nutshell

In an earlier section, I defined "entropy" as a measure of the energy that cannot be converted into additional work but is instead transformed into heat. This notion of entropy was originally defined by Rudolph Clausius in 1865. At the time of Clausius, heat was believed to be a kind of fluid that could move from one system to another, and temperature was a property of a system that affected the flow of heat.

In the next few decades, a different view of heat emerged in the scientific community: systems are made up of molecules, and heat is a result of the motion, or kinetic energy, of those molecules. This new view was largely a result of the work of Ludwig Boltzmann, who developed what is now called *statistical mechanics*.

Ludwig Boltzmann, 1844–1906 (AIP Emilio Segre Visual Archives, Segre Collection)

Statistical mechanics proposes that large-scale properties (e.g., heat) emerge from microscopic properties (e.g., the motions of trillions of molecules). For example, think about a room full of moving air molecules. A *classical* mechanics analysis would determine the position and velocity of each molecule, as well as all the forces acting on that molecule, and would use this information to determine the future position and velocity of that molecule. Of course, if there are fifty quadrillion molecules, this approach would take rather a long time—in fact it always would be impossible, both in practice and, as quantum mechanics has shown, in principle. A *statistical* mechanics approach gives up on determining the exact position, velocity, and future behavior of each molecule and instead tries to predict the *average* positions and velocities of large *ensembles* of molecules.

In short, classical mechanics attempts to say something about every single microscopic entity (e.g., molecule) by using Newton's laws. Thermodynamics gives laws of macroscopic entities—heat, energy, and entropy—without acknowledging that any microscopic molecules are the source of these macroscopic entities. Statistical mechanics is a bridge between these two extremes, in that it explains how the behavior of the macroscopic entities arises from *statistics* of large ensembles of microscopic entities.

There is one problem with the statistical approach—it gives only the *probable* behavior of the system. For example, if all the air molecules in a room are flying around randomly, they are most likely to be spread out all over the room, and all of us will have enough air to breathe. This is what we predict and depend on, and it has never failed us yet. However, according to statistical mechanics, since the molecules are flying around randomly, there is some very small chance that at some point they will all fly over to the same corner at the same time. Then any person who happened to be in that corner would be crushed by the huge air pressure, and the rest of us would suffocate from lack of air. As far as I know, such an event has never happened in any room anywhere. However, there is nothing in Newton's laws that says it can't happen; it's just incredibly unlikely. Boltzmann reasoned that if there are enough microscopic entities to average over, his statistical approach will give the right answer virtually all the time, and indeed, in practice it does so. But at the time Boltzmann was formulating his new science, the suggestion that a physical law could apply only "virtually all of the time" rather than *exactly* all of the time was repellent to many other scientists. Furthermore, Boltzmann's insistence on the reality of microscopic entities such as molecules and atoms was also at odds with his colleagues. Some have speculated that the rejection of his ideas by most of his fellow scientists contributed to his suicide in 1906, at the age of 62. Only years after his death were his ideas generally accepted; he is now considered to be one of the most important scientists in history.

Microstates and Macrostates

Given a room full of air, at a given instant in time each molecule has a certain position and velocity, even if it is impossible to actually measure all of them. In statistical mechanics terminology, the particular collection of exact molecule positions and velocities at a given instant is called the *microstate* of the whole room at that instant. For a room full of air molecules randomly flying around, the most probable type of microstate at a given time is that the air molecules are spread uniformly around the room. The least probable type of microstate is that the air molecules are all clumped together as closely as possible in a single location, for example, the corner of the room. This seems simply obvious, but Boltzmann noted that the reason for this is that there are many more possible microstates of the system in which the air molecules are spread around uniformly than there are microstates in which they all are clumped together.

The situation is analogous to a slot machine with three rotating pictures (figure 3.2). Suppose each of the three pictures can come up "apple," "orange," "cherry," "pear," or "lemon." Imagine you put in a quarter, and pull the handle

FIGURE 3.2. Slot machine with three rotating fruit pictures, illustrating the concepts *microstate* and *macrostate*. (Drawing by David Moser.)

to spin the pictures. It is much more likely that the pictures will all be different (i.e., you lose your money) than that the pictures will all be the same (i.e., you win a jackpot). Now imagine such a slot machine with fifty quadrillion pictures, and you can see that the probability of all coming up the same is very close to zero, just like the probability of the air molecules ending up all clumped together in the same location.

A *type of microstate*, for example, "pictures all the same—you win" versus "pictures not all the same—you lose" or "molecules clumped together—we can't breathe" versus "molecules uniformly spread out—we can breathe," is called a *macrostate* of the system. A macrostate can correspond to many different microstates. In the slot machine, there are many different microstates consisting of three nonidentical pictures, each of which corresponds to the single "you lose" macrostate, and only a few microstates that correspond to the "you win" macrostate. This is how casinos are sure to make money. *Temperature* is a macrostate—it corresponds to many different possible microstates of molecules at different velocities that happen to average to the same temperature.

Using these ideas, Boltzmann interpreted the second law of thermodynamics as simply saying that an isolated system will more likely be in a more probable macrostate than in a less probable one. To our ears this sounds like a tautology but it was a rather revolutionary way of thinking about the point back then, since it included the notion of probability. Boltzmann defined the *entropy* of a macrostate as a function of the number of microstates that

FIGURE 3.3. Boltzmann's tombstone, in Vienna. (Photograph courtesy of Martin Roell.)

could give rise to that macrostate. For example, on the slot machine of figure 3.2, where each picture can come up "apple," "orange," "cherry," "pear," or "lemon," it turns out that there are a total of 125 possible combinations (microstates), out of which five correspond to the macrostate "pictures all the same—you win" and 120 correspond to the macrostate "pictures not all the same—you lose." The latter macrostate clearly has a higher Boltzmann entropy than the former.

Boltzmann's entropy obeys the second law of thermodynamics. Unless *work* is done, Boltzmann's entropy will always increase until it gets to a macrostate with highest possible entropy. Boltzmann was able to show that, under many conditions, his simple and intuitive definition of entropy is equivalent to the original definition of Clausius.

The actual equation for Boltzmann's entropy, now so fundamental to physics, appears on Boltzmann's tombstone in Vienna (figure 3.3).

Shannon Information

Many of the most basic scientific ideas are spurred by advances in technology. The nineteenth-century studies of thermodynamics were inspired and driven by the challenge of improving steam engines. The studies of information by mathematician Claude Shannon were likewise driven by the

Claude Shannon, 1916–2001. (Reprinted with permission of Alcatel-Lucent USA, Inc.)

twentieth-century revolution in communications—particularly the development of the telegraph and telephone. In the 1940s, Shannon adapted Boltzmann's ideas to the more abstract realm of communications. Shannon worked at Bell Labs, a part of the American Telephone and Telegraph Company (AT&T). One of the most important problems for AT&T was to figure out how to transmit signals more quickly and reliably over telegraph and telephone wires.

Shannon's mathematical solution to this problem was the beginning of what is now called *information theory*. In his 1948 paper "A Mathematical Theory of Communication," Shannon gave a narrow definition of information and proved a very important theorem, which gave the maximum possible transmission rate of information over a given channel (wire or other medium), even if there are errors in transmission caused by noise on the channel. This maximum transmission rate is called the *channel capacity*.

Shannon's definition of information involves a *source* that sends *messages* to a *receiver*. For example, figure 3.4 shows two examples of a source talking to a receiver on the phone. Each word the source says can be considered a *message* in the Shannon sense. Just as the telephone doesn't understand the words being said on it but only transmits the electrical pulses used to encode the voice, Shannon's definition of information completely ignores the *meaning* of the messages and takes into account only how often the source sends each of the possible different messages to the receiver.

52 | BACKGROUND AND HISTORY

FIGURE 3.4. Top: Information content (zero) of Nicky's conversation with Grandma. Bottom: Higher information content of Jake's conversation with Grandma. (Drawings by David Moser.)

Shannon asked, "How much information is transmitted by a source sending messages to a receiver?" In analogy with Boltzmann's ideas, Shannon defined the information of a macrostate (here, a source) as a function of the number of possible microstates (here, ensembles of possible messages) that could be sent by that source. When my son Nicky was barely a toddler, I would put him on the phone to talk with Grandma. He loved to talk on the phone, but could say only one word—"da." His messages to Grandma were "da da da da da...." In other words, the Nicky-macrostate had only one possible microstate (sequences of "da"s), and although the macrostate was cute, the information content was, well, zero. Grandma knew just what to expect. My son Jake, two years older, also loved to talk on the phone but had a much bigger vocabulary and would tell Grandma all about his activities, projects, and adventures, constantly surprising her with his command of language. Clearly the information content of the Jake-source was much higher, since so many microstates—i.e., more different collections of messages—could be produced.

Shannon's definition of information content was nearly identical to Boltzmann's more general definition of entropy. In his classic 1948 paper, Shannon defined the information content in terms of the *entropy* of the message source. (This notion of entropy is often called *Shannon entropy* to distinguish it from the related definition of entropy given by Boltzmann.)

People have sometimes characterized Shannon's definition of information content as the "average amount of surprise" a receiver experiences on receiving a message, in which "surprise" means something like the "degree of uncertainty" the receiver had about what the source would send next. Grandma is clearly more surprised at each word Jake says than at each word Nicky says, since she already knows exactly what Nicky will say next but can't as easily predict what Jake will say next. Thus each word Jake says gives her a higher average "information content" than each word Nicky says.

In general, in Shannon's theory, a message can be any unit of communication, be it a letter, a word, a sentence, or even a single bit (a zero or a one). Once again, the entropy (and thus information content) of a source is defined in terms of message probabilities and is not concerned with the "meaning" of a message.

Shannon's results set the stage for applications in many different fields. The best-known applications are in the field of coding theory, which deals with both data compression and the way codes need to be structured to be reliably transmitted. Coding theory affects nearly all of our electronic communications; cell phones, computer networks, and the worldwide global positioning system are a few examples.

Information theory is also central in cryptography and in the relatively new field of bioinformatics, in which entropy and other information theory measures are used to analyze patterns in gene sequences. It has also been applied to analysis of language and music and in psychology, statistical inference, and artificial intelligence, among many other fields. Although information theory was inspired by notions of entropy in thermodynamics and statistical mechanics, it is controversial whether or not information theory has had much of a reverse impact on those and other fields of physics. In 1961, communications engineer and writer John Pierce quipped that "efforts to marry communication theory and physics have been more interesting than fruitful." Some physicists would still agree with him. However, there are a number of new approaches to physics based on concepts related to Shannon's information theory (e.g., quantum information theory and the physics of information) that are beginning to be fruitful as well as interesting.

As you will see in subsequent chapters, information theoretic notions such as entropy, information content, mutual information, information dynamics, and others have played central though controversial roles in attempts to define the notion of complexity and in characterizing different types of complex systems.

CHAPTER 4 | Computation

Quo facto, quando orientur controversiae, non magis disputatione opus erit inter duos philosophos, quam inter duos Computistas. Sufficiet enim calamos in manus sumere sedereque ad abacos, et sibi mutuo dicere: Calculemus!

[*If controversies were to arise, there would be no more need of disputation between two philosophers than between two accountants. For it would suffice to take their pencils in their hands, to sit down to their slates, and say to each other, "Let us calculate."*]

—G. Leibniz (Trans. B. Russell)

PEOPLE USUALLY THINK of computation as the thing that a computer does, such as calculations in spreadsheets, word processing, e-mail, and the like. And they usually think of a computer as the machine on one's desk (or lap) that has electronic circuits inside, and usually a color monitor and a mouse, and that in the distant past used old-fashioned technology such as vacuum tubes. We also have a vague idea that our brains themselves are roughly like computers, with logic, memory, input, and output.

However, if you peruse some of the scholarly literature or seminar titles in complex systems, you will find the word *computation* used in some rather unfamiliar contexts: a biology book on "computation in cells and tissues"; a keynote lecture on "immune system computation"; an economics lecture concerning "the nature and limits of distributed computation in markets"; an article in a prestigious science journal on "emergent computation in plants." And this is just a small sampling of such usage.

The notion of computation has come a long way since the early days of computers, and many scientists now view the phenomenon of computation as being widespread in nature. It is clear that cells, tissues, plants, immune systems, and financial markets do not work anything like the computer on your desk, so what exactly do these people mean by *computation*, and why do they call it that?

In order to set the stage for addressing this question in chapter 12, this chapter gives an overview of the history of ideas about computation and what can be computed, and describes basics of computational concepts used by scientists to understand natural complex systems.

What Is Computation and What Can Be Computed?

Information, as narrowly defined by Shannon, concerns the predictability of a message source. In the real world, however, information is something that is analyzed for meaning, that is remembered and combined with other information, and that produces results or actions. In short, information is *processed* via computation.

The meaning of *computation* has changed dramatically over the years. Before the late 1940s, *computing* meant performing mathematical calculations by hand (what nineteenth-century British schoolboys would have called "doing sums"). *Computers* were people who did such calculations. One of my former professors, Art Burks, used to tell us how he had married a "computer"—the term used for women who were enlisted during World War II to hand-calculate ballistic trajectories. Alice Burks was working as such a computer when she met Art.

Nowadays computation is what computers of the electronic variety do and what natural complex systems seem to do as well. But what exactly *is* computation, and how much can it accomplish? Can a computer compute anything, in principle, or does it have any limits? These are questions that were answered only in the mid-twentieth century.

Hilbert's Problems and Gödel's Theorem

The study of the foundations and limitations of computation, which led to the invention of electronic computers, was developed in response to a set of seemingly abstract (and abstruse) math problems. These problems were posed in the year 1900 at the International Congress of Mathematicians in Paris by the German mathematician David Hilbert.

David Hilbert, 1862–1943
(AIP Emilio Segre Visual
Archives, Lande Collection)

Hilbert's lecture at this congress set out a list of mathematical New Year's resolutions for the new century in the form of twenty-three of the most important unsolved problems in mathematics. Problems 2 and 10 ended up making the biggest splash. Actually, these were not just problems in mathematics; they were problems *about* mathematics itself and what can be proved by using mathematics. Taken together, these problems can be stated in three parts as follows:

1. *Is mathematics complete?* That is, can every mathematical statement be proved or disproved from a given finite set of axioms?
   For example, remember Euclid's axioms from high-school geometry? Remember using these axioms to prove statements such as "the angles in a triangle sum to 180 degrees"? Hilbert's question was this: Given some fixed set of axioms, is there a proof for every true statement?
2. *Is mathematics consistent?* In other words, can only the *true* statements be proved? The notion of what is meant by "true statement" is technical, but I'll leave it as intuitive. For example, if we could prove some false statement, such as $1 + 1 = 3$, mathematics would be inconsistent and in big trouble.
3. *Is every statement in mathematics decidable?* That is, is there a definite procedure that can be applied to every statement that will tell us in finite time whether or not the statement is true or false? The idea here is that you could come up with a mathematical statement such as, "Every even integer greater than 2 can be expressed as the sum of two prime

Kurt Gödel, 1906–1978
(Photograph courtesy of
Princeton University Library)

numbers," hand it to a mathematician (or a computer), who would apply a precise recipe (a "definite procedure"), which would yield the correct answer "true" or "false" in finite time.

The last question is known by its German name as the *Entscheidungsproblem* ("decision problem"), and goes back to the seventeenth-century mathematician Gottfried Leibniz. Leibniz actually built his own calculating machine, and believed that humans could eventually build a machine that could determine the truth or falsity of any mathematical statement.

Up until 1930, these three problems remained unsolved, but Hilbert expressed confidence that the answer would be "yes" in each case, and asserted that "there is no such thing as an unsolvable problem."

However, his optimism turned out to be short-lived. Very short-lived. At the same meeting in 1930 at which Hilbert made his confident assertion, a twenty-five-year-old mathematician named Kurt Gödel astounded the mathematical world by presenting a proof of the so-called incompleteness theorem. This theorem stated that if the answer to question 2 above is "yes" (i.e., mathematics is consistent), then the answer to question 1 (is mathematics complete?) has to be "no."

Gödel's incompleteness theorem looked at arithmetic. He showed that if arithmetic is consistent, then there are true statements in arithmetic that cannot be proved—that is, arithmetic is incomplete. If arithmetic were inconsistent, then there would be false statements that could be proved, and all of mathematics would come crashing down.

Gödel's proof is complicated. However, intuitively, it can be explained very easily. Gödel gave an example of a mathematical statement that can be translated into English as: "This statement is not provable."

Think about it for a minute. It's a strange statement, since it talks about itself—in fact, it asserts that it is not provable. Let's call this statement "Statement A." Now, suppose Statement A could indeed be proved. But then it would be false (since it states that it cannot be proved). That would mean a false statement could be proved—arithmetic would be inconsistent. Okay, let's assume the opposite, that Statement A cannot be proved. That would mean that Statement A is true (because it asserts that it cannot be proved), but then there is a true statement that cannot be proved—arithmetic would be incomplete. Ergo, arithmetic is either inconsistent or incomplete.

It's not easy to imagine how this statement gets translated into mathematics, but Gödel did it—therein lies the complexity and brilliance of Gödel's proof, which I won't cover here.

This was a big blow for the large number of mathematicians and philosophers who strongly believed that Hilbert's questions would be answered affirmatively. As the mathematician and writer Andrew Hodges notes: "This was an amazing new turn in the enquiry, for Hilbert had thought of his programme as one of tidying up loose ends. It was upsetting for those who wanted to find in mathematics something that was perfect and unassailable...."

Turing Machines and Uncomputability

While Gödel dispatched the first and second of Hilbert's questions, the British mathematician Alan Turing killed off the third.

In 1935 Alan Turing was a twenty-three-year-old graduate student at Cambridge studying under the logician Max Newman. Newman introduced Turing to Gödel's recent incompleteness theorem. When he understood Gödel's result, Turing was able to see how to answer Hilbert's third question, the *Entscheidungsproblem*, and his answer, again, was "no."

How did Turing show this? Remember that the *Entscheidungsproblem* asks if there is always a "definite procedure" for deciding whether a statement is

Alan Turing, 1912–1954
(Photograph copyright ©2003 by
Photo Researchers Inc. Reproduced
by permission.)

provable. What does "definite procedure" mean? Turing's first step was to define this notion. Following the intuition of Leibniz of more than two centuries earlier, Turing formulated his definition by thinking about a powerful calculating machine—one that could not only perform arithmetic but also could manipulate symbols in order to prove mathematical statements. By thinking about how humans might calculate, he constructed a mental design of such a machine, which is now called a *Turing machine*. The Turing machine turned out to be a blueprint for the invention of the electronic programmable computer.

A QUICK INTRODUCTION TO TURING MACHINES

As illustrated in figure 4.1, a Turing machine consists of three parts: (1) A tape, divided into squares (or "cells"), on which symbols can be written and from which symbols can be read. The tape is infinitely long in both directions. (2) A movable read/write tape head that reads symbols from the tape and writes symbols to the tape. At any time, the head is in one of a number of *states*. (3) A set of rules that tell the head what to do next.

The head starts out at a particular tape cell and in a special **start** state. At each time step, the head reads the symbol at its current tape cell. The head then follows the rule that corresponds to that symbol and the head's current

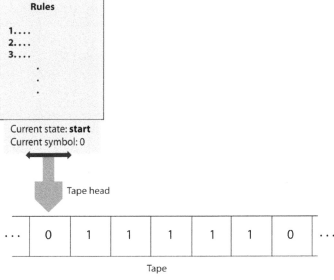

FIGURE 4.1. Illustration of a Turing machine.

state. The rule tells the head what symbol to write on the current tape cell (replacing the previous symbol); whether the head should move to the right, move to the left, or stay put; and what the head's new state is. When the head goes into a special **halt** state, the machine is done and stops.

The input to the machine is the set of symbols written on the tape before the machine starts. The output from the machine is the set of symbols written on the tape after the machine halts.

A SIMPLE EXAMPLE

A simple example will make all this clearer. To make things really simple, assume that (like a real computer) the only possible symbols that can be on a tape are 0, 1, and a *blank* symbol indicating a blank tape cell. Let's design a Turing Machine that reads a tape containing all blank cells except for some number of 1s sandwiched between exactly two 0s (e.g., 011110) and determines whether the number of 1s is even or odd. If even, then the final output of the machine will be a single 0 (and all other cells blank) on the tape. If odd, the final output will be a single 1 (and all other cells blank). Assume the input always has exactly two 0s on the tape, with zero or more 1s between them, and all blanks on either side of them.

Our Turing machine's head will have four possible states: **start**, **even**, **odd**, and **halt**. The head will start on the left-most 0 and in the **start** state. We will write rules that cause the head to move to the right, one cell at a time, replacing the 0s and 1s it encounters with blanks. If the head reads a 1 in the current cell, the head goes into the **odd** state and moves one cell to the right. If it reads a 1 again it goes into the **even** state and moves one cell to the right, and so on, switching between the **even** and **odd** states as 1s are read.

When the head reads a 0, it has come to the end of the input, and whatever state it is in, **odd** or **even**, is the correct one. The machine will then write a corresponding 0 or 1 in its current cell and go into the **halt** state.

Here are the rules for the tape head that implement this algorithm:

1. If you are in the **start** state and read a 0, then change to the **even** state, replace the 0 with a *blank* (i.e., erase the 0), and move one cell to the right.
2. If you are in the **even** state and read a 1, change to the **odd** state, replace the 1 with a *blank*, and move one cell to the right.
3. If you are in the **odd** state and read a 1, change to the **even** state, replace the 1 with a *blank*, and move one cell to the right.
4. If you are in the **odd** state and read a 0, replace that 0 with a 1 and change to the **halt** state.
5. If you are in the **even** state and read a 0, replace that 0 with a 0 (i.e., don't change it) and change to the **halt** state.

The process of starting with input symbols on the tape and letting the Turing machine serially process that input by using such rules is called "running the Turing machine on the input."

Definite Procedures Defined as Turing Machines

In the example above, assuming that the input is in the correct format of zero or more 1s sandwiched between two 0s, running this Turing machine on the input is guaranteed to produce a correct answer for any input (including the special case of zero 1s, which is considered to be an even number). Even though it seems kind of clunky, you have to admit that this process is a "definite procedure"—a precise set of steps—for solving the even/odd problem. Turing's first goal was to make very concrete this notion of *definite procedure*. The idea is that, given a particular problem to solve, you can construct a definite procedure for solving it by designing a Turing machine that solves it. Turing machines were put forth as the *definition* of "definite procedure," hitherto a vague and ill-defined notion.

When formulating these ideas, Turing didn't build any actual machines (though he built significant ones later on). Instead, all his thinking about Turing machines was done with pencil and paper alone.

Universal Turing Machines

Next, Turing proved an amazing fact about Turing machines: one can design a special *universal* Turing machine (let's call it U) that can emulate the workings of any other Turing machine. For U to emulate a Turing machine M running on an input I, U starts with part of its tape containing a sequence of 0s, 1s, and *blank*s that encodes input I, and part of its tape containing a sequence of 0s, 1s, and *blank*s that encodes machine M. The concept of *encoding* a machine M might not be familiar to you, but it is not hard. First, we recognize that all rules (like the five given in the "simple example" above) can be written in a shorthand that looks like

—Current State—Current Symbol—New State—New Symbol—Motion—

In this shorthand, Rule 1 above would be written as:

—**start**—0—**even**—*blank*—right—

(The separator '—' is not actually needed, but makes the rule easier for us to read.) Now to encode a rule, we simply assign different three-digit binary numbers to each of the possible states: for example, **start** = 000, **even** = 001, **odd** = 010, and **halt** = 100. Similarly, we can assign three-digit binary numbers to each of the possible symbols: for example, symbol '0' = 000, symbol '1' = 001, and symbol *blank* = 100. Any such binary numbers will do, as long as each one represents only one symbol, and we are consistent in using them. It's okay that the same binary number is used for, say, **start** and '0'; since we know the structure of the rule shorthand above, we will be able to tell what is being encoded from the context.

Similarly, we could encode "move right" by 000, and "move left" by 111. Finally, we could encode the '—' separator symbol above as 111. Then we could, for example, encode Rule 1 as the string

111000111000111001111100111000111

which just substitutes our codes for the states and symbols in the shorthand above. The other rules would also be encoded in this way. Put the encoding strings all together, one after another, to form a single, long string, called the *code* of Turing machine M. To have U emulate M running on input I, U's initial tape contains both input I and M's code. At each step U reads the current symbol in I from the input part of the tape, decodes the appropriate rule from the M part of the tape, and carries it out on the input part, all the while keeping track (on some other part of its tape) of what state M would be in if M were actually running on the given input.

When M would have reached the **halt** state, U also halts, with the input (now output) part of its tape now containing the symbols M would have on its tape after it was run on the given input I. Thus we can say that "U runs M on I." I am leaving out the actual states and rules of U, since they are pretty complicated, but I assure you that such a Turing machine can be designed. Moreover, what U does is precisely what a modern programmable computer does: it takes a program M that is stored in one part of memory, and runs it on an input I that is stored in another part of memory. It is notable that the first programmable computers were developed only ten years or so after Turing proved the existence of a universal Turing machine.

Turing's Solution to the Entscheidungsproblem

Recall the question of the *Entscheidungsproblem*: Is there always a definite procedure that can decide whether or not a statement is true?

It was the existence of a universal Turing machine that enabled Turing to prove that the answer is "no." He realized that the input I to a Turing machine M could be the code of another Turing machine. This is equivalent to the input of a computer program being the lines of another computer program. For example, you might write a program using a word processor such as Microsoft Word, save it to a file, and then run the "Word Count" program on the file. Here, your program is an input to another program (Word Count), and the output is the number of words in your program. The Word Count program doesn't run your program; it simply counts the words in it as it would for any file of text.

Analogously, you could, without too much difficulty, design a Turing machine M that counted the 1s in its input, and then run M on the code for a second Turing machine M'. M would simply count the 1s in M''s code. Of course, the Universal Turing Machine U could have the code for M in the "program" part of its tape, have the code for M' in the "input" part of its tape, and run M on M'. Just to be perverse, we could put the code for M in both the "program" and "input" parts of U's tape, and have M run on its own code! This would be like having the Word Count program run on its own lines of code, counting the number of words it contains itself. No problem!

All this may seem fairly pedestrian to your computer-sophisticated mind, but at the time Turing was developing his proof, his essential insight—that the very same string of 0s and 1s on a tape could be interpreted as either a program or as input to another program—was truly novel.

Now we're ready to outline Turing's proof.

Turing proved that the answer to the *Entscheidungsproblem* is "no" by using a mathematical technique called "proof by contradiction." In this technique, you first assume that the answer is "yes" and use further reasoning to show that this assumption leads to a contradiction, and so cannot be the case.

Turing first assumed that the answer is "yes," that is, there is always a definite procedure that can decide whether or not a given statement is true. Turing then proposed the following statement:

> **Turing Statement**: Turing machine M, given input I, will reach the **halt** state after a finite number of time steps.

By the first assumption, there is some definite procedure that, given M and I as input, will decide whether or not this particular statement is true. Turing's proof shows that this assumption leads to a contradiction.

It is important to note that there are some Turing machines that never reach the **halt** state. For instance, consider a Turing machine similar to the one in our example above but with only two rules:

1. If you are in the **start** state and read a 0 or a 1, change to the **even** state and move one cell to the right.
2. If you are in the **even** state and read a 0 or a 1, change to the **start** state and move one cell to the left.

This is a perfectly valid Turing machine, but it is one that will never halt. In modern language we would describe its action as an "infinite loop"—behavior that is usually the result of a programming bug. Infinite loops don't have to be so obvious; in fact, there are many very subtle ways to create an infinite loop, whose presence is then very hard to detect.

Our assumption that there is a definite procedure that will decide the Turing Statement is equivalent to saying that we can design a Turing machine that is an infinite loop detector.

More specifically, our assumption says that we can design a Turing machine H that, given the code for any machine M and any input I on its tape, will, in finite time, answer "yes" if M would halt on input I, and "no" if M would go into an infinite loop and never halt.

The problem of designing H is called the "Halting problem." Note that H itself must always halt with either a "yes" or "no" answer. So H can't simply run M on I, since there would be some possibility that M (and thus H) would never halt. H has to do its job in some less straightforward way.

It's not immediately clear how it would work, but nonetheless we have assumed that our desired H exists. Let's denote $H(M, I)$ as the action of running H on input M and I. The output is "yes" (e.g., a single 1 on the tape) if M would halt on I and "no" (e.g., a single 0 on the tape) if M would not halt on I.

Now, said Turing, we create a modified version of H, called H', that takes as input the code for some Turing machine M, and calculates $H(M, M)$. That is, H' performs the same steps that H would to determine whether M would halt on its own code M. However, make H' different from H in what it does after obtaining a "yes" or a "no". H just halts, with the answer on its tape. H' halts only if the answer is "No, M does not halt on code M." If the answer is "Yes, M does halt on code M," H' goes into an infinite loop and never halts.

Whew, this might be getting a bit confusing. I hope you are following me so far. This is the point in every Theory of Computation course at which students either throw up their hands and say "I can't get my mind around this stuff!" or clap their hands and say "I *love* this stuff!"

Needless to say, I was the second kind of student, even though I shared the confusion of the first kind of student. Maybe you are the same. So take a deep breath, and let's go on.

Now for Turing's big question:

What does H' do when given as input the code for H', namely, its own code? Does it halt?

At this point, even the second kind of student starts to get a headache. This is really hard to get your head around. But let's try anyway.

First, suppose that H' does not halt on input H'. But then we have a problem. Recall that H', given as input the code for some Turing machine M, goes into an infinite loop if and only if M *does* halt on M. So H' not halting on input M implies that M *does* halt on input M. See what's coming? H' not halting on input H' implies that H' does halt on input H'. But H' can't both halt and not halt, so we have a contradiction.

Therefore, our supposition that H' does not halt on H' was wrong, so H' must halt on input H'. But now we have the opposite problem. H' only halts on its input M if M *does not* halt on M. So H' only halts on H' if H' does not halt on H'. Another contradiction!

These contradictions show that H' can neither halt on input H' nor go into an infinite loop on input H'. An actual machine has to do one or the other, so H' can't exist. Since everything that H' does is allowed for a Turing machine, except possibly running H, we have shown that H itself cannot exist.

Thus—and this is Turing's big result—**there can be no definite procedure for solving the Halting problem.** The Halting problem is an example that proves that the answer to the *Entscheidungsproblem* is "no"; not every mathematical statement has a definite procedure that can decide its truth or falsity. With this, Turing put the final nail in the coffin for Hilbert's questions.

Turing's proof of the uncomputability of the Halting problem, sketched above, uses precisely the same core idea as Gödel's incompleteness proof. Gödel figured out a way to encode mathematical statements so that they could talk about themselves. Turing figured out a way to encode mathematical statements as Turing machines and thus run on one another.

At this point, I should summarize Turing's momentous accomplishments. First, he rigorously *defined* the notion of "definite procedure." Second, his definition, in the form of Turing machines, laid the groundwork for the invention of electronic programmable computers. Third, he showed what few people ever expected: there are limits to what can be computed.

The Paths of Gödel and Turing

The nineteenth century was a time of belief in infinite possibility in mathematics and science. Hilbert and others believed they were on the verge of realizing Leibniz's dream: discovering an automatic way to prove or disprove any statement, thus showing that there is nothing mathematics could not conquer. Similarly, as we saw in chapter 2, Laplace and others believed that, using Newton's laws, scientists could in principle predict everything in the universe.

In contrast, the discoveries in mathematics and physics of the early to middle twentieth century showed that such infinite possibility did not in fact exist. Just as quantum mechanics and chaos together quashed the hope of perfect prediction, Gödel's and Turing's results quashed the hope of the unlimited power of mathematics and computing. However, as a direct consequence of his negative answer to the *Entscheidungsproblem*, Turing set the stage for the next great discovery, electronic programmable computers, which have since changed almost everything about the way science is done and the way our lives are lived.

After publishing their complementary proofs in the 1930s, Turing and Gödel took rather different paths, though, like everyone else at that time, their lives were deeply affected by the rise of Hitler and the Third Reich. Gödel, in spite of suffering from on-and-off mental health problems, continued his work on the foundations of mathematics in Vienna until 1940, when he moved to the United States to avoid serving in the German army. (According to his biographer Hao Wang, while preparing for American citizenship Gödel found a logical inconsistency in the U.S. Constitution, and his friend Albert Einstein had to talk him out of discussing it at length during his official citizenship interview.)

Gödel, like Einstein, was made a member of the prestigious Institute for Advanced Study in Princeton and continued to make important contributions to mathematical logic. However, in the 1960s and 1970s, his mental health deteriorated further. Toward the end of his life he became seriously paranoid and was convinced that he was being poisoned. As a result, he refused to eat and eventually starved to death.

Turing also visited the Institute for Advanced Study and was offered a membership but decided to return to England. During World War II, he became part of a top-secret effort by the British government to break the so-called Enigma cipher that was being used by the German navy to encrypt communications. Using his expertise in logic and statistics, as well as progress in electronic computing, he took the lead in developing code-breaking machines that were eventually able to decrypt almost all Enigma communications. This gave Britain a great advantage in its fight against Germany and arguably was a key factor in the eventual defeat of the Nazis.

After the war, Turing participated in the development of one of the first programmable electronic computers (stemming from the idea of his universal Turing machine), at Manchester University. His interests returned to questions about how the brain and body "compute," and he studied neurology and physiology, did influential work on the theory of developmental biology, and wrote about the possibility of intelligent computers. However, his personal life presented a problem to the mores of the day: he did not attempt to hide his homosexuality. Homosexuality was illegal in 1950s Britain; Turing was arrested for actively pursuing relationships with men and was sentenced to drug "therapy" to treat his "condition." He also lost his government security clearance. These events may have contributed to his probable suicide in 1954. Ironically, whereas Gödel starved himself to avoid being (as he believed) poisoned, Turing died from eating a poisoned (cyanide-laced) apple. He was only 41.

CHAPTER 5 | Evolution

All great truths begin as blasphemies.
—George Bernard Shaw, *Annajanska, The Bolshevik Empress*

THE SECOND LAW OF THERMODYNAMICS states that the total entropy of an isolated system will always increase until it reaches its maximum value. Everyone knows this instinctively—it happens not only in our understanding of science, but also in our daily lives, and is ingrained in humans' conceptions of history and in our art, literature, and religions. The Buddha tells us that "Subject to decay are all compounded things." The Old Testament prophet Isaiah foretells that "The earth shall wax old like a garment." Shakespeare asks,

> O! how shall summer's honey breath hold out,
> Against the wrackful siege of battering days,
> When rocks impregnable are not so stout,
> Nor gates of steel so strong but Time decays?

It is a gloomy message, this inexorable march toward maximum entropy. But nature gives us a singular counterexample: Life. By anyone's measure, living systems are *complex*—they exist somewhere in the middle ground between order and disorder. According to our intuitions, over the long history of life, living systems have become vastly more complex and intricate rather than more disordered and entropic.

We know that to decrease entropy, *work* must be done. Who or what is doing the work of creating and maintaining living systems and making them more complex? Some of the world's religions propose that a deity is responsible, but in the mid-1800s, Charles Darwin proposed that instead, the history of life has resulted from the invisible hand of evolution via natural selection.

No idea in science has been more threatening to humans' conceptions about themselves than Darwin's theory of evolution; it arguably has been *the* most controversial idea in the history of science. But it is also one of the best ideas. The philosopher Daniel Dennett strongly affirms this:

> If I were to give an award for the single best idea anyone has ever had, I'd give it to Darwin, ahead of Newton and Einstein and everyone else. In a single stroke, the idea of evolution by natural selection unifies the realm of life, meaning, and purpose with the realm of space and time, cause and effect, mechanism and physical law.

This chapter sketches the history and main ideas of Darwinian evolution and how it produces organization and adaptation. Concepts from evolutionary theory will come up again and again in the remainder of the book. In chapter 18, I describe how some of these concepts are being radically modified in light of the unexpected results coming out of the molecular revolution in biology and the results of complex systems ideas as applied to evolution.

Pre-Darwinian Notions of Evolution

The word *evolution* means "gradual change." Biological evolution is the process of gradual (and sometimes rapid) change in biological forms over the history of life. Until the eighteenth century, the prevailing opinion was that biological forms do not change over time; rather, all organisms were created by a deity and have largely remained in their original form since their creation. Although some ancient Greek and Indian philosophers had proposed that humans arose via transmutation from other species, in the West the conception of divine creation began to be widely questioned only in the eighteenth century.

In the mid-1700s, 100 years before Darwin proposed his theory, a French zoologist named George Louis Leclerc de Buffon published a many-volume work entitled *Historie Naturelle*, in which he described the similarities between different species. Buffon suggested that the earth is much older than the Biblical 6,000 years and that all modern organisms evolved from a single ancestor, though he did not propose a mechanism for this evolution. Buffon's work in biology and geology was a significant break from the prevailing creationist viewpoint. Not surprising, the Catholic Church in France burned copies of his books.

Charles Darwin's grandfather, Erasmus Darwin, was another prominent eighteenth-century scientist who believed in the evolution of all species from a single ancient ancestor. He proposed mechanisms for evolution that were precursors to his grandson's theory of natural selection. Erasmus Darwin expressed his ideas both in scientific writing and in poetry:

> Organic life beneath the shoreless waves
> Was born and nurs'd in ocean's pearly caves;
> First forms minute, unseen by spheric glass,
> Move on the mud, or pierce the watery mass;
> These, as successive generations bloom,
> New powers acquire and larger limbs assume;
> Whence countless groups of vegetation spring,
> And breathing realms of fin and feet and wing.

If only modern-day scientists were so eloquent! However, like the Catholics in France, the Anglican Church didn't much like these ideas.

The most famous pre-Darwinian evolutionist is Jean-Baptiste Lamarck. A French aristocrat and botanist, Lamarck published a book in 1809, *Philosophie Zoologique*, in which he proposed his theory of evolution: new types of organisms are spontaneously generated from inanimate matter, and these species evolve via the "inheritance of acquired characteristics." The idea was that organisms adapted to their environment during their lifetimes, and that these *acquired* adaptations were then passed directly to the organisms' offspring. One example in Lamarck's book was the acquisition of long legs by wading birds, such as storks. Such birds, he believed, originally had to stretch their legs in order to keep their bodies out of the water. This continual stretching made their legs longer, and the acquired trait of longer legs was passed on to the birds' offspring, who stretched their legs even longer, passing this trait on to their own offspring, and so on. The result is the very long legs we now see on wading birds.

Lamarck gave many other such examples. He also asserted that evolution entails a "tendency to progression," in which organisms evolve to be increasingly "advanced," with humans at the pinnacle of this process. Thus, changes in organisms are predominately changes for the better, or at least, for the more complex.

Lamarck's ideas were rejected by almost all of his contemporaries—not only by proponents of divine creation but also by people who believed in

Jean-Baptiste Lamarck, 1744–1829 (Illustration from *LES CONTEMPORAINS N 554: Lamarck, naturaliste {1744–1829}*, by Louis Théret, Bonne Press, 1903. Photograph copyright © by Scientia Digital [http://www.scientiadigital.com]. Reprinted by perimission.)

evolution. The evolutionists were not at all convinced by Lamarck's examples of evolution via inheritance of acquired characteristics, and indeed, his empirical data were weak and were generally limited to his own speculations on how certain traits of organisms came about.

However, it seems that Charles Darwin himself was, at least at first, favorably impressed by Lamarck: "Lamarck…had few clear facts, but so bold and many such profound judgment that he foreseeing consequence was endowed with what may be called the prophetic spirit in science. The highest endowment of lofty genius." Darwin also believed that, in addition to natural selection, the inheritance of acquired characteristics was one of the mechanisms of evolution (though this belief did not survive as part of what we now call "Darwinism").

Neither Lamarck nor Darwin had a good theory of how such inheritance could take place. However, as the science of genetics became better understood in the years after Darwin, the inheritance of acquired characteristics seemed almost certain to be impossible. By the beginning of the twentieth century, Lamarck's theories were no longer taken seriously in evolutionary biology, though several prominent psychologists still believed in them as an explanation of some aspects of the mind, such as instinct. For example, Sigmund Freud expressed the view that "if [the] instinctual life of animals permits of any explanation at all, it can only be this: that they carry over into their new existence the experience of their kind; that is to say, that they have preserved in their minds memories of what their ancestors experienced." I don't think these beliefs remained in psychology much beyond the time of Freud.

Origins of Darwin's Theory

Charles Darwin should be an inspiration to youthful underachievers everywhere. As a child, he was a mediocre student in an overachieving family. (His usually loving father, a successful country doctor, in a moment of frustration complained bitterly to the teenaged Charles: "You care for nothing but shooting, dogs, and rat-catching, and you will be a disgrace to yourself and your family!") Underachieving as he might have been then, he went on to be the most famous, and most important, biologist of all time.

In 1831, while trying to decide on his future career (country doctor or country parson seemed to be the choices), Darwin was offered a dual job as both "naturalist" and "captain's dining companion" on a survey ship, the H.M.S. *Beagle*. The ship's captain was a "gentleman," and a bit lonely, so he wanted to dine with another gentleman rather than with the riff-raff of the ship's crew. Darwin was his man.

Darwin spent almost five years on the *Beagle* (1831–1836), much of the time in South America, where, in addition to his dining duties, he collected plants, animals, and fossils and did a lot of reading, thinking, and writing.

Charles Darwin, 1809–1882. Photograph taken in 1854, a few years before he published *Origin of Species*. (Reproduced with permission from John van Wyhe, ed., The Complete Work of Charles Darwin Online [http://darwin-online.org.uk/].)

Fortunately he wrote many letters and kept extensive notebooks full of his observations, ideas, opinions, reactions to books, et cetera; his detailed recording of his thoughts went on for the rest of his life. If Darwin were alive today, he clearly would have been an obsessive blogger.

During and after the *Beagle* voyage, Darwin got a lot of ideas from his reading of scientific books and articles from various disciplines. He was convinced by Charles Lyell's *Principles of Geology* (1830) that geological features (mountains, canyons, rock formations) arise from gradual processes of erosion, wind, myriad floods, volcanic eruptions, and earthquakes, rather than from catastrophic events such as the biblical Noah's flood. Such a view of gradualism—that small causes, taken over long periods, can have very large effects—was anathema to religious fundamentalists of the day, but Lyell's evidence was compelling to Darwin, especially as, on his voyage, he could see for himself the results of different kinds of geological processes.

Thomas Malthus's *Essay on the Principle of Population* (1798) drew Darwin's attention to the fact that population growth leads to competition for food and other resources. Malthus's essay was about *human* population growth, but Darwin would adapt these ideas to explain the evolution of all living organisms via a continual "struggle for existence."

Darwin also read Adam Smith's free-market manifesto, *The Wealth of Nations* (1776). This book exposed him to Smith's notion of the *invisible hand* in economics, whereby a collection of individuals acting in their own self-interest produces maximum benefit for the entire community.

From his own observations in South America and elsewhere, Darwin was acutely struck by the tremendous variation among living beings and by the apparent adaptation of different species to their environments. One of his most famous examples is the finches of the Galápagos Islands, 600 miles off the coast of Ecuador. Darwin observed that different species of these small birds, although otherwise quite similar to one another, have wide variations in beak size and shape. Darwin was eventually able to show that different species of finches had common ancestors who had evidently migrated to individual islands in the Galápagos chain. He also showed that the type of beak was adapted to the individual species' food sources, which differed among the islands. Darwin hypothesized that the geographical isolation imposed by the different islands, as well as the local environmental conditions, led to the evolution of these many different species from a small number of ancestors.

We can imagine Darwin with these ideas swirling in his head during his voyage and afterward, back in England, trying to make sense of the data he had collected. Gradual change over long periods can produce very large effects.

Population growth combined with limited resources creates a struggle for existence. Collections of individuals acting in self-interested ways produce global benefit. Life seems to allow almost infinite variation, and a species' particular traits seem designed for the very environment in which the species lives. Species branch out from common ancestors.

Over the years, it all came together in his mind as a coherent theory. Individual organisms have more offspring than can survive, given limited food resources. The offspring are not exact copies of the parents but have some small amount of random variation in their traits. The traits that allow some offspring to survive and reproduce will be passed on to further offspring, thus spreading in the population. Very gradually, through reproduction with random variation and individual struggles for existence, new species will be formed with traits ideally adapted to their environments. Darwin called this process *evolution by natural selection*.

For years after the development of his theories, Darwin shared his ideas with only a few people (Charles Lyell and some others). In part, his reticence was due to a desire for additional data to bolster his conclusions, but also contributing was a deep concern that his theories would bring unhappiness to religious people, in particular to his own wife, who was deeply religious. Having once considered becoming a country parson himself, he expressed discomfort with his main conclusion: "I am almost convinced (quite contrary to the opinion I started with) that species are not (it is like confessing a murder) immutable."

However, Darwin's notebooks of the time also revealed his understanding of the philosophical implications of his work for the status of humans. He wrote, "Plato...says in Phaedo that our 'necessary ideas' arise from the pre-existence of the soul, are not derivable from experience—read monkeys for preexistence."

Competition is not only the centerpiece of evolution, but is also a great motivator in science itself. Darwin's hesitation to publish his work quickly melted away when he discovered that he was about to be scooped. In 1858, Darwin received a manuscript from another English naturalist, Alfred Russell Wallace, entitled *On the Tendency of Varieties to Depart Indefinitely from the Original Type*. Darwin was alarmed to find that Wallace had independently come up with the same basic ideas of evolution by natural selection. Darwin expressed his dismay in a letter to Lyell: "[A]ll my originality, whatever it may amount to, will be smashed." However, he generously offered to help Wallace publish his essay, but requested that his own work also be published at the same time, in spite of his worries about this request being "base and paltry."

Lyell agreed that, in order to solve the priority problem, Darwin and Wallace should publish their work together. This joint work was read to the Linnean Society in the summer of 1858. By the end of 1859, Darwin had published his 400-plus-page book *On the Origin of Species*.

It turns out that the priority issue was not fully solved. Unbeknown to Darwin, twenty-eight years before the publication of the *Origin*, a little-known Scot named Patrick Matthew had published an obscure book with an equally obscure title, *On Naval Timber and Arboriculture*, in whose appendix he proposed something very much like Darwin's evolution by natural selection. In 1860, Matthew read about Darwin's ideas in the periodical *Gardiner's Chronicle* and wrote a letter to the publication citing his priority. Darwin, ever anxious to do the right thing, responded with his own letter: "I freely acknowledge that Mr. Matthew has anticipated by many years the explanation which I have offered of the origin of species, under the name of natural selection...I can do no more than offer my apologies to Mr. Matthew for my entire ignorance of his publication."

So who actually is responsible for the idea of evolution by natural selection? Evidently, this is another example of an idea that was "in the air" at the time, an idea that someone would inevitably come up with. Darwin's colleague Thomas Huxley realized this and chided himself: "How extremely stupid not to have thought of that!"

Why does Darwin get all the credit? There are several reasons, including the fact that he was at that time a more famous and respected scientist than the others, but the most important reason is that Darwin's book, unlike the works of Wallace and Matthew, contained a more coherent set of ideas and a tremendous amount of evidence supporting those ideas. Darwin was the person who turned natural selection from an interesting and plausible speculation into an extremely well-supported theory.

To summarize the major ideas of Darwin's theory:

- Evolution has occurred; that is, all species descend from a common ancestor. The history of life is a branching tree of species.
- Natural selection occurs when the number of births is greater than existing resources can support so that individuals undergo competition for resources.
- Traits of organisms are inherited with variation. The variation is in some sense *random*—that is, there is no force or bias leading to variations that increase fitness (though, as I mentioned previously, Darwin himself accepted Lamarck's view that there are such forces). Variations that turn out to be adaptive in the current environment are

likely to be *selected*, meaning that organisms with those variations are more likely to survive and thus pass on the new traits to their offspring, causing the number of organisms with those traits to increase over subsequent generations.
- Evolutionary change is constant and gradual via the accumulation of small, favorable variations.

According to this view, the result of evolution by natural selection is the appearance of "design" but with no designer. The appearance of design comes from chance, natural selection, and long periods of time. Entropy decreases (living systems become more organized, seemingly more *designed*) as a result of the work done by natural selection. The energy for this work comes from the ability of individual organisms to metabolize energy from their environments (e.g., sunlight and food).

Mendel and the Mechanism of Heredity

A major issue not explained by Darwin's theory was exactly how traits are passed on from parent to offspring, and how variation in those traits—upon which natural selection acts—comes about. The discovery that DNA is the carrier of hereditary information did not take place until the 1940s. Many theories of heredity were proposed in the 1800s, but none was widely accepted until the "rediscovery" in 1900 of the work of Gregor Mendel.

Mendel was an Austrian monk and physics teacher with a strong interest in nature. Having studied the theories of Lamarck on the inheritance of acquired traits, Mendel performed a sequence of experiments, over a period of eight years, on generations of related pea plants to see whether he could verify Lamarck's claims. His results not only disconfirmed Lamarck's speculations but also revealed some surprising facts about the nature of heredity.

Mendel looked at several different traits of pea plants: smoothness and color of seeds; shape of pea pod; color of pods and flowers; locations of flowers on the plants; and height of stems. Each of these traits (or "characters") could have one of two distinct forms (e.g., the pod color could be *green* or *yellow*; the stem height could be *tall* or *dwarf*).

Mendel's long years of experiments revealed several things that are still considered roughly valid in modern-day genetics. First, he found that the plants' offspring did not take on any traits that were acquired by the parents during their lifetimes. Thus, Lamarckian inheritance did not take place.

Gregor Mendel, 1822–1884
(From the National Library of Medicine)
[http://wwwils.nlm.nih.gov/visibleproofs/galleries/technologies/dna.html].

Second, he found that heredity took place via discrete "factors" that are contributed by the parents, one factor being contributed by each parent for each trait (e.g., each parent contributes either a factor for tall stems or dwarf stems). These factors roughly correspond to what we would call *genes*. Thus, the medium of inheritance, whatever it was, seemed to be discrete, not continuous as was proposed by Darwin and others. (Note that pea plants reproduce via either self-pollination or cross-pollination with other pea plants.)

For each trait he was studying, Mendel found that each plant has a pair of genes responsible for the trait. (For simplicity, I am using more modern terminology; the term "gene" was not used in Mendel's time.) Each gene of the pair encodes a "value" for the trait—for example, *tall* vs. *dwarf*. This value is called an *allele*. For stem height there are three possibilities for the allele pairs encoded by these two genes: both alleles the same (*tall/tall* or *dwarf/dwarf*) or different (*tall/dwarf*, which is equivalent to *dwarf/tall*).

Moreover, Mendel found that, for each trait, one allele is *dominant* (e.g., *tall* is dominant for stem height) and the other *recessive* (e.g., *dwarf* is recessive for stem height). A *tall/tall* individual will always be tall. A *tall/dwarf* individual will also be tall since tall is dominant; only one copy of the dominant allele is needed. Only a *dwarf/dwarf* individual—with two copies of the recessive allele—will be dwarf.

As an example, suppose you have two *tall/dwarf* individuals that cross-pollinate. Both the parents are tall, but there is a 25% chance that both will pass on their *dwarf* gene to the child, making it *dwarf/dwarf*.

Mendel used such reasoning and the laws of probability to predict, very successfully, how many plants in a given generation will display the dominant or recessive version of a given trait, respectively. Mendel's experiments contradicted the widely believed notion of "blending inheritance"—that the offspring's traits typically will be an average of the parents' traits.

Mendel's work was the first to explain and quantitatively predict the results of inheritance, even though Mendel did not know what substance his "factors" were made out of, or how they recombined as a result of mating. Unfortunately, his 1865 paper, "Experiments in Plant Hybridization," was published in a rather obscure journal and was not appreciated as being of great importance until 1900, after which several scientists had obtained similar results in experiments.

The Modern Synthesis

You would think that the dissemination of Mendel's results would be a big boost for Darwinism, since it provided Darwin's theory with an experimentally tested mechanism of inheritance. But for decades, Mendel's ideas were considered to be opposed to Darwin's. Darwin's theory asserted that evolution, and therefore variation, is continuous (i.e., organisms can differ from one another in arbitrarily minute ways) and Mendel's theory proposed that variation is discrete (a pea plant is either tall or dwarf, but nothing in between). Many early adherents to Mendel's theories believed in *mutation theory*—a proposal that variation in organisms is due to mutations in offspring, possibly very large, which themselves drive evolution, with natural selection only a secondary mechanism for preserving (or deleting) such mutations in a population. Darwin and his early followers were completely against this idea; the cornerstones of Darwin's theory were that individual variations must be very small, natural selection on these tiny variations is what drives evolution, and evolution is gradual. "Natura non facit saltum" (Nature does not make leaps) was Darwin's famous dismissal of mutation theory.

After many bitter arguments between the early Darwinists and Mendelians, this false opposition was cleared up by the 1920s when it was discovered that, unlike the traits of Mendel's pea plants, most traits in organisms are determined by many genes, each with several different alleles. The huge number of possible combinations of these many different alleles can result in seemingly continuous variation in an organism. Discrete variation in the *genes* of an organism can result in continuous-seeming variation in the organism's *phenotype*—the physical traits (e.g., height, skin color, etc.) resulting from these genes. Darwinism and Mendelism were finally recognized as being complementary, not opposed.

One reason the early Darwinists and Mendelians disagreed so strongly is that, although both sides had experimental evidence supporting their position, neither side had the appropriate conceptual framework (i.e., multiple genes controlling traits) or mathematics to understand how their respective theories fit together. A whole new set of mathematical tools had to be developed to analyze the results of Mendelian inheritance with many interacting genes operating under natural selection in a mating population. The necessary tools were developed in the 1920s and 1930s, largely as a result of the work of the mathematical biologist Ronald Fisher.

Fisher, along with Francis Galton, was a founder of the field of modern statistics. He was originally spurred by real-world problems in agriculture and animal breeding. Fisher's work, along with that of J.B.S. Haldane and Sewall Wright, showed that Darwin's theories were indeed compatible with Mendel's. Moreover, the combined work of Fisher, Haldane, and Wright provided a mathematical framework—*population genetics*—for understanding the dynamics of alleles in an evolving population undergoing Mendelian inheritance and natural selection. This unification of Darwinism and Mendelism, along with the framework of population genetics, was later called "the Modern Synthesis."

Fisher, Wright, and Haldane are known as the three founders of the Modern Synthesis. There were many strong disagreements among the three, particularly a bitter fight between Fisher and Wright over the relative roles of natural selection and "random genetic drift." In the latter process, certain alleles become dominant in a population merely as a chance event. For instance, suppose that in a population of pea plants, neither the *dwarf* nor *tall* alleles really affect the fitness of the plants as a whole. Also suppose that at some point the *dwarf* allele, just by chance, appears in a higher fraction of plants than the *tall* allele. Then, if each *dwarf* and *tall* plant has about the same number of offspring plants, the *dwarf* allele will likely be even more frequent in the next generation, simply because there were more parent plants with the *dwarf* allele. In general, if there is no selective advantage of either trait, one or the other trait will eventually be found in 100% of the individuals in the population. Drift is a stronger force in small rather than large populations, because in large populations, the small fluctuations that eventually result in drift tend to cancel one another out.

Wright believed that random genetic drift played a significant role in evolutionary change and the origin of new species, whereas in Fisher's view, drift played only an insignificant role at best.

These are both reasonable and interesting speculations. One would think that Fisher and Wright would have had lots of heated but friendly discussions about it over beer (that is, when the Briton, Fisher, and the American, Wright, were on the same continent). However, what started as a very productive and stimulating interchange between them ended up with Fisher and Wright each publishing papers that offended the other, to the point that communication between them basically ended by 1934. The debate over the respective roles of natural selection versus drift was almost as bitter as the earlier one between the Mendelians and the Darwinists—ironic, since it was largely the work of Fisher and Wright that showed that these two sides actually need not have disagreed.

The Modern Synthesis was further developed in the 1930s and 1940s and was solidified into a set of principles of evolution that were almost universally accepted by biologists for the following fifty years:

- Natural selection is the major mechanism of evolutionary change and adaptation.
- Evolution is a gradual process, occurring via natural selection on very small random variations in individuals. Variation of this sort is highly abundant in populations and is not biased in any direction (e.g., it does not intrinsically lead to "improvement," as believed by Lamarck). The source of individual variation is random genetic mutations and recombinations.
- Macroscale phenomena, such as the origin of new species, can be explained by the microscopic process of gene variation and natural selection.

The original architects of the Modern Synthesis believed they had solved the major problems of explaining evolution, even though they still did not know the molecular basis of genes or by what mechanism variation arises. As the evolutionist Ian Tattersall relates, "Nobody could ever again look at the evolutionary process without very consciously standing on the edifice of the Synthesis. And this edifice was not only one of magnificent elegance and persuasiveness; it had also brought together practitioners of all the major branches of organismic biology, ending decades of infighting, mutual incomprehension, and wasted energies."

Challenges to the Modern Synthesis

Serious challenges to the validity of the Modern Synthesis began brewing in the 1960s and 1970s. Perhaps the most prominent of the challengers were paleontologists Stephen Jay Gould and Niles Eldredge, who pointed out some discrepancies between what the Modern Synthesis predicted and what the actual fossil record showed. Gould went on to be simultaneously the best-known proponent and expositor of Darwinian evolution (through his many books and articles for nonscientists) and the most vociferous critic of the tenets of the Synthesis.

One major discrepancy is the prediction of the Modern Synthesis for gradual change in the morphology of organisms as compared with what Gould, Eldredge, and others claimed was the actual pattern in the fossil record: long periods of no change in the morphology of organisms (and no new species

Stephen Jay Gould, 1941–2002. (Jon Chase/Harvard News Office, © 1997 President and Fellows of Harvard College, reproduced by permission.)

Niles Eldredge (Courtesy of Niles Eldredge.)

emerging) punctuated by (relatively) short periods of large change in morphology, resulting in the emergence of new species. This pattern became labeled *punctuated equilibria*. Others defended the Modern Synthesis, asserting that the fossil record was too incomplete for scientists to make such an inference. (Some detractors of punctuated equilibria nicknamed the theory "evolution by jerks." Gould countered that the proponents of gradualism supported "evolution by creeps.") Punctuated equilibria have also been widely observed in laboratory experiments that test evolution and in simplified computer simulations of evolution.

Thus Gould and his collaborators asserted that the "gradualism" pillar of the Modern Synthesis is wrong. They also believed that the other two pillars—the primacy of natural selection and small gene variations to explain the history of life—were not backed up by evidence.

Although Gould agreed that natural selection is an important mechanism of evolutionary change, he asserted that the roles of *historical contingency* and *biological constraints* are at least as important as that of natural selection.

Historical contingency refers to all the random accidents, large and small, that have contributed to the shaping of organisms. One example is the impact of large meteors wiping out habitats and causing extinction of groups of species, thus allowing other species to emerge. Other examples are the unknown quirks of fate that gave carnivorous mammals an advantage over the carnivorous birds that once rivaled them in numbers but which are now extinct.

Gould's metaphor for the role of contingency is an imaginary "tape of life"—a kind of time-lapse movie covering all of evolution since the beginning of life on Earth. Gould asks, What would happen if the tape were restarted with slightly different initial conditions? Would we see anything like the array of organisms that evolved during the first playing of the tape? The answer of the Modern Synthesis would presumably be "yes"—natural selection would again shape organisms to be optimally adapted to the environment, so they would look much the same as what we have now. Gould's answer is that the role played by historical contingency would make the replayed tape much different.

Biological constraints refer to the limitations on what natural selection can create. Clearly natural selection can't defy the laws of physics—it can't create a flying creature that does not obey the laws of gravity or a perpetual-motion animal that needs no food. Gould and many others have argued that there are biological constraints as well as physical constraints that limit what kind of organisms can evolve.

This view naturally leads to the conclusion that not all traits of organisms are explainable as "adaptations." Clearly traits such as hunger and sex drive lead us to survival and reproduction. But some traits may have arisen by accident, or as side effects of adaptive traits or developmental constraints. Gould has been quite critical of evolutionists he calls "strict adaptationists"—those who insist that natural selection is the only possible explanation for complex organization in biology.

Furthermore, Gould and his colleagues attacked the third pillar of the Synthesis by proposing that some of the large-scale phenomena of evolution cannot be explained in terms of the microscopic process of gene variation and natural selection, but instead require natural selection to work on levels higher than genes and individuals—perhaps entire species.

Some evidence for Gould's doubts about the Modern Synthesis came from work in molecular evolution. In the 1960s, Motoo Kimura proposed a theory of "neutral evolution," based on observations of protein evolution, that challenged the central role of natural selection in evolutionary change. In the 1970s, chemists Manfred Eigen and Peter Schuster observed behavior analogous to punctuated equilibria in evolution of viruses made up of RNA, and developed an explanatory theory in which the unit of evolution was not an individual virus, but a collective of viruses—a *quasi-species*—that consisted of mutated copies of an original virus.

These, and other challenges to the Modern Synthesis were by no means accepted by all evolutionists, and, as in the early days of Darwinism, debates among rival views often became rancorous. In 1980, Gould wrote that "[T]he synthetic theory...is effectively dead, despite its persistence as textbook orthodoxy." Going even further, Niles Eldredge and Ian Tattersall contended that the view of evolution due to the Modern Synthesis "is one of the greatest myths of twentieth-century biology." On the other side, the eminent evolutionary biologists Ernst Mayr and Richard Dawkins strongly defended the tenets of the Synthesis. Mayr wrote, "I am of the opinion that nothing is seriously wrong with the achievements of the evolutionary synthesis and that it does not need to be replaced." Dawkins wrote, "The theory of evolution by cumulative natural selection is the only theory we know of that is in principle capable of explaining the existence of organized complexity." Many people still hold to this view, but, as I describe in chapter 18, the idea that gradual change via natural selection is the major, if not the only force in shaping life is coming under increasing skepticism as new technologies have allowed the field of genetics to explode with unexpected discoveries, profoundly changing how people think about evolution.

It must be said that although Gould, Eldredge, and others have challenged the tenets of the Modern Synthesis, they, like virtually all biologists, still strongly embrace the basic ideas of Darwinism: that evolution has occurred over the last four billion years of life and continues to occur; that all modern species have originated from a single ancestor; that natural selection has played an important role in evolution; and that there is no "intelligent" force directing evolution or the design of organisms.

CHAPTER 6 | Genetics, Simplified

SOME OF THE CHALLENGES to the Modern Synthesis have found support in the last several decades in results coming from molecular biology, which have changed most biologists' views of how evolution takes place.

In chapter 18, I describe some of these results and their impact on genetics and evolutionary theory. As background for this and other discussions throughout the book, I give here a brief review of the basics of genetics. If you are already familiar with this subject, this chapter can be skipped.

It has been known since the early 1800s that all living organisms are composed of tiny cells. In the later 1800s, it was discovered that the nucleus of every cell contains large, elongated molecules that were dubbed *chromosomes* ("colored bodies," since they could be stained so easily in experiments), but their function was not known. It also was discovered that an individual cell reproduces itself by dividing into two identical cells, during which process (dubbed *mitosis*) the chromosomes make identical copies of themselves. Many cells in our bodies undergo mitosis every few hours or so—it is an integral process of growth, repair, and general maintenance of the body.

Meiosis, discovered about the same time, is the process in *diploid* organisms by which eggs and sperm are created. Diploid organisms, including most mammals and many other classes of organisms, are those in which chromosomes in all cells (except sperm and egg, or *germ* cells) are found in pairs (twenty-three pairs in humans). During meiosis, one diploid cell becomes four germ cells, each of which has half the number of chromosomes as the original cell. Each chromosome pair in the original cell is cut into parts, which recombine to form chromosomes for the four new germ cells. During fertilization, the chromosomes in two germ cells fuse together to create the correct number of chromosome pairs.

The result is that the genes on a child's chromosome are a mixed-up version of its parents' chromosomes. This is a major source of variation in organisms with sexual reproduction. In organisms with no sexual reproduction the child looks pretty identical to the parent.

All this is quite complicated, so it is no surprise that biologists took a long time to unravel how it all works. But this was just the beginning.

The first suggestion that chromosomes are the carriers of heredity was made by Walter Sutton in 1902, two years after Mendel's work came to be widely known. Sutton hypothesized that chromosomes are composed of units ("genes") that correspond to Mendelian factors, and showed that meiosis gives a mechanism for Mendelian inheritance. Sutton's hypothesis was verified a few years later by Thomas Hunt Morgan via experiments on that hero of genetics, the fruit fly. However, the molecular makeup of genes, or how they produced physical traits in organisms, was still not known.

By the late 1920s, chemists had discovered both ribonucleic acid (RNA) and deoxyribonucleic acid (DNA), but the connection with genes was not discovered for several more years. It became known that chromosomes contained DNA, and some people suspected that this DNA might be the substrate of genes. Others thought that the substrate consisted of proteins found in the cell nucleus. DNA of course turned out to be the right answer, and this was finally determined experimentally by the mid-1940s.

But several big questions remained. How exactly does an organism's DNA cause the organism to have particular traits, such as *tall* or *dwarf* stems? How does DNA create a near-exact copy of itself during cell division (mitosis)? And how does the variation, on which natural selection works, come about at the DNA level?

These questions were all answered, at least in part, within the next ten years. The biggest break came when, in 1953, James Watson and Francis Crick figured out that the structure of DNA is a double helix. In the early 1960s, the combined work of several scientists succeeded in breaking the *genetic code*—how the parts of DNA encode the amino acids that make up proteins. A *gene*—a concept that had been around since Mendel without any understanding of its molecular substrate—could now be defined as a substring of DNA that codes for a particular protein. Soon after this, it was worked out how the code was translated by the cell into proteins, how DNA makes copies of itself, and how variation arises via copying errors, externally caused *mutations*, and sexual recombination. This was clearly a "tipping point" in genetics research. The science of genetics was on a roll, and hasn't stopped rolling yet.

The Mechanics of DNA

The collection of all of an organism's physical traits—its *phenotype*—comes about largely due to the character of and interactions between proteins in cells. Proteins are long chains of molecules called *amino acids*.

Every cell in your body contains almost exactly the same complete DNA sequence, which is made up of a string of chemicals called *nucleotides*. Nucleotides contain chemicals called *bases*, which come in four varieties, called (for short) A, C, G, and T. In humans, strings of DNA are actually double strands of paired A, C, G, and T molecules. Due to chemical affinities, A always pairs with T, and C always pairs with G.

Sequences are usually written with one line of letters on the top, and the paired letters (*base pairs*) on the bottom, for example,

$$T C C G A T T \ldots$$

$$A G G C T A A \ldots$$

In a DNA molecule, these double strands weave around one another in a double helix (figure 6.1).

Subsequences of DNA form *genes*. Roughly, each gene codes for a particular protein. It does that by coding for each of the amino acids that make up the protein. The way amino acids are coded is called the *genetic code*. The code is the same for almost every organism on Earth. Each amino acid corresponds to a triple of nucleotide bases. For example, the DNA triplet A A G corresponds to the amino acid *phenylalanine*, and the DNA triplet C A C corresponds to the amino acid *valine*. These triplets are called *codons*.

So how do proteins actually get formed by genes? Each cell has a complex set of molecular machinery that performs this task. The first step is *transcription* (figure 6.2), which happens in the cell nucleus. From a single strand of the DNA, an enzyme (an active protein) called *RNA polymerase* unwinds a small part of the DNA from its double helix. This enzyme then uses one of the DNA strands to create a *messenger* RNA (or mRNA) molecule that is a letter-for-letter copy of the section of DNA. Actually, it is an *anticopy*: in every place where the gene has C, the mRNA has G, and in every place where the gene has A, the mRNA has U (its version of T). The original can be reconstructed from the anticopy.

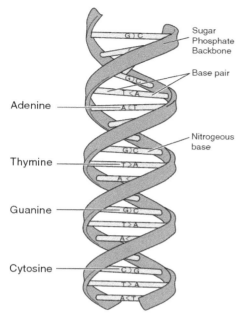

FIGURE 6.1. Illustration of the double helix structure of DNA. (From the National Human Genome Research Institute, Talking Glossary of Genetic Terms [http://www.genome.gov/glossary.cfm.])

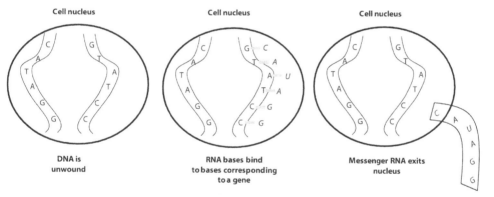

FIGURE 6.2. Illustration of transcription of DNA into messenger RNA. Note that the letter *U* is RNA's version of DNA's letter *T*.

The process of transcription continues until the gene is completely transcribed as mRNA.

The second step is *translation* (figure 6.3), which happens in the cell cytoplasm. The newly created mRNA strand moves from the nucleus to the cytoplasm, where it is read, one codon at a time, by a cytoplasmic structure called a *ribosome*. In the ribosome, each codon is brought together with a corresponding anticodon residing on a molecule of *transfer* RNA (tRNA). The anticodon consists of the complementary bases. For example, in figure 6.3,

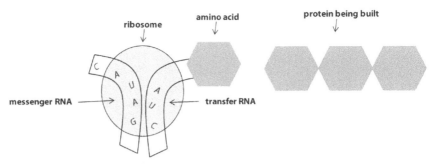

FIGURE 6.3. Illustration of translation of messenger RNA into amino acids.

the mRNA codon being translated is U A G, and the anticodon is the complementary bases A U C. A tRNA molecule that has that anticodon will attach to the mRNA codon, as shown in the figure. It just so happens that every tRNA molecule has attached to it both an anticodon *and* the corresponding amino acid (the codon A U C happens to code for the amino acid *isoleucine* in case you were interested). Douglas Hofstadter has called tRNA "the cell's flash cards."

The ribosome cuts off the amino acids from the tRNA molecules and hooks them up into a protein. When a stop-codon is read, the ribosome gets the signal to stop, and releases the protein into the cytoplasm, where it will go off and perform whatever function it is supposed to do.

The transcription and translation of a gene is called the gene's *expression* and a gene is *being expressed* at a given time if it is being transcribed and translated.

All this happens continually and simultaneously in thousands of sites in each cell, and in all of the trillions of cells in your body. It's amazing how little energy this takes—if you sit around watching TV, say, all this subcellular activity will burn up fewer than 100 calories per hour. That's because these processes are in part fueled by the random motion and collisions of huge numbers of molecules, which get their energy from the "ambient heat bath" (e.g., your warm living room).

The paired nature of nucleotide bases, A with T and C with G, is also the key to the replication of DNA. Before mitosis, enzymes unwind and separate strands of DNA. For each strand, other enzymes read the nucleotides in the DNA strand, and to each one attach a new nucleotide (new nucleotides are continually manufactured in chemical processes going on in the cell), with A attached to T, and C attached to G, as usual. In this way, each strand of the original two-stranded piece of DNA becomes a new two-stranded piece of DNA, and each cell that is the product of mitosis gets one of these complete two-stranded DNA molecules. There are many complicated processes in the cell that keep this replication process on track. Occasionally (about once every 100 billion nucleotides), errors will occur (e.g., a wrong base will be attached), resulting in mutations.

It is important to note that there is a wonderful self-reference here: All this complex cellular machinery—the mRNA, tRNA, ribosomes, polymerases, and so forth—that effect the transcription, translation, and replication of DNA are themselves encoded in that very DNA. As Hofstadter remarks: "The DNA contains coded versions of its own decoders!" It also contains coded versions of all the proteins that go into synthesizing the nucleotides the DNA is made up of. It's a self-referential circularity that would no doubt have pleased Turing, had he lived to see it explained.

The processes sketched above were understood in their basic form by the mid-1960s in a heroic effort by geneticists to make sense of this incredibly complex system. The effort also brought about a new understanding of evolution at the molecular level.

In 1962, Crick, Watson, and biologist Maurice Wilkins jointly received the Nobel prize in medicine for their discoveries about the structure of DNA. In 1968, Har Gobind Korana, Robert Holley, and Marshall Nirenberg received the same prize for their work on cracking the genetic code. By this time, it finally seemed that the major mysteries of evolution and inheritance had been mostly worked out. However, as we see in chapter 18, it is turning out to be a lot more complicated than anyone ever thought.

CHAPTER 7 | Defining and Measuring Complexity

THIS BOOK IS ABOUT COMPLEXITY, but so far I haven't defined this term rigorously or given any clear way to answer questions such as these: Is a human brain more complex than an ant brain? Is the human genome more complex than the genome of yeast? Did complexity in biological organisms increase over the last four billion years of evolution? Intuitively, the answer to these questions would seem to be "of course." However, it has been surprisingly difficult to come up with a universally accepted definition of complexity that can help answer these kinds of questions.

In 2004 I organized a panel discussion on complexity at the Santa Fe Institute's annual Complex Systems Summer School. It was a special year: 2004 marked the twentieth anniversary of the founding of the institute. The panel consisted of some of the most prominent members of the SFI faculty, including Doyne Farmer, Jim Crutchfield, Stephanie Forrest, Eric Smith, John Miller, Alfred Hübler, and Bob Eisenstein—all well-known scientists in fields such as physics, computer science, biology, economics, and decision theory. The students at the school—young scientists at the graduate or postdoctoral level—were given the opportunity to ask any question of the panel. The first question was, "How do you define *complexity*?" Everyone on the panel laughed, because the question was at once so straightforward, so expected, and yet so difficult to answer. Each panel member then proceeded to give a different definition of the term. A few arguments even broke out between members of the faculty over their respective definitions. The students were a bit shocked and frustrated. If the faculty of the Santa Fe Institute—the most famous institution in the world devoted to research on complex systems—could not agree on what was meant by *complexity,* then how can there even begin to be a science of complexity?

The answer is that there is not yet a single science of complexity but rather several different sciences of complexity with different notions of what complexity means. Some of these notions are quite formal, and some are still very informal. If the *sciences* of complexity are to become a unified *science* of complexity, then people are going to have to figure out how these diverse notions—formal and informal—are related to one another, and how to most usefully refine the overly complex notion of *complexity*. This is work that largely remains to be done, perhaps by those shocked and frustrated students as they take over from the older generation of scientists.

I don't think the students should have been shocked and frustrated. Any perusal of the history of science will show that the lack of a universally accepted definition of a central term is more common than not. Isaac Newton did not have a good definition of force, and in fact, was not happy about the concept since it seemed to require a kind of magical "action at a distance," which was not allowed in mechanistic explanations of nature. While genetics is one of the largest and fastest growing fields of biology, geneticists still do not agree on precisely what the term *gene* refers to at the molecular level. Astronomers have discovered that about 95% of the universe is made up of "dark matter" and "dark energy" but have no clear idea what these two things actually consist of. Psychologists don't have precise definitions for *idea* or *concept*, or know what these correspond to in the brain. These are just a few examples. Science often makes progress by inventing new terms to describe incompletely understood phenomena; these terms are gradually refined as the science matures and the phenomena become more completely understood. For example, physicists now understand all forces in nature to be combinations of four different kinds of fundamental forces: electromagnetic, strong, weak, and gravitational. Physicists have also theorized that the seeming "action at a distance" arises from the interaction of elementary particles. Developing a single theory that describes these four fundamental forces in terms of quantum mechanics remains one of the biggest open problems in all of physics. Perhaps in the future we will be able to isolate the different fundamental aspects of "complexity" and eventually unify all these aspects in some overall understanding of what we now call complex phenomena.

The physicist Seth Lloyd published a paper in 2001 proposing three different dimensions along which to measure the complexity of an object or process:

How hard is it to describe?

How hard is it to create?

What is its degree of organization?

Lloyd then listed about forty measures of complexity that had been proposed by different people, each of which addressed one or more of these three questions using concepts from dynamical systems, thermodynamics, information theory, and computation. Now that we have covered the background for these concepts, I can sketch some of these proposed definitions.

To illustrate these definitions, let's use the example of comparing the complexity of the human genome with the yeast genome. The human genome contains approximately three billion base pairs (i.e., pairs of nucleotides). It has been estimated that humans have about 25,000 genes—that is, regions that code for proteins. Surprisingly, only about 2% of base pairs are actually parts of genes; the nongene parts of the genome are called *noncoding regions*. The noncoding regions have several functions: some of them help keep their chromosomes from falling apart; some help control the workings of actual genes; some may just be "junk" that doesn't really serve any purpose, or has some function yet to be discovered.

I'm sure you've heard of the Human Genome project, but you may not know that there was also a Yeast Genome Project, in which the complete DNA sequences of several varieties of yeast were determined. The first variety that was sequenced turned out to have approximately twelve million base pairs and six thousand genes.

Complexity as Size

One simple measure of complexity is size. By this measure, humans are about 250 times as complex as yeast if we compare the number of base pairs, but only about four times as complex if we count genes.

Since 250 is a pretty big number, you may now be feeling rather complex, at least as compared with yeast. However, disappointingly, it turns out that the amoeba, another type of single-celled microorganism, has about 225 times as many base pairs as humans do, and a mustard plant called *Arabidopsis* has about the same number of genes that we do.

Humans are obviously more complex than amoebae or mustard plants, or at least I would like to think so. This means that genome size is not a very good measure of complexity; our complexity must come from something deeper than our absolute number of base pairs or genes (See figure 7.1).

Complexity as Entropy

Another proposed measure of the complexity of an object is simply its Shannon entropy, defined in chapter 3 to be the average information

FIGURE 7.1. Clockwise from top left: Yeast, an amoeba, a human, and *Arabidopsis*. Which is the most complex? If you used genome length as the measure of complexity, then the amoeba would win hands down (if only it had hands). (Yeast photograph from NASA, [http://www.nasa.gov/mission_pages/station/science/experiments/Yeast-GAP.html]; amoeba photograph from NASA [http://ares.jsc.nasa.gov/astrobiology/biomarkers/_images/amoeba.jpg]; *Arabidopsis* photograph courtesy of Kirsten Bomblies; Darwin photograph reproduced with permission from John van Wyhe, ed., The Complete Work of Charles Darwin Online [http://darwin-online.org.uk/].)

content or "amount of surprise" a message source has for a receiver. In our example, we could define a *message* to be one of the symbols A, C, G, or T. A highly ordered and very easy-to-describe sequence such as "A A A A A A A ... A" has entropy equal to zero. A completely random sequence has the maximum possible entropy.

There are a few problems with using Shannon entropy as a measure of complexity. First, the object or process in question has to be put in the form of "messages" of some kind, as we did above. This isn't always easy or straightforward—how, for example, would we measure the entropy of the human brain? Second, the highest entropy is achieved by a random set of messages. We could make up an artificial genome by choosing a bunch of random As, Cs, Gs, and Ts. Using entropy as the measure of complexity, this random, almost certainly nonfunctional genome would be considered more complex than the human genome. Of course one of the things that makes humans complex, in the intuitive sense, is precisely that our genomes aren't random but have been evolved over long periods to encode genes *useful* to our survival, such as the ones that control the development of eyes and muscles. The most complex entities are not the most ordered or random ones but somewhere in between. Simple Shannon entropy doesn't capture our intuitive concept of complexity.

Complexity as Algorithmic Information Content

Many people have proposed alternatives to simple entropy as a measure of complexity. Most notably Andrey Kolmogorov, and independently both Gregory Chaitin and Ray Solomonoff, proposed that the complexity of an object is the size of the shortest computer program that could generate a complete description of the object. This is called the *algorithmic information content* of the object. For example, think of a very short (artificial) string of DNA:

A C A C A C A C A C A C A C A C A C A C (string 1).

A very short computer program, "Print A C ten times," would spit out this pattern. Thus the string has low algorithmic information content. In contrast, here is a string I generated using a pseudo-random number generator:

A T C T G T C A A G A C G G A A C A T (string 2)

Assuming my random number generator is a good one, this string has no discernible overall pattern to it, and would require a longer program, namely "Print the exact string A T C T G T C A A A A C G G A A C A T." The idea is that string 1 is compressible, but string 2 is not, so contains more algorithmic information. Like entropy, algorithmic information content assigns higher information content to random objects than ones we would intuitively consider to be complex.

The physicist Murray Gell-Mann proposed a related measure he called "effective complexity" that accords better with our intuitions about complexity. Gell-Mann proposed that any given entity is composed of a combination of regularity and randomness. For example, string 1 above has a very simple regularity: the repeating A C motif. String 2 has no regularities, since it was generated at random. In contrast, the DNA of a living organism has some regularities (e.g., important correlations among different parts of the genome) probably combined with some randomness (e.g., true junk DNA).

To calculate the effective complexity, first one figures out the best description of the regularities of the entity; the effective complexity is defined as the amount of information contained in that description, or equivalently, the algorithmic information content of the set of regularities.

String 1 above has the regularity that it is A C repeated over and over. The amount of information needed to describe this regularity is the algorithmic information content of this regularity: the length of the program "Print A C some number of times." Thus, entities with very predictable structure have low effective complexity.

In the other extreme, string 2, being random, has no regularities. Thus there is no information needed to describe its regularities, and while the algorithmic information content of the string itself is maximal, the algorithmic information content of the string's *regularities*—its effective complexity—is zero. In short, as we would wish, both very ordered and very random entities have low effective complexity.

The DNA of a viable organism, having many independent and interdependent regularities, would have high effective complexity because its regularities presumably require considerable information to describe.

The problem here, of course, is how do we figure out what the regularities are? And what happens if, for a given system, various observers do not agree on what the regularities are?

Gell-Mann makes an analogy with scientific theory formation, which is, in fact, a process of finding regularities about natural phenomena. For any given phenomenon, there are many possible theories that express its regularities, but clearly some theories—the simpler and more elegant ones—are better than others. Gell-Mann knows a lot about this kind of thing—he shared the 1969 Nobel prize in Physics for his wonderfully elegant theory that finally made sense of the (then) confusing mess of elementary particle types and their interactions.

In a similar way, given different proposed sets of regularities that fit an entity, we can determine which is best by using the test called Occam's Razor. The best set of regularities is the smallest one that describes the entity in question and at the same time minimizes the remaining random component of that entity. For example, biologists today have found many regularities in the human genome, such as genes, regulatory interactions among genes, and so on, but these regularities still leave a lot of seemingly random aspects that don't obey any regularities—namely, all that so-called junk DNA. If the

Murray Gell-Mann of biology were to come along, he or she might find a better set of regularities that is simpler than that which biologists have so far identified and that is obeyed by more of the genome.

Effective complexity is a compelling idea, though like most of the proposed measures of complexity, it is hard to actually measure. Critics also have pointed out that the subjectivity of its definition remains a problem.

Complexity as Logical Depth

In order to get closer to our intuitions about complexity, in the early 1980s the mathematician Charles Bennett proposed the notion of *logical depth*. The logical depth of an object is a measure of how difficult that object is to construct. A highly ordered sequence of A, C, G, T (e.g., string 1, mentioned previously) is obviously easy to construct. Likewise, if I asked you to give me a random sequence of A, C, G, and T, that would be pretty easy for you to do, especially with the help of a coin you could flip or dice you could roll. But if I asked you to give me a DNA sequence that would produce a viable organism, you (or any biologist) would be very hard-pressed to do so without cheating by looking up already-sequenced genomes.

In Bennett's words, "Logically deep objects ... contain internal evidence of having been the result of a long computation or slow-to-simulate dynamical process, and could not plausibly have originated otherwise." Or as Seth Lloyd says, "It is an appealing idea to identify the complexity of a thing with the amount of information processed in the most plausible method of its creation."

To define logical depth more precisely, Bennett equated the *construction of an object* with the computation of a string of 0s and 1s encoding that object. For our example, we could assign to each nucleotide letter a two-digit code: $A = 00, C = 01, G = 10$, and $T = 11$. Using this code, we could turn any sequence of A, C, G, and T into a string of 0s and 1s. The logical depth is then defined as the number of steps that it would take for a properly programmed Turing machine, starting from a blank tape, to construct the desired sequence as its output.

Since, in general, there are different "properly programmed" Turing machines that could all produce the desired sequence in different amounts of time, Bennett had to specify which Turing machine should be used. He proposed that the shortest of these (i.e., the one with the least number of states and rules) should be chosen, in accordance with the above-mentioned Occam's Razor.

Logical depth has very nice theoretical properties that match our intuitions, but it does not give a practical way of measuring the complexity of any natural object of interest, since there is typically no practical way of finding the smallest Turing machine that could have generated a given object, not to mention determining how long that machine would take to generate it. And this doesn't even take into account the difficulty, in general, of describing a given object as a string of 0s and 1s.

Complexity as Thermodynamic Depth

In the late 1980s, Seth Lloyd and Heinz Pagels proposed a new measure of complexity, *thermodynamic depth*. Lloyd and Pagels' intuition was similar to Bennett's: more complex objects are harder to construct. However, instead of measuring the number of steps of the Turing machine needed to construct the description of an object, thermodynamic depth starts by determining "the most plausible scientifically determined sequence of events that lead to the thing itself," and measures "the total amount of thermodynamic and informational resources required by the physical construction process."

For example, to determine the thermodynamic depth of the human genome, we might start with the genome of the very first creature that ever lived and list all the evolutionary genetic events (random mutations, recombinations, gene duplications, etc.) that led to modern humans. Presumably, since humans evolved billions of years later than amoebas, their thermodynamic depth is much greater.

Like logical depth, thermodynamic depth is appealing in theory, but in practice has some problems as a method for measuring complexity. First, there is the assumption that we can, in practice, list all the events that lead to the creation of a particular object. Second, as pointed out by some critics, it's not clear from Seth Lloyd and Heinz Pagels' definition just how to define "an event." Should a genetic mutation be considered a single event or a group of millions of events involving all the interactions between atoms and subatomic particles that cause the molecular-level event to occur? Should a genetic recombination between two ancestor organisms be considered a single event, or should we include all the microscale events that cause the two organisms to end up meeting, mating, and forming offspring? In more technical language, it's not clear how to "coarse-grain" the states of the system— that is, how to determine what are the relevant *macrostates* when listing events.

Complexity as Computational Capacity

If complex systems—both natural and human-constructed—can perform computation, then we might want to measure their complexity in terms of the sophistication of what they can compute. The physicist Stephen Wolfram, for example, has proposed that systems are complex if their computational abilities are equivalent to those of a universal Turing machine. However, as Charles Bennett and others have argued, the ability to perform universal computation doesn't mean that a system by itself is complex; rather, we should measure the complexity of the behavior of the system coupled with its inputs. For example, a universal Turing machine alone isn't complex, but together with a machine code and input that produces a sophisticated computation, it creates complex behavior.

Statistical Complexity

Physicists Jim Crutchfield and Karl Young defined a different quantity, called *statistical complexity*, which measures the minimum amount of information about the past behavior of a system that is needed to optimally predict the *statistical behavior* of the system in the future. (The physicist Peter Grassberger independently defined a closely related concept called *effective measure complexity*.) *Statistical complexity* is related to Shannon's entropy in that a system is thought of as a "message source" and its behavior is somehow quantified as discrete "messages." Here, predicting the statistical behavior consists of constructing a model of the system, based on observations of the messages the system produces, such that the model's behavior is *statistically* indistinguishable from the behavior of the system itself.

For example, a model of the message source of string 1 above could be very simple: "repeat A C"; thus its statistical complexity is low. However, in contrast to what could be done with entropy or algorithmic information content, a simple model could also be built of the message source that generates string 2: "choose at random from A, C, G, or T." The latter is possible because models of statistical complexity are permitted to include random choices. The quantitative value of statistical complexity is the information content of the simplest such model that predicts the system's behavior. Thus, like effective complexity, statistical complexity is low for both highly ordered and random systems, and is high for systems in between—those that we would intuitively consider to be complex.

Like the other measures described above, it is typically not easy to measure statistical complexity if the system in question does not have a ready interpretation as a message source. However, Crutchfield, Young, and their colleagues have actually measured the statistical complexity of a number of real-world phenomena, such as the atomic structure of complicated crystals and the firing patterns of neurons.

Complexity as Fractal Dimension

So far all the complexity measures I have discussed have been based on information or computation theoretic concepts. However, these are not the only possible sources of measures of complexity. Other people have proposed concepts from dynamical systems theory to measure the complexity of an object or process. One such measure is the *fractal dimension* of an object. To explain this measure, I must first explain what a *fractal* is.

The classic example of a fractal is a coastline. If you view a coastline from an airplane, it typically looks rugged rather than straight, with many inlets, bays, prominences, and peninsulas (Figure 7.2, top). If you then view the same coastline from your car on the coast highway, it still appears to have the exact same kind of ruggedness, but on a smaller scale (Figure 7.2, bottom). Ditto for the close-up view when you stand on the beach and even for the ultra close-up view of a snail as it crawls on individual rocks. The similarity of the shape of the coastline at different scales is called "self-similarity."

The term *fractal* was coined by the French mathematician Benoit Mandelbrot, who was one of the first people to point out that the world is full of fractals—that is, many real-world objects have a rugged self-similar structure. Coastlines, mountain ranges, snowflakes, and trees are often-cited examples. Mandelbrot even proposed that the universe is fractal-like in terms of the distribution of galaxies, clusters of galaxies, clusters of clusters, et cetera. Figure 7.3 illustrates some examples of self-similarity in nature.

Although the term *fractal* is sometimes used to mean different things by different people, in general a fractal is a geometric shape that has "fine structure at every scale." Many fractals of interest have the self-similarity property seen in the coastline example given above. The logistic-map bifurcation diagram from chapter 2 (figure 2.6) also has some degree of self-similarity; in fact the chaotic region of this (R greater than 3.57 or so) and many other systems are sometimes called *fractal attractors*.

Mandelbrot and other mathematicians have designed many different mathematical models of fractals in nature. One famous model is the so-called Koch

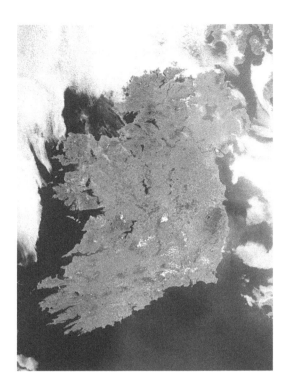

FIGURE 7.2. Top: Large-scale aerial view of Ireland, whose coastline has self-similar (fractal) properties. Bottom: Smaller-scale view of part of the Irish coastline. Its rugged structure at this scale resembles the rugged structure at the larger scale. (Top photograph from NASA Visible Earth [http://visibleearth.nasa.gov/]. Bottom photograph by Andreas Borchert, licensed under Creative Commons [http://creativecommons.org/licenses/by/3.0/].)

FIGURE 7.3. Other examples of fractal-like structures in nature: A tree, a snowflake (microscopically enlarged), a cluster of galaxies. (Tree photograph from the National Oceanic and Atmospheric Administration Photo Library. Snowflake photograph from [http://www.SnowCrystals.com], courtesy of Kenneth Libbrecht. Galaxy cluster photograph from NASA Space Telescope Science Institute.)

curve (Koch, pronounced "Coke," is the name of the Swedish mathematician who proposed this fractal). The Koch curve is created by repeated application of a rule, as follows.

1. Start with a single line.

———————————————

2. Apply the Koch curve rule: "For each line segment, replace its middle third by two sides of a triangle, each of length 1/3 of the original segment." Here there is only one line segment; applying the rule to it yields:

3. Apply the Koch curve rule to the resulting figure. Keep doing this forever. For example, here are the results from a second, third, and fourth application of the rule:

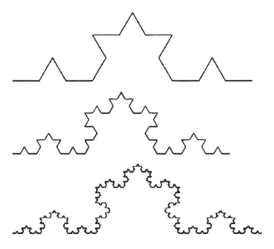

This last figure looks a bit like an idealized coastline. (In fact, if you turn the page 90 degrees to the left and squint really hard, it looks just like the west coast of Alaska.) Notice that it has true self-similarity: all of the subshapes, and their subshapes, and so on, have the same shape as the overall curve. If we applied the Koch curve rule an infinite number of times, the figure would be self-similar at an infinite number of scales—a perfect fractal. A real coastline of course does not have true self-similarity. If you look at a small section of the coastline, it does not have exactly the same shape as the entire coastline, but is visually similar in many ways (e.g., curved and rugged). Furthermore, in real-world objects, self-similarity does not go all the way to infinitely small scales. Real-world structures such as coastlines are often called "fractal" as a shorthand, but it is more accurate to call them "fractal-like," especially if a mathematician is in hearing range.

Fractals wreak havoc with our familiar notion of spatial dimension. A line is one-dimensional, a surface is two-dimensional, and a solid is three-dimensional. What about the Koch curve?

First, let's look at what exactly *dimension* means for regular geometric objects such as lines, squares, and cubes.

Start with our familiar line segment. Bisect it (i.e., cut it in half). Then bisect the resulting line segments, continuing at each level to bisect each line segment:

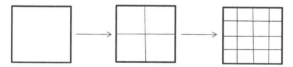

Each level is made up of two half-sized copies of the previous level.

Now start with a square. Bisect each side. Then bisect the sides of the resulting squares, continuing at each level to bisect every side:

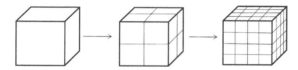

Each level is made up of four one-quarter-sized copies of the previous level.

Now, you guessed it, take a cube and bisect all the sides. Keep bisecting the sides of the resulting cubes:

Each level is made up of eight one-eighth-sized copies of the previous level.

This sequence gives a meaning of the term *dimension*. In general, each level is made up of smaller copies of the previous level, where the number of copies is 2 raised to the power of the dimension ($2^{dimension}$). For the line, we get $2^1 = 2$ copies at each level; for the square we get $2^2 = 4$ copies at each level, and for the cube we get $2^3 = 8$ copies at each level. Similarly, if you trisect instead of bisect the lengths of the line segments at each level, then each level is made up of $3^{dimension}$ copies of the previous level. I'll state this as a general formula:

> Create a geometric structure from an original object by repeatedly dividing the length of its sides by a number x. Then each level is made up of $x^{dimension}$ copies of the previous level.

Indeed, according to this definition of dimension, a line is one-dimensional, a square two-dimensional and a cube three-dimensional. All good and well.

Let's apply an analogous definition to the object created by the Koch rule. At each level, the line segments of the object are three times smaller than before, and each level consists of four copies of the previous level. By our definition above, it must be true that $3^{dimension}$ is equal to 4. What is the dimension? To figure it out, I'll do a calculation out of your sight (but detailed in the notes), and attest that according to our formula, the dimension is approximately 1.26. That is, the Koch curve is neither one- nor two-dimensional, but in between. Amazingly enough, fractal dimensions are not integers. That's what makes fractals so strange.

In short, the fractal dimension quantifies the number of copies of a self-similar object at each level of magnification of that object. Equivalently, fractal dimension quantifies how the total size (or area, or volume) of an object will change as the magnification level changes. For example, if you measure the total length of the Koch curve each time the rule is applied, you will find that each time the length has increased by 4/3. Only perfect fractals—those whose levels of magnification extend to infinity—have precise fractal dimension. For real-world finite fractal-like objects such as coastlines, we can measure only an approximate fractal dimension.

I have seen many attempts at intuitive descriptions of what fractal dimension means. For example, it has been said that fractal dimension represents the "roughness," "ruggedness," "jaggedness," or "complicatedness" of an object; an object's degree of "fragmentation"; and how "dense the structure" of the object is. As an example, compare the coastline of Ireland (figure 7.2) with that of South Africa (figure 7.4). The former has higher fractal dimension than the latter.

One description I like a lot is the rather poetic notion that fractal dimension "quantifies the cascade of detail" in an object. That is, it quantifies how much detail you see at all scales as you dive deeper and deeper into the infinite cascade of self-similarity. For structures that aren't fractals, such as a smooth round marble, if you keep looking at the structure with increasing magnification, eventually there is a level with no interesting details. Fractals, on the other hand, have interesting details at all levels, and fractal dimension in some sense quantifies how interesting that detail is as a function of how much magnification you have to do at each level to see it.

This is why people have been attracted to fractal dimension as a way of measuring complexity, and many scientists have applied this measure to

FIGURE 7.4. Coastline of South Africa. (Photograph from NASA Visible Earth [http://visibleearth.nasa.gov].)

real-world phenomena. However, ruggedness or cascades of detail are far from the only kind of complexity we would like to measure.

Complexity as Degree of Hierarchy

In Herbert Simon's famous 1962 paper "The Architecture of Complexity" Simon proposed that the complexity of a system can be characterized in terms of its degree of *hierarchy*: "the complex system being composed of subsystems that, in turn, have their own subsystems, and so on." Simon was a distinguished political scientist, economist, and psychologist (among other things); in short, a brilliant polymath who probably deserves a chapter of his own in this book.

Simon proposed that the most important common attributes of complex systems are *hierarchy* and *near-decomposibility*. Simon lists a number of complex systems that are structured hierarchically—e.g., the body is composed of organs, which are in turn composed of cells, which are in turn composed of celluar subsystems, and so on. In a way, this notion is similar to fractals in the idea that there are self-similar patterns at all scales.

Near-decomposibility refers to the fact that, in hierarchical complex systems, there are many more strong interactions within a subsystem than between subsystems. As an example, each cell in a living organism has a metabolic network that consists of a huge number of interactions among substrates, many more than take place between two different cells.

Simon contends that evolution can design complex systems in nature only if they can be put together like building blocks—that is, only if they are hierachical and nearly decomposible; a cell can evolve and then become a building block for a higher-level organ, which itself can become a building block for an even higher-level organ, and so forth. Simon suggests that what the study of complex systems needs is "a theory of hierarchy."

Many others have explored the notion of hierarchy as a possible way to measure complexity. As one example, the evolutionary biologist Daniel McShea, who has long been trying to make sense of the notion that the complexity of organisms increases over evolutionary time, has proposed a hierarchy scale that can be used to measure the degree of hierarchy of biological organisms. McShea's scale is defined in terms of levels of *nestedness*: a higher-level entity contains as parts entities from the next lower level. McShea proposes the following biological example of nestedness:

Level 1: *Prokaryotic* cells (the simplest cells, such as bacteria)

Level 2: Aggregates of level 1 organisms, such as *eukaryotic* cells (more complex cells whose evolutionary ancestors originated from the fusion of prokaryotic cells)

Level 3: Aggregates of level 2 organisms, namely all multicellular organisms

Level 4: Aggregates of level 3 organisms, such as insect colonies and "colonial organisms" such as the Portuguese man o' war.

Each level can be said to be more complex than the previous level, at least as far as nestedness goes. Of course, as McShea points out, nestedness only describes the *structure* of an organism, not any of its *functions*.

McShea used data both from fossils and modern organisms to show that the maximum hierarchy seen in organisms increases over evolutionary time. Thus this is one way in which complexity seems to have quantifiably increased with evolution, although measuring the degree of hierarchy in actual organisms can involve some subjectivity in determining what counts as a "part" or even a "level."

There are many other measures of complexity that I don't have space to cover here. Each of these measures captures something about our notion of complexity but all have both theoretical and practical limitations, and have so far rarely been useful for characterizing any real-world system. The diversity of measures that have been proposed indicates that the notions of complexity that we're trying to get at have many different interacting dimensions and probably can't be captured by a single measurement scale.

PART II | Life and Evolution in Computers

Nature proceeds little by little from things lifeless to animal life in such a way that it is impossible to determine the exact line of demarcation.

—Aristotle, *History of Animals*

[W]e *all know intuitively what life is. It is edible, lovable, or lethal.*

—James Lovelock, *The Ages of Gaia*

CHAPTER 8 | Self-Reproducing Computer Programs

What Is Life?

CHAPTER 5 DESCRIBED SOME OF THE HISTORY of ideas about how life has evolved. But a couple of things were missing, such as, how did life originate in the first place? And what exactly constitutes being alive? As you can imagine, both questions are highly contentious in the scientific world, and no one yet has definitive answers. Although I do not address the first question here, there has been some fascinating research on it in the complex systems community.

The second question—what is life, exactly?—has been on the minds of people probably for as long as "people" have existed. There is still no good agreement among either scientists or the general public on the definition of life. Questions such as "When does life begin?" or "What form could life take on other planets?" are still the subject of lively, and sometimes vitriolic, debate.

The idea of creating artificial life is also very old, going back at least two millennia to legends of the Golem and of Ovid's Pygmalion, continuing in the nineteenth-century story of Frankenstein's monster, all the way to the present era of movies such as *Blade Runner* and *The Matrix*, and computer games such as "Sim Life."

These works of fiction both presage and celebrate a new, technological version of the "What is life?" question: Is it possible for computers or robots to be considered "alive"? This question links the previously separate topics of computation and of life and evolution.

You can ask ten biologists what are the ten key requisites for life and you'll get a different list each time. Most are likely to include autonomy, metabolism, self-reproduction, survival instinct, and evolution and adaptation. As a start, can we understand these processes mechanistically and capture them in computers?

Many people have argued a vehement "no" for the following reasons:

Autonomy: A computer can't do anything on its own; it can do only what humans program it to do.

Metabolism: Computers can't create or gather their own energy from their environment like living organisms do; they have to be fed energy (e.g., electricity) by humans.

Self-reproduction: A computer can't reproduce itself; to do so it would have to contain a description of itself, and that description would have to contain a description of itself, and so on *ad infinitum*.

Survival instinct: Computers don't care whether they survive or not and they don't care how successful they are. (For example, in a lecture I attended by a prominent psychologist, the speaker, discussing the success of computer chess programs, asserted that "Deep Blue may have beat Kasparov, but it didn't get any joy out of it.")

Evolution and adaptation: A computer can't evolve or adapt on its own; it is restricted to change only in ways specified ahead of time by its programmer.

Although these arguments are still believed by many people, all of them have been claimed to be disproven in one way or another in the field of *artificial life*, whose purview is the simulation and "creation" of life on computers. In this chapter and the next I focus on those issues most closely related to Darwinism—self-reproduction and evolution.

Self-Reproduction in Computers

The self-reproduction argument is the most mathematical one: it states that self-reproduction in a computer would lead to an infinite regress.

Let's investigate this issue via the simplest version of the computer self-reproduction problem: write a computer program that prints out an exact copy of itself and nothing else.

I've written the following programs in a simple computer language that even nonprogrammers should be able to understand. (It's actually a pseudo language, with a few commands that most real languages wouldn't have, but still plausible and thrown in to make things simpler.)

Here's a first attempt. I start out with the name of the program:

```
program copy
```

Now I need to add an instruction to print out the name of the program:

```
program copy
    print("program copy")
```

The print command simply prints on the computer screen the characters between the first and last quotation marks, followed by a carriage return. Now I need to add an instruction to print out that second line:

```
program copy
    print("program copy")
    print("    print("program copy")")
```

Note that, since I want the program to print out an *exact* copy of itself, the second print command has to include the four spaces of indentation I put before the first print command, plus the two quotation marks, which are now themselves being quoted (the print command prints anything, including quotation marks, between the outermost quotation marks). Now I need another line to print out that third line:

```
program copy
    print("program copy")
    print("    print("program copy")")
    print("    print("    print("program copy")")")
```

and so forth. By now you can probably see how this strategy, in which each command prints an exact copy of the command preceding it, leads to an infinite regress. How can this be avoided? Before reading on, you might spend a few moments trying to solve this puzzle.

This simple-sounding problem turns out to have echos in the work of Kurt Gödel and Alan Turing, which I described in chapter 4. The solution also contains an essential means by which biological systems themselves get around the infinite regress. The solution was originally found, in the context of a more complicated problem, by the twentieth-century Hungarian mathematician John von Neumann.

Von Neumann was a pioneer in fields ranging from quantum mechanics to economics and a designer of one of the earliest electronic computers. His design consisted of a central processing unit that communicates with a random

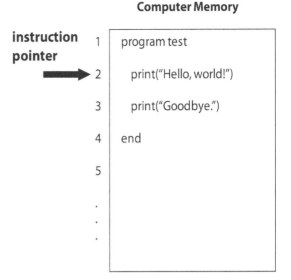

FIGURE 8.1. A simplified picture of computer memory, with numbered locations 1–5 and beyond, four of which contain lines of a program. The instruction pointer points to the instruction currently being executed by the computer. The lines sometimes contain leading spaces, which are ignored when the instruction is executed.

access memory in which both programs and data can be stored. It remains the basic design of all standard computers today. Von Neumann was also one of the first scientists who thought deeply about connections between computation and biology. He dedicated the last years of his life to solving the problem of how a machine might be able to reproduce itself; his solution was the first complete design for a self-reproducing machine. The self-copying computer program I will show you was inspired by his "self-reproducing automaton" and illustrates its fundamental principle in a simplified way.

Before showing you the self-copying program, I need to explain a few more things about the programming language I will be using.

Consider the picture of computer memory given in figure 8.1. Computer memory, in our highly simplified case, consists of numbered locations or "addresses," here numbered 1–5 and beyond. Each location contains some text. These lines of text can be interpreted by the computer as commands in a program or as data to be used by a program. The program currently stored in memory, when executed, will print

```
Hello, world!
Goodbye.
```

To accomplish this, the computer has an "instruction pointer"—a number also stored in memory, which always is equal to the memory location of the instruction currently being executed by the computer. The instruction pointer—let's call it ip for short—is initially set to the memory location containing the program's first line. We say it "points to" that instruction. At each step in the computation the instruction pointed to by ip is carried out, and ip is increased by 1.

For example, in figure 8.1, the value of ip is 2, and we say that ip is pointing to the line print("Hello, world!").

We call ip a *variable* since its value changes ("varies") as the computation is carried out.

We also can define another variable, line[n], as equal to the string of characters in memory location *n*. For example, the command print(line[2]) will print

```
print("Hello, world!")
```

Finally, our language contains a loop command. For example, the following lines of computer code,

```
x = 0
loop until x = 4
{
print("Hello, world!")
x = x + 1
}
```

will print

```
Hello, world!
Hello, world!
Hello, world!
Hello, world!
```

The commands inside the two curly brackets get repeated until the loop condition (here x = 4) is true. The variable x is used as a counter—it starts off at zero and is increased by 1 each time the loop is performed. When it gets to 4, the loop stops.

Now we are ready for the self-copying program, which appears in its entirety in figure 8.2. The best way to understand a computer program is to hand-simulate it; that is, to go through it line by line keeping track of what it does.

Suppose this program has been loaded into computer memory as shown in figure 8.2, and suppose a user comes along and types selfcopy on the computer's command line. This signals the computer to start interpreting the program called selfcopy. The interpreter—part of the computer's

FIGURE 8.2. A self-copying program.

Computer Memory

1	program selfcopy
2	L = ip − 1
3	loop until line[L] = "end"
4	{
5	print(line[L])
6	L = L + 1
7	}
8	print("end")
9	end

operating system—sets the instruction pointer to 1, which points to the name of the program. The ip then moves down, line by line, executing each instruction.

In memory location 2 a variable L is set to ip−1. Recall that ip is the location of the instruction currently being executed. So when line 2 is executed, ip is set to 2 and L is set to $2-1 = 1$. (Note that L will now stay equal to 1 until it is reset, even though ip changes as each instruction is executed.)

Next, a loop is entered, which will be iterated until line[L] is equal to the character string end. Remember that line[L] is equal to the string located in memory location L. Right now, L is set to 1, so line[L] is equal to the string program selfcopy. This is not equal to the string end, so the loop is continued. In the loop, line[L] is printed and L is incremented. First, with L = 1, program selfcopy is printed; then L is set to 2.

Now, line[L] is the second line of the program, namely L = ip - 1. Again, this string is not equal to end, so the loop is continued. In this way, each line of the program is printed out. A particularly interesting line is line 5: when line 5 is being executed with L = 5, the instruction print(line[L]) prints itself out. When L = 9 and line[L] is equal to end, the loop ends. At this point, lines 1–8 have been printed. The instruction pointer moves to line 8 (the instruction immediately following the loop), which, when executed, prints out the string "end," completing the self-copying.

The essence of self-copying in this program is to use the same information stored in memory in two ways: first as instructions to be executed, and second as data to be used (i.e., printed) by those instructions. This dual use of information is what allows us to avoid an infinite regress of the kind illustrated earlier by my first attempt at a self-copying program.

The Deeper Meaning of the Self-Reproducing Computer Program

The dual use of information is also at the heart of Gödel's paradox, embodied by his self-referential sentence "This statement is not provable."

This is a bit tricky to understand. First, let's note that this sentence, like any other English sentence, can be looked at in two ways: (1) as the literal string of letters, spaces, and punctuation contained in the sentence, and (2) as the *meaning* of that literal string, as interpreted by an English speaker.

To be very clear, let's call the literal string of characters S. That is, $S =$ "This statement is not provable." We can now state facts about S: for example, it contains twenty-six letters, four spaces, and one period.

Let's call the meaning of the sentence M. We can rewrite M as follows: "Statement S is not provable." In a way, you can think of M as a "command" and of S as the "data" for that command. The weird (and wonderful) thing is that the data S is the same thing as the command M. The chief reason Gödel was able to translate his English sentence into a paradox in mathematics was that he was able to phrase M as a mathematical statement and S as a number that encoded the string of characters of that mathematical statement.

This is all very tricky. This kind of distinction between literal strings of characters and the meaning of those strings, and the paradoxes that self-reference can produce, are discussed in a detailed and very entertaining way in Douglas Hofstadter's book *Gödel, Escher, Bach: an Eternal Golden Braid*.

Similarly, this kind of dual use of information is key to Turing's proof of the undecidability of the Halting problem. Remember H and H' from chapter 4? Do you recall how H' ran on its own code? That is, just like our self-reproducing computer program above, H' was used in two ways: as an interpreted program and as the input for that program.

Self-Replication in DNA

At this point you may be groaning that we're back in the abstract realm of logical headaches. But give me a minute to bring us back to reality. The really amazing thing is that this dual use of information is key to the way DNA replicates itself. As I described in chapter 6, DNA is made up of strings of nucleotides. Certain substrings (*genes*) encode amino acids making up proteins, including the enzymes (special kinds of proteins) that effect the splitting of the double helix and the copying of each strand via messenger RNA, transfer RNA, ribosomes, et cetera. In a very crude analogy, the DNA strings encoding the enzymes that perform the copying roughly correspond to the lines of code in the self-copying program. These "lines of code" in DNA are executed when the enzymes are created and act on the DNA itself, interpreting it as data to be split up and copied.

However, you may have noticed something I have so far swept under the rug. There is a major difference between my self-copying program and DNA self-reproduction. The self-copying program required an interpreter to execute it: an instruction pointer to move down the lines of computer code and a computer operating system to carry them out (e.g., actually perform the storing and retrieving of internal variables such as ip and L, actually print strings of characters, and so on). The interpreter is completely external to the program itself.

In contrast, in the case of DNA, the instructions for building the "interpreter"—the messenger RNA, transfer RNA, ribosomes, and all the other machinery of protein synthesis—are encoded along with everything else in the DNA. That is, DNA not only contains the code for its self-replicating "program" (i.e., the enzymes that perform the splitting and copying of DNA) but also it encodes its own interpreter (the cellular machinery that translates DNA into those very enzymes).

Von Neumann's Self-Reproducing Automaton

Von Neumann's original self-reproducing automaton (described mathematically but not actually built by von Neumann) similarly contained not only a self-copying program but also the machinery needed for its own interpretation. Thus it was truly a *self*-reproducing machine. This explains why von Neumann's construction was considerably more complicated than my simple self-copying program. That it was formulated in the 1950s, before the details of biological self-reproduction were well understood, is testament

John von Neumann, 1903–1957
(AIP Emilio Segre Visual Archives)

to von Neumann's insight. Von Neumann's design of this automaton and mathematical proof of its correctness were mostly completed when he died in 1957, at the age of 53, of cancer possibly caused by his exposure to radiation during his work on the atomic bomb. The proof was completed by von Neumann's colleague, Arthur Burks. The complete work was eventually published in 1966 as a book, *Theory of Self-Reproducing Automata*, edited by Burks.

Von Neumann's design for a self-reproducing automaton was one of the first real advances in the science of artificial life, demonstrating that self-reproduction by machine was indeed possible in principle, and providing a "logic" of self-reproduction that turned out to have some remarkable similarities to the one used by living systems.

Von Neumann recognized that these results could have profound consequences. He worried about public perception of the possibilities of self-reproducing machines, and said that he did not want any mention of the "reproductive potentialities of the machines of the future" made to the mass media. It took a while, but the mass media eventually caught up. In 1999, computer scientists Ray Kurzweil and Hans Moravec celebrated the possibility of super-intelligent self-reproducing robots, which they believe will be built in the near future, in their respective nonfiction (but rather far-fetched) books *The Age of Spiritual Machines* and *Robot*. In 2000 some of the possible perils of self-reproducing nano-machines were decried by Bill Joy, one of the founders of Sun Microsystems, in a now famous article in *Wired* magazine called "Why the Future Doesn't Need Us." So far none of these predictions has come to pass. However, complex self-reproducing machines may soon be a reality: some simple self-reproducing robots have already been constructed by roboticist Hod Lipson and his colleagues at Cornell University.

John von Neumann

It is worth saying a few words about von Neumann himself, one of the most important and interesting figures in science and mathematics in the twentieth century. He is someone you should know about if you don't already. Von Neumann was, by anyone's measure, a true genius. During his relatively short life he made fundamental contributions to at least six fields: mathematics, physics, computer science, economics, biology, and neuroscience. He is the type of genius whom people tell stories about, shaking their heads and wondering whether someone that smart could really be a member of the human species. I like these stories so much, I want to retell a few of them here.

Unlike Einstein and Darwin, whose genius took a while to develop, Hungarian born "Johnny" von Neumann was a child prodigy. He supposedly could divide eight-digit numbers in his head at the age of six. (It evidently took him a while to notice that not everyone could do this; as reported in one of his biographies, "When his mother once stared rather aimlessly in front of her, six-year-old Johnny asked: 'What are you calculating?'") At the same age he also could converse with his father in ancient Greek.

At the age of eighteen von Neumann went to university, first in Budapest, then in Germany and Switzerland. He first took the "practical" course of studying chemical engineering but couldn't be kept away from mathematics. He received a doctorate in math at the age of twenty-three, after doing fundamental work in both mathematical logic and quantum mechanics. His work was so good that just five years later he was given the best academic job in the world—a professorship (with Einstein and Gödel) at the newly formed Institute for Advanced Study (IAS) in Princeton.

The institute didn't go wrong in their bet on von Neumann. During the next ten years, von Neumann went on to invent the field of game theory (producing what has been called "the greatest paper on mathematical economics ever written"), design the conceptual framework of one of the first programmable computers (the EDVAC, for which he wrote what has been called "the most important document ever written about computing and computers"), and make central contributions to the development of the first atomic and hydrogen bombs. This was all before his work on self-reproducing automata and his exploration of the relationships between the logic of computers and the workings of the brain. Von Neumann also was active in politics (his positions were very conservative, driven by strong anti-communist views) and eventually became a member of the Atomic Energy Commission, which advised the U.S. president on nuclear weapons policy.

Von Neumann was part of what has been called the "Hungarian phenomenon," a group of several Hungarians of similar age who went on to become world-famous scientists. This group also included Leo Szilard, whom we heard about in chapter 3, the physicists Eugene Wigner, Edward Teller, and Denis Gabor, and the mathematicians Paul Erdös, John Kemeny, and Peter Lax. Many people have speculated on the causes of this improbable cluster of incredible talent. But as related by von Neumann biographer Norman MacRae, "Five of Hungary's six Nobel Prize winners were Jews born between 1875 and 1905, and one was asked why Hungary in his generation had brought forth so many geniuses. Nobel laureate Wigner replied that he did not understand the question. Hungary in that time had produced only one genius, Johnny von Neumann."

Von Neumann was in many ways ahead of his time. His goal was, like Turing's, to develop a general theory of information processing that would encompass both biology and technology. His work on self-reproducing automata was part of this program. Von Neumann also was closely linked to the so-called cybernetics community—an interdisciplinary group of scientists and engineers seeking commonalities among complex, adaptive systems in both natural and artificial realms. What we now call "complex systems" can trace its ancestry to cybernetics and the related field of systems science. I explore these connections further in the final chapter.

Von Neumann's interest in computation was not always well received at the elite Institute for Advanced Study. After completing his work on the EDVAC, von Neumann brought several computing experts to the IAS to work with him on designing and building an improved successor to EDVAC. This system was called the "IAS computer"; its design was a basis for the early computers built by IBM. Some of the "pure" scientists and mathematicians at IAS were uncomfortable with so practical a project taking place in their ivory tower, and perhaps even more uncomfortable with von Neumann's first application of this computer, namely weather prediction, for which he brought a team of meteorologists to the IAS. Some of the purists didn't think this kind of activity fit in with the institute's theoretical charter. As IAS physicist Freeman Dyson put it, "The [IAS] School of Mathematics has a permanent establishment which is divided into three groups, one consisting of pure mathematics, one consisting of theoretical physicists, and one consisting of Professor von Neumann." After von Neumann's death, the IAS computer project was shut down, and the IAS faculty passed a motion "to have no experimental science, no laboratories of any kind at the Institute." Freeman Dyson described this as, "The snobs took revenge."

CHAPTER 9 | Genetic Algorithms

AFTER HE ANSWERED THE QUESTION "Can a machine reproduce itself?" in the affirmative, von Neumann wanted to take the next logical step and have computers (or computer programs) reproduce themselves with mutations and compete for resources to survive in some environment. This would counter the "survival instinct" and "evolution and adaptation" arguments mentioned above. However, von Neumann died before he was able to work on the evolution problem.

Others quickly took up where he left off. By the early 1960s, several groups of researchers were experimenting with evolution in computers. Such work has come to be known collectively as *evolutionary computation*. The most widely known of these efforts today is the work on *genetic algorithms* done by John Holland and his students and colleagues at the University of Michigan.

John Holland is, in some sense, the academic grandchild of John von Neumann. Holland's own Ph.D. advisor was Arthur Burks, the philosopher, logician, and computer engineer who assisted von Neumann on the EDVAC computer and who completed von Neumann's unfinished work on self-reproducing automata. After his work on the EDVAC, Burks obtained a faculty position in philosophy at the University of Michigan and started the Logic of Computers group, a loose-knit collection of faculty and students who were interested in the foundations of computers and of information processing in general. Holland joined the University of Michigan as a Ph.D. student, starting in mathematics and later switching to a brand-new program called "communication sciences" (later "computer and communication sciences"), which was arguably the first real computer science department in the world. A few years later, Holland became the program's first Ph.D.

John Holland. (Photograph copyright © by the Santa Fe Institute. Reprinted by permission.)

recipient, giving him the distinction of having received the world's first Ph.D. in computer science. He was quickly hired as a professor in that same department.

Holland got hooked on Darwinian evolution when he read Ronald Fisher's famous book, *The Genetical Theory of Natural Selection*. Like Fisher (and Darwin), Holland was struck by analogies between evolution and animal breeding. But he looked at the analogy from his own computer science perspective: "That's where genetic algorithms came from. I began to wonder if you could breed programs the way people would say, breed good horses and breed good corn."

Holland's major interest was in the phenomenon of adaptation—how living systems evolve or otherwise change in response to other organisms or to a changing environment, and how computer systems might use similar principles to be adaptive as well. His 1975 book, *Adaptation in Natural and Artificial Systems*, laid out a set of general principles for adaptation, including a proposal for genetic algorithms.

My own first exposure to genetic algorithms was in graduate school at Michigan, when I took a class taught by Holland that was based on his book. I was instantly enthralled by the idea of "evolving" computer programs. (Like Thomas Huxley, my reaction was, "How extremely stupid not to have thought of that!")

A Recipe for a Genetic Algorithm

The term *algorithm* is used these days to mean what Turing meant by *definite procedure* and what cooks mean by *recipe*: a series of steps by which an input is transformed to an output.

In a *genetic* algorithm (GA), the desired output is a solution to some problem. Say, for example, that you are assigned to write a computer program that controls a robot janitor that picks up trash around your office building. You decide that this assignment will take up too much of your time, so you want to employ a genetic algorithm to evolve the program for you. Thus, the desired output from the GA is a robot-janitor control program that allows the robot to do a good job of collecting trash.

The input to the GA has two parts: a *population* of candidate programs, and a *fitness function* that takes a candidate program and assigns to it a *fitness* value that measures how well that program works on the desired task.

Candidate programs can be represented as strings of bits, numbers, or symbols. Later in this chapter I give an example of representing a robot-control program as a string of numbers.

In the case of the robot janitor, the fitness of a candidate program could be defined as the square footage of the building that is covered by the robot, when controlled by that program, in a set amount of time. The more the better.

Here is the recipe for the GA.

Repeat the following steps for some number of *generations*:

1. Generate an initial population of candidate solutions. The simplest way to create the initial population is just to generate a bunch of random programs (strings), called "individuals."
2. Calculate the fitness of each individual in the current population.
3. Select some number of the individuals with highest fitness to be the *parents* of the next generation.
4. Pair up the selected parents. Each pair produces offspring by recombining parts of the parents, with some chance of random mutations, and the offspring enter the new population. The selected parents continue creating offspring until the new population is full (i.e., has the same number of individuals as the initial population). The new population now becomes the current population.
5. Go to step 2.

Genetic Algorithms in the Real World

The GA described above is simple indeed, but versions of it have been used to solve hard problems in many scientific and engineering areas, as well as in art, architecture, and music.

Just to give you a flavor of these problems: GAs have been used at the General Electric Company for automating parts of aircraft design, Los Alamos National Lab for analyzing satellite images, the John Deere company for automating assembly line scheduling, and Texas Instruments for computer chip design. GAs were used for generating realistic computer-animated horses in the 2003 movie *The Lord of the Rings: The Return of the King*, and realistic computer-animated stunt doubles for actors in the movie *Troy*. A number of pharmaceutical companies are using GAs to aid in the discovery of new drugs. GAs have been used by several financial organizations for various tasks: detecting fraudulent trades (London Stock Exchange), analysis of credit card data (Capital One), and forecasting financial markets and portfolio optimization (First Quadrant). In the 1990s, collections of artwork created by an interactive genetic algorithm were exhibited at several museums, including the Georges Pompidou Center in Paris. These examples are just a small sampling of ways in which GAs are being used.

Evolving Robby, the Soda-Can-Collecting Robot

To introduce you in more detail to the main ideas of GAs, I take you through a simple extended example. I have a robot named "Robby" who lives in a (computer simulated, but messy) two-dimensional world that is strewn with empty soda cans. I am going to use a genetic algorithm to evolve a "brain" (that is, a control strategy) for Robby.

Robby's job is to clean up his world by collecting the empty soda cans. Robby's world, illustrated in figure 9.1, consists of 100 squares (sites) laid out in a 10 × 10 grid. You can see Robby in site 0,0. Let's imagine that there is a wall around the boundary of the entire grid. Various sites have been littered with soda cans (but with no more than one can per site).

Robby isn't very intelligent, and his eyesight isn't that great. From wherever he is, he can see the contents of one adjacent site in the north, south, east, and west directions, as well as the contents of the site he occupies. A site can be empty, contain a can, or be a wall. For example, in figure 9.1, Robby, at site 0,0, sees that his current site is empty (i.e., contains no soda cans), the "sites" to the north and west are walls, the site to the south is empty, and the site to the east contains a can.

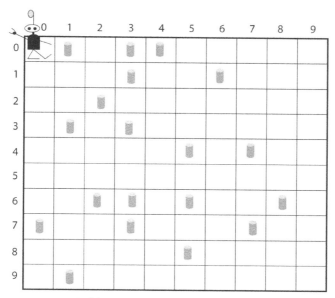

FIGURE 9.1. Robby's world. A 10 x 10 array, strewn with soda cans.

For each cleaning session, Robby can perform exactly 200 actions. Each action consists of one of the following seven choices: move to the north, move to the south, move to the east, move to the west, choose a random direction to move in, stay put, or bend down to pick up a can. Each action may generate a reward or a punishment. If Robby is in the same site as a can and picks it up, he gets a reward of ten points. However, if he bends down to pick up a can in a site where there is no can, he is fined one point. If he crashes into a wall, he is fined five points and bounces back into the current site.

Clearly, Robby's reward is maximized when he picks up as many cans as possible, without crashing into any walls or bending down to pick up a can if no can is there.

Since this is a simple problem, it would probably be pretty easy for a human to figure out a good strategy for Robby to follow. However, the point of genetic algorithms is that humans, being intrinsically lazy, don't have to figure out anything; we just let computer evolution figure it out for us. Let's use a genetic algorithm to evolve a good strategy for Robby.

The first step is to figure out exactly what we are evolving; that is, what exactly constitutes a *strategy*? In general, a strategy is a set of rules that gives, for any situation, the action you should take in that situation. For Robby, a "situation" is simply what he can see: the contents of his current site plus the contents of the north, south, east, and west sites. For the question "what to do in each situation," Robby has seven possible things he can do: move north, south, east, or west; move in a random direction; stay put; or pick up a can.

Therefore, a strategy for Robby can be written simply as a list of all the possible situations he could encounter, and for each possible situation, which of the seven possible actions he should perform.

How many possible situations are there? Robby looks at five different sites (current, north, south, east, west), and each of those sites can be labeled as empty, contains can, or wall. This means that there are 243 different possible situations (see the notes for an explanation of how I calculated this). Actually, there aren't really that many, since Robby will never face a situation in which his current site is a wall, or one in which north, south, east, and west are all walls. There are other "impossible" situations as well. Again, being lazy, we don't want to figure out what all the impossible situations are, so we'll just list all 243 situations, and know that some of them will never be encountered.

Table 9.1 is an example of a strategy—actually, only part of a strategy, since an entire strategy would be too long to list here.

Robby's situation in figure 9.1 is

North	South	East	West	Current Site
Wall	Empty	Can	Wall	Empty

To decide what to do next, Robby simply looks up this situation in his strategy table, and finds that the corresponding action is MoveWest. So he moves west. And crashes into a wall.

I never said this was a *good* strategy. Finding a good strategy isn't our job; it's the job of the genetic algorithm.

TABLE 9-1

Situation					Action
North	*South*	*East*	*West*	*Current Site*	
Empty	Empty	Empty	Empty	Empty	MoveNorth
Empty	Empty	Empty	Empty	Can	MoveEast
Empty	Empty	Empty	Empty	Wall	MoveRandom
Empty	Empty	Empty	Can	Empty	PickUpCan
⋮	⋮	⋮	⋮	⋮	⋮
Wall	Empty	Can	Wall	Empty	MoveWest
⋮	⋮	⋮	⋮	⋮	⋮
Wall	Wall	Wall	Wall	Wall	StayPut

I wrote the code for a genetic algorithm to evolve strategies for Robby. In my GA, each individual in the population is a strategy—a listing of the actions that correspond to each possible situation. That is, given a strategy such as the one in table 9.1, an individual to be evolved by the GA is just a listing of the 243 actions in the rightmost column, in the order given:

MoveNorth MoveEast MoveRandom PickUpCan ... MoveWest ... StayPut

The GA remembers that the first action in the string (here MoveNorth) goes with the first situation ("Empty Empty Empty Empty Empty"), the second action (here MoveEast) goes with the second situation ("Empty Empty Empty Empty Can"), and so on. In other words, I don't have to explicitly list the situations corresponding to these actions; instead the GA remembers the order in which they are listed. For example, suppose Robby happened to observe that he was in the following situation:

North	South	East	West	Current Site
Empty	Empty	Empty	Empty	Can

I build into the GA the knowledge that this is situation number 2. It would look at the strategy table and see that the action in position 2 is **MoveEast**. Robby moves east, and then observes his next situation; the GA again looks up the corresponding action in the table, and so forth.

My GA is written in the programming language C. I won't include the actual program here, but this is how it works.

1. **Generate the initial population.** The GA starts with an initial population of 200 *random* individuals (strategies).

   A random population is illustrated in figure 9.2. Each individual strategy is a list of 243 "genes." Each gene is a number between 0 and 6, which stands for an action (0 = *MoveNorth*, 1 = *MoveSouth*, 2 = *MoveEast*, 3 = *MoveWest*, 4 = *StayPut*, 5 = *PickUp*, and 6 = *RandomMove*). In the initial population, these numbers are filled in at random. For this (and all other probabilistic or random choices), the GA uses a pseudo-random-number generator.
   Repeat the following for 1,000 generations:
2. **Calculate the *fitness* of each individual in the population.** In my program, the fitness of a strategy is determined by seeing how well the strategy lets Robby do on 100 different cleaning sessions. A cleaning session consists of putting Robby at site 0, 0, and throwing down a bunch of cans at random (each site can contain at most one can; the

Individual 1:
233003234216303435305460061025625151141622604356543340665115141565022064064205100664321616152165202236443336334601332650300040622050243165006111305146664232401245633345524126143441361020150630642551654043264463156164510543665346310551646005164

Individual 2:
164113431210253603403612414312011042354625253042020445164336656103532215310513144062212061463143215461025652364442202534034530502005620634026331002453416430151631210012214400664012665246351650154123113132453304433212634555005314213064423311000

Individual 3:
204233444024112261321364526324642122061221222526606261444361253251266406133534015341111020616422665314552254023405115503130222020065445125062206631426135532010000400031640130154160162006134440626160505641421553133236021503355131253632642630551

. . .

Individual 200:
346325251360010122256121060433011352051553201306560053222350433242506412425526553463534552305332661201063212455442344061365430246240160663016464641103026540006334126150352262106063624260550616616344255124354464110023463330440102533212142402251

FIGURE 9.2. Random initial population. Each individual consists of 243 numbers, each of which is between 0 and 6, and each of which encodes an action. The location of a number in a string indicates to which situation the action corresponds.

probability of a given site containing a can is 50%). Robby then follows the strategy for 200 actions in each session. The score of the strategy in each session is the number of reward points Robby accumulates minus the total fines he incurs. The strategy's *fitness* is its average score over 100 different cleaning sessions, each of which has a different configuration of cans.

3. **Apply evolution** to the current population of strategies to create a new population. That is, repeat the following until the new population has 200 individuals:

   (a) Choose two parent individuals from the current population probabilistically based on fitness. That is, the higher a strategy's fitness, the more chance it has to be chosen as a parent.

(b) Mate the two parents to create two children. That is, randomly choose a position at which to split the two number strings; form one child by taking the numbers before that position from parent A and after that position from parent B, and vice versa to form the second child.

(c) With a small probability, mutate numbers in each child. That is, with a small probability, choose one or more numbers and replace them each with a randomly generated number between 0 and 6.

(d) Put the two children in the new population.

4. Once the new population has 200 individuals, return to step 2 with this new generation.

The magic is that, starting from a set of 200 random strategies, the genetic algorithm creates strategies that allow Robby to perform very well on his cleaning task.

The numbers I used for the population size (200), the number of generations (1,000), the number of actions Robby can take in a session (200), and the number of cleaning sessions to calculate fitness (100) were chosen by me, somewhat arbitrarily. Other numbers can be used and can also produce good strategies.

I'm sure you are now on the edge of your seat waiting to find out what happened when I ran this genetic algorithm. But first, I have to admit that before I ran it, I overcame my laziness and constructed my own "smart" strategy, so I could see how well the GA could do compared with me. My strategy for Robby is: "If there is a can in the current site, pick it up. Otherwise, if there is a can in one of the adjacent sites, move to that site. (If there are multiple adjacent sites with cans, I just specify the one to which Robby moves.) Otherwise, choose a random direction to move in."

This strategy actually isn't as smart as it could be; in fact, it can make Robby get stuck cycling around empty sites and never making it to some of the sites with cans.

I tested my strategy on 10,000 cleaning sessions, and found that its average (per-session) score was approximately 346. Given that at the beginning of each session, about 50%, or 50, of the sites contain cans, the maximum possible score for any strategy is approximately 500, so my strategy is not very close to optimal.

Can the GA do as well or better than this? I ran it to see. I took the highest-fitness individual in the final generation, and also tested it on 10,000 new and different cleaning sessions. Its average (per-session) score was approximately 483—that is, nearly optimal!

How Does the GA-Evolved Strategy Solve the Problem?

Now the question is, what is this strategy doing, and why does it do better than my strategy? Also, how did the GA evolve it?

Let's call my strategy M and the GA's strategy G. Below is each strategy's genome.

M: 6563536562523532526563536561513531512523532521513531516563 5365
    6252353252656353656050353050252353252050353050151353151252 3532
    5215135315105035305025235325205035305065635365625235325265 6353
    6561513531512523532521513531516563536562523532526563 53454

G: 2543551532562352510563554611513361541510341561105501500520 3025
    6256132252350325112052333054055231255051336154150665264150 2665
    0601226445360563152025643105435463240435033415325025325135 2352
    0451501301562134362523532231350512605133562015245143 43432

Staring at the genome of a strategy doesn't help us too much in understanding how that strategy works. We can see a few genes that make sense, such as the important situations in which Robby's current site contains a can, such as the second situation ("Empty Empty Empty Empty Can"), which has action 5 (*PickUp*) in both strategies. Such situations always have action 5 in M, but only most of the time in G. For example, I managed to determine that the following situation

North	South	East	West	Current Site
Empty	Can	Empty	Can	Can

has action 3 (*MoveWest*), which means Robby doesn't pick up the can in his current site. This seems like a bad idea, yet G does better than M overall! The key, it turns out, is not these isolated genes, but the way different genes interact, just as has been found in real genetics. And just as in real genetics, it's very difficult to figure out how these various interactions lead to the overall behavior or fitness.

It makes more sense to look at the actual behavior of each strategy—its *phenotype*—rather than its genome. I wrote a graphics program to display Robby's moves when using a given strategy, and spent some time watching the behavior of Robby when he used strategy M and when he used strategy G. Although the two strategies behave similarly in many situations, I found that strategy G employs two tricks that cause it to perform better than strategy M.

First, consider a situation in which Robby does not sense a can in his current site or in any of his neighboring sites. If Robby is following strategy

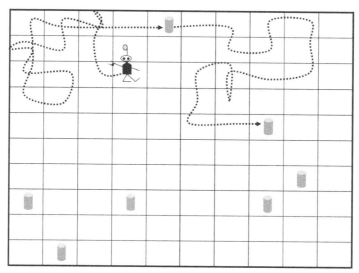

Strategy M

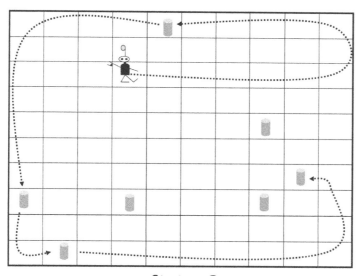

Strategy G

FIGURE 9.3. Robby in a "no-can" wilderness. The dotted lines show the paths he took in my simulation when he was following strategies M (top) and G (bottom).

M, he chooses a random move to make. However, if he is following strategy G, Robby moves to the east until he either finds a can or reaches a wall. He then moves north, and continues to circle the edge of the grid in a counterclockwise direction until a can is encountered or sensed. This is illustrated in figure 9.3 by the path Robby takes under each strategy (dotted line).

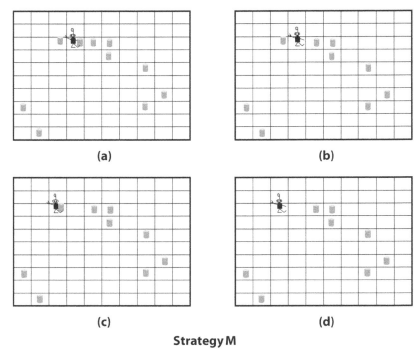

Strategy M

FIGURE 9.4. Robby in a cluster of cans, using strategy M over four time steps.

Not only does this circle-the-perimeter strategy prevent Robby from crashing into walls (a possibility under *M* whenever a random move is made), but it also turns out that circling the perimeter is a more efficient way to encounter cans than simply moving at random.

Second, with *G* the genetic algorithm discovered a neat trick by having Robby not pick up a can in his current site in certain situations.

For example, consider the situation illustrated in figure 9.4a. Given this situation, if Robby is following *M*, he will pick up the can in his current site, move west, and then pick up the can in his new site (pictures b–d). Because Robby can see only immediately adjacent sites, he now cannot see the remaining cluster of cans. He will have to move around at random until he encounters another can by chance.

In contrast, consider the same starting situation with *G*, illustrated in figure 9.5a. Robby doesn't pick up the can in his current site; instead he moves west (figure 9.5b). He then picks up the western-most can of the cluster (figure 9.5c). The can he didn't pick up on the last move acts as a marker so Robby can "remember" that there are cans on the other side of it. He goes on to pick up all of the remaining cans in the cluster (figure 9.5d–9.5k).

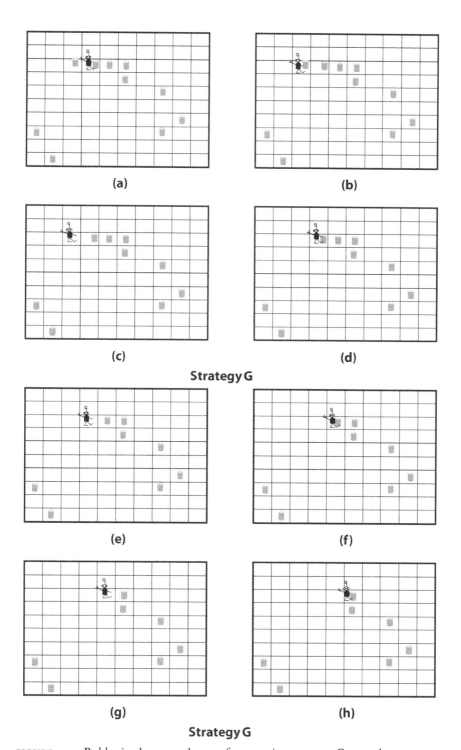

Strategy G

(a) (b) (c) (d) (e) (f) (g) (h)

Strategy G

FIGURE 9.5. Robby in the same cluster of cans, using strategy G over eleven time steps. (*Continued on next page*)

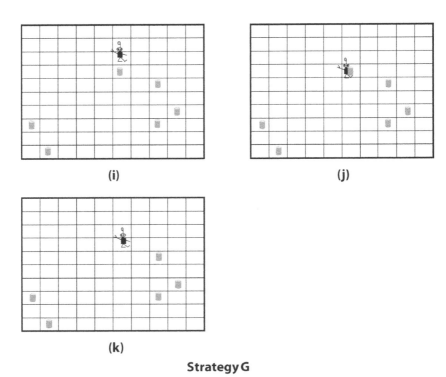

Strategy G

FIGURE 9.5. (*Continued*)

I knew that my strategy wasn't perfect, but this little trick never occurred to me. Evolution can be pretty clever. GAs often come up with things we humans don't consider.

Geneticists often test their theories about gene function by doing "knock-out mutations," in which they use genetic engineering techniques to prevent the genes in question from being transcribed and see what effect that has on the organism. I can do the same thing here. In particular, I did an experiment in which I "knocked out" the genes in G that made this trick possible: I changed genes such that each gene that corresponds to a "can in current site" situation has the action *PickUp*. This lowered the average score of G from its original 483 to 443, which supports my hypothesis that this trick is partly responsible for G's success.

How Did the GA Evolve a Good Strategy?

The next question is, how did the GA, starting with a random population, manage to evolve such a good strategy as G?

To answer this question, let's look at how strategies improved over generations. In figure 9.6, I plot the fitness of the best strategy in each generation

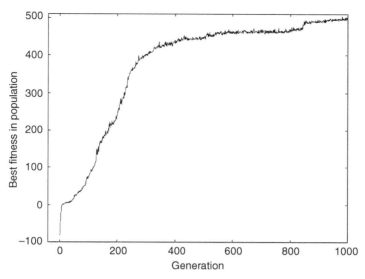

FIGURE 9.6. Plot of best fitness in the population versus generation for the run of the GA in which strategy G was evolved.

in my run of the GA. You can see that the best fitness starts out way below zero, rises very quickly until about generation 300, and then improves more slowly for the rest of the run.

The first generation consists of 200 randomly generated strategies. As you might expect, all of them are very, very bad. The best one has fitness of only −81 and the worst one has fitness −825.

I looked at the behavior of Robby when using the worst strategy of this generation, on several sessions, each starting with a different environment (configuration of cans). In some environments, Robby makes a few moves, then gets stuck, executing action *StayPut* again and again, for the entire session. In others he spends the session crashing into a wall over and over again. In others he spends his whole time trying to pick up a nonexistent can in his current site. No surprise that evolution weeded out this strategy quite early on.

I also looked at the behavior of Robby using the best strategy of this generation, which is still a pretty bad one that gets stuck in ways similar to those in the worst strategy. However, it has a couple of advantages over the worst one: it is less likely to continually crash into a wall, and it occasionally moves into a site with a can and actually picks up the can! This being the best strategy of its generation, it has an excellent chance of being selected to reproduce. When it indeed is selected, its children inherit these good traits (along with lots of bad traits).

By the tenth generation, the fitness of the best strategy in the population has risen all the way to zero. This strategy usually gets stuck in a *StayPut* loop, occasionally getting stuck in a cycle moving back and forth between two sites. Very occasionally it crashes into walls. And like its ancestor from the first generation, it very occasionally picks up cans.

The GA continues its gradual improvement in best fitness. By generation 200 the best strategy has discovered the all-important trait of moving to sites with cans and then picking up those cans—at least a lot of the time. However, when stranded in a no-can wilderness, it wastes a lot of time by making random moves, similar to strategy M. By generation 250 a strategy equal in quality to M has been found, and by generation 400, the fitness is up beyond the 400 level, with a strategy that would be as good as G if only it made fewer random moves. By generation 800 the GA has discovered the trick of leaving cans as markers for adjacent cans, and by generation 900 the trick of finding and then moving around the perimeter of the world has been nearly perfected, requiring only a few tweaks to get it right by generation 1,000.

Although Robby the robot is a relatively simple example for teaching people about GAs, it is not all that different from the way GAs are used in the real world. And as in the example of Robby, in real-world applications, the GA will often evolve a solution that works, but it's hard to see *why* it works. That is often because GAs find good solutions that are quite different from the ones humans would come up with. Jason Lohn, a genetic algorithms expert from the National Astronautical and Space Administration (NASA), emphasizes this point: "Evolutionary algorithms are a great tool for exploring the dark corners of design space. You show [your designs] to people with 25 years' experience in the industry and they say 'Wow, does that really work?'.... We frequently see evolved designs that are completely unintelligible."

In Lohn's case, unintelligible as it might be, it does indeed work. In 2005 Lohn and his colleagues won a "Human Competitive" award for their GA's design of a novel antenna for NASA spacecraft, reflecting the fact that the GA's design was an improvement over that of human engineers.

PART III | Computation Writ Large

The proper domain of computer science is information processing writ large across all of nature.

—Chris Langton (Quoted in Roger Lewin, *Complexity: Life at the Edge of Chaos*)

CHAPTER 10

Cellular Automata, Life, and the Universe

Computation in Nature

A recent article in *Science* magazine, called "Getting the Behavior of Social Insects to Compute," described the work of a group of entomologists who characterize the behavior of ant colonies as "computer algorithms," with each individual ant running a simple program that allows the colony as a whole to perform a complex computation, such as reaching a consensus on when and where to move the colony's nest.

This would be an easy computation for me to program on my computer: I could just appoint one (virtual) ant as leader and decision maker. All the other ants would simply observe the leader's decision and follow it. However, as we have seen, in ant colonies there is no leader; the ant-colony "computer" consists of millions of autonomous ants, each of which can base its decisions and actions only on the small fraction of other ants it happens to interact with. This leads to a kind of computation very different from the kind our desktop computers perform with a central processing unit and random-access memory.

Along the same lines, a 1994 article by three prominent brain researchers asked, "Is the brain a computer?" and answered, "If we embrace a broader concept of computation, then the answer is a definite Yes." Like ant colonies, it is clear that the way the brain computes—with billions of neurons working in parallel without central control—is very different from the way current-day digital computers work.

In the previous two chapters we explored the notion of life and evolution occurring *in* computers. In this part of the book, we look at the opposite notion; the extent to which computation itself occurs in nature. In what sense do natural systems "compute"? At a very general level, one might say that *computation* is what a complex system does with *information* in order to succeed or adapt in its environment. But can we make this statement more precise? Where is the information, and what exactly does the complex system do with it?

In order to make questions like this more amenable to study, scientists generally will *idealize* the problem—that is, simplify it as much as possible while still retaining the features that make the problem interesting.

In this spirit of simplification, many people have studied computation in nature via an idealized model of a complex system called a *cellular automaton*.

Cellular Automata

Recall from chapter 4 that Turing machines provide a way of formalizing the notion of "definite procedure"—that is, computation. A computation is the transformation of the input initially on a Turing machine's tape, via the machine's set of rules, to the output on its tape after the **halt** state is reached. This abstract machine inspired the design of all subsequent digital computers. Because of John von Neumann's contribution to this design, our current-day computers are called "von-Neumann-style architectures."

The von-Neumann-style architecture consists of a random access memory (RAM) that stores both program instructions and data, and a central processing unit (CPU) that fetches instructions and data from memory and executes the instructions on the data. As you probably know, although programmers write instructions in high-level programming languages, instructions and data are actually stored in the computer as strings of 1s and 0s. Instructions are executed by translating such bit strings into simple logic operations applied to the data, which are then computed by the CPU. Only a few types of simple logic operators are needed to perform any computation, and today's CPUs can compute billions of these logic operations per second.

A cellular automaton, being an idealized version of a complex system, has a very different kind of architecture. Imagine a grid of battery-powered lightbulbs, as shown in figure 10.1. Each lightbulb is connected to all of its neighboring lightbulbs in the north, south, east, west, and diagonal directions. In the figure, these connections are shown for only one of the lightbulbs, but imagine that all the other ones have corresponding connections.

FIGURE 10.1. An array of lightbulbs, each of which is connected to its neighbors in the north, south, east, west, and diagonal directions, as is illustrated for one of the lightbulbs. Each lightbulb can either be in state **on** or state **off**. Imagine that all four edges wrap around in a circular fashion—for example, the upper left bulb has the upper right bulb as its western neighbor and the lower left bulb as its northern neighbor.

In figure 10.2 (left box), some of the lightbulbs have been turned on (to make the figure simpler, I didn't draw the connections). After this initial configuration of on and off lightbulbs has been set up, each lightbulb will run a clock that tells it when to "update its state"— that is, turn on or off; and all the clocks are synchronized so all lightbulbs update their states at the same time, over and over again. You can think of the grid as a model of fireflies flashing or turning off in response to the flashes of nearby fireflies, or of neurons firing or being inhibited by the actions of close-by neurons, or, if you prefer, simply as a work of abstract art.

How does a lightbulb "decide" whether to turn on or off at each time step? Each bulb follows a rule that is a function of the states in its *neighborhood*—that is, its own state (i.e., **on** or **off**) and those of the eight neighbors to which it is connected.

For example, let's say that the rule followed by each lightbulb is, "If the majority of bulbs in my neighborhood (including myself) are on, turn on (or stay on, if already on), otherwise turn off (or stay off, if already off)." That is, for each neighborhood of nine bulbs, if five or more of them are on, then the middle one is on at the next time step. Let's look at what the lightbulb grid does after one time step.

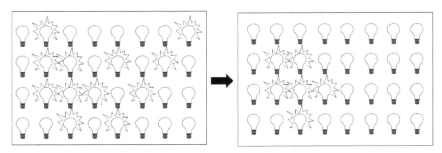

FIGURE 10.2. Left: The same array of lightbulbs as in figure 10.1, set up in an initial configuration of on and off states. Connections between lightbulbs are not shown. Right: each bulb's state has been updated according to the rule "take on whichever state is a majority in my local neighborhood."

As explained in the caption to figure 10.1, to make sure that each lightbulb indeed has eight neighbors, we will give the grid circular boundaries. Imagine that the top edge is folded over and touches the bottom edge, and the left edge is folded over and touches the right edge, forming a donut shape. This gives every lightbulb eight neighbors.

Now let's go back to the rule defined above. Figure 10.2 shows the initial grid and its configuration after following the rule for one time step.

I could have defined a more complicated rule, such as, "If at least two but no more than seven bulbs in my neighborhood are on, then turn on, otherwise turn off," and the updated grid would have looked different. Or "if exactly one bulb is off or exactly four bulbs are on in my neighborhood, turn off, otherwise turn on." There are lots of possibilities.

Exactly how many possible different rules could be defined? "Lots" is really understating it. The answer is "two raised to the power 512" (2^{512}), a huge number, many times larger than the number of atoms in the universe. (See the notes to find out how this answer was derived.)

This grid of lightbulbs is a cellular automaton. More generally, a cellular automaton is a grid (or *lattice*) of *cells*, where a cell is a simple unit that turns on or off in response to the states in its local neighborhood. (In general, cells can be defined with any number of states, but here we'll just talk about the on/off kind.) A cellular automaton *rule*—also called a *cell update rule*—is simply the identical rule followed by each cell, which tells the cell what its state should be at the next time step as a function of the current states in its local neighborhood.

Why do I say that such a simple system is an idealized model of a complex system? Like complex systems in nature, cellular automata are composed of large numbers of simple components (i.e., cells), with no central controller, each of which communicates with only a small fraction of the other components. Moreover, cellular automata can exhibit very complex behavior that is difficult or impossible to predict from the cell update rule.

Cellular automata were invented—like so many other good ideas—by John von Neumann, back in the 1940s, based on a suggestion by his colleague, the mathematician Stan Ulam. (This is a great irony of computer science, since cellular automata are often referred to as **non**-*von-Neumann-style* architectures, to contrast with the *von-Neumann-style* architectures that von Neumann also invented.) As I described in chapter 8, von Neumann was trying to formalize the *logic* of self-reproduction in machines, and he chose cellular automata as a way to approach this problem. In short, he was able to design a cellular automaton rule, in his case with twenty-nine states per cell instead of just two, that would create a perfect reproduction of any initial pattern placed on the cellular automaton lattice.

Von Neumann also was able to show that his cellular automaton was equivalent to a universal Turing machine (cf. chapter 4). The cell update rule plays the role of the rules for the Turing machine tape head, and the configuration of states plays the role of the Turing machine tape—that is, it encodes the *program* and *data* for the universal machine to run. The step-by-step updates of the cells correspond to the step-by-step iteration of the universal Turing machine. Systems that are equivalent in power to universal Turing machines (i.e., can compute anything that a universal Turing machine can) are more generally called *universal computers*, or are said to be *capable of universal computation* or to support *universal computation*.

The Game of Life

Von Neumann's cellular automaton rule was rather complicated; a much simpler, two-state cellular automaton also capable of universal computation was invented in 1970 by the mathematician John Conway. He called his invention the "Game of Life." I'm not sure where the "game" part comes in, but the "life" part comes from the way in which Conway phrased the rule. Denoting **on** cells as *alive* and **off** cells as *dead*, Conway defined the rule in terms of four life processes: *birth*, a dead cell with exactly three live neighbors becomes alive at the next time step; *survival*, a live cell with exactly two or three live neighbors stays alive; *loneliness*, a live cell with fewer than two neighbors dies and a dead cell with fewer than three neighbors stays dead;

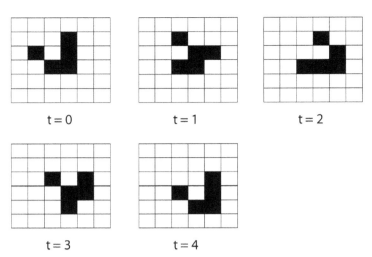

FIGURE 10.3. Close-up picture of a glider in the Game of Life. After four time steps, the original glider configuration has moved in the southeast direction.

and *overcrowding*, a live or dead cell with more than three live neighbors dies or stays dead.

Conway came up with this cellular automaton rule when looking for a rule that would create *interesting* (or perhaps life-like) behavior. Indeed the Game of Life has plenty of interesting behavior and there is a whole community of Life aficionados whose main hobby is to discover initial patterns that will create such behavior.

One simple pattern with interesting behavior is the *glider*, illustrated in figure 10.3. Here, instead of lightbulbs, I simply represent **on** (live) states by black squares and **off** (dead) states by white squares. The figure shows a glider "moving" in a southeast direction from its initial position. Of course, it's not the cells that move; they are all fixed in place. The moving entities are **on** states that form a coherent, persisting shape. Since, as I mentioned earlier, the cellular automaton's boundaries wrap around to create a donut shape, the glider will continue moving around and around the lattice forever.

Other intricate patterns that have been discovered by enthusiasts include the *spaceship*, a fancier type of glider, and the *glider gun*, which continually shoots out new gliders. Conway showed how to simulate Turing machines in Life by having the changing **on/off** patterns of states simulate a tape head that reads and writes on a simulated tape.

John Conway also sketched a proof (later refined by others) that Life could simulate a universal computer. This means that given an initial configuration of on and off states that encodes a program and the input data for that program, Life will run that program on that data, producing a pattern that represents the program's output.

Conway's proof consisted of showing how glider guns, gliders, and other structures could be assembled so as to carry out the *logical operations* **and**, **or**, and **not**. It has long been known that any machine that has the capacity to put together all possible combinations of these logic operations is capable of universal computation. Conway's proof demonstrated that, in principle, all such combinations of logical operations are possible in the Game of Life.

It's fascinating to see that something as simple to define as the Life cellular automaton can, in principle, run any program that can be run on a standard computer. However, in practice, any nontrivial computation will require a large collection of logic operations, interacting in specific ways, and it is very difficult, if not impossible, to design initial configurations that will achieve nontrivial computations. And even if it were possible, the ensuing computation would be achingly slow, not to mention wasteful, since the huge parallel, non-von-Neumann-style computing resources of the cellular automaton would be used to simulate, in a very slow manner, a traditional von-Neumann-style computer.

For these reasons, people don't use Life (or other "universal" cellular automata) to perform real-world computations or even to model natural systems. What we really want from cellular automata is to harness their parallelism and ability to form complex patterns in order to achieve computations in a nontraditional way. The first step is to characterize the kinds of patterns that cellular automata can form.

The Four Classes

In the early 1980s, Stephen Wolfram, a physicist working at the Institute for Advanced Study in Princeton, became fascinated with cellular automata and the patterns they make. Wolfram is one of those legendary former child prodigies whom people like to tell stories about. Born in London in 1959, Wolfram published his first physics paper at age 15. Two years later, in the summer after his first year at Oxford, a time when typical college students get jobs as lifeguards or hitchhike around Europe with a backpack, Wolfram wrote a paper in the field of "quantum chromodynamics" that caught the attention of Nobel prize–winning physicist Murray Gell-Mann, who invited Wolfram to join his group at Caltech (California Institute of Technology). Two years later, at age twenty, Wolfram received a Ph.D. in theoretical physics.

Stephen Wolfram. (Photograph courtesy of Wolfram Research, Inc.)

(Most students take at least five years to get a Ph.D., *after* graduating from college.) He then joined the Caltech faculty, and was soon awarded one of the first MacArthur Foundation "genius" grants. A couple of years later, he was invited to join the faculty at the Institute for Advanced Study in Princeton.

Whew. With all that fame, funding, and the freedom to do whatever he wanted, Wolfram chose to study the dynamics of cellular automata.

In the spirit of good theoretical physics, Wolfram set out to study the behavior of cellular automata in the simplest form possible—using one-dimensional, two-state cellular automata in which each cell is connected only to its two nearest neighbors (figure 10.4a). Wolfram termed these "elementary cellular automata." He figured that if he couldn't understand what was going on in these seemingly ultra-simple systems, there was no chance of understanding more complex (e.g., two-dimensional or multistate) cellular automata.

Figure 10.4 illustrates one particular elementary cellular automaton rule. Figure 10.4a shows the lattice—now just a line of cells, each connected to its nearest neighbor on either side. As before, a cell is represented by a square—black for **on**, and white for **off**. The edges of the lattice wrap around to make a circle. Figure 10.4b shows the rule that each cell follows: for each of the eight possible state configurations for a three-cell neighborhood, the update state for the center cell is given. For example, whenever a three-cell neighborhood consists of all **off** states, the center cell should stay **off** at the next time step. Likewise, whenever a three-cell neighborhood has the configuration **off-off-on**, the center cell should change its state to **on** at the next time step. Note that

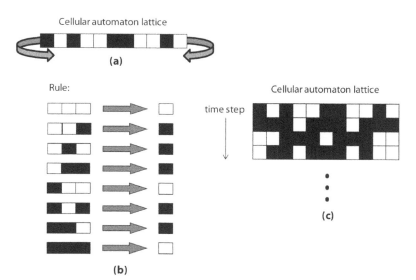

FIGURE 10.4. (a) An illustration of a one-dimensional lattice whose ends wrap around in a circle; (b) A particular elementary cellular automaton rule (Rule 110) expressed as a list of three-cell configurations and corresponding update states for the configuration's center cell; (c) A space-time diagram, showing four successive configurations of the cellular automaton.

the term *rule* refers to the entire list of configurations and update states, not to the individual lines in the list. Figure 10.4c shows a space-time diagram for this cellular automaton. The top row of cells is the one-dimensional lattice set up with a particular initial configuration of **on** and **off** states. Each successive row going down is the updated lattice at the next time step. Such plots are called space-time diagrams because they track the spatial configuration of a cellular automaton over a number of time steps.

Since there are only eight possible configurations of states for a three-cell neighborhood (cf. figure 10.4b) and only two possible ways to fill in the update state (**on** or **off**) for each of these eight configurations, there are only 256 (2^8) possible rules for elementary cellular automata. By the 1980s, computers were powerful enough for Wolfram to thoroughly investigate every single one of them by looking at their behavior starting from many different initial lattice configurations.

Wolfram assigned an identifying number to each elementary cellular automaton rule as illustrated in figure 10.5. He called the **on** state "1" and the **off** state "0," and wrote the rule's update states as a string of 1s and 0s, starting with the update state for the all **on** neighborhood and ending with the update state for the all **off** neighborhood. As shown, the rule given in

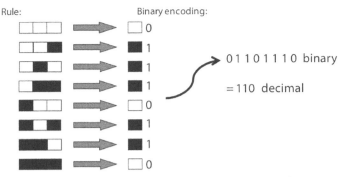

FIGURE 10.5. An illustration of the numbering system for elementary cellular automata used by Stephen Wolfram.

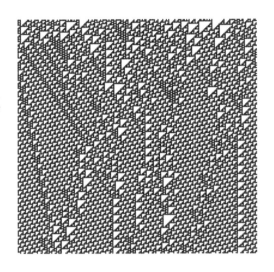

FIGURE 10.6. Rule 110 space-time diagram. The one-dimensional cellular automaton lattice has 200 cells, which are shown starting with a random initial configuration of states, and updating over 200 time steps.

figure 10.4 is written as 0 1 1 0 1 1 1 0. Wolfram then interpreted this string as a number written in binary (i.e., base 2). The string 0 1 1 0 1 1 1 0 in binary is equal to the number 110 in decimal. This rule is thus called "Rule 110." As another example, the rule with update states 0 0 0 1 1 1 1 0 is "Rule 30." (See the notes for a review on how to convert base 2 numbers to decimal.)

Wolfram and his colleagues developed a special programming language, called Mathematica, designed in part to make it easy to simulate cellular automata. Using Mathematica, Wolfram programmed his computer to run elementary cellular automata and to display space-time diagrams that show their behavior. For example, figures 10.6 and 10.7 are plots like that given in figure 10.4, just on a larger scale. The top horizontal row of figure 10.6 is a random initial configuration of 200 black and white cells, and each successive row is the result of applying Rule 110 to each cell in the previous row, for

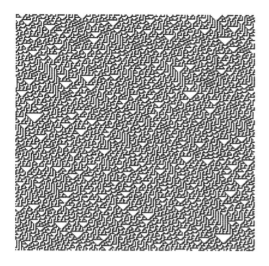

FIGURE 10.7. Rule 30 space-time diagram, with an initial configuration of random states.

200 time steps. The plot in figure 10.7 shows the pattern created by Rule 30, starting from a random initial configuration.

Looking at figures 10.6 and 10.7, perhaps you can sense why Wolfram got so excited about elementary cellular automata. How, exactly, did these complex patterns emerge from the very simple cellular automaton rules that created them?

Seeing such complexity emerge from simple rules was evidently an epiphany for Wolfram. He later said, "The Rule 30 automaton is the most surprising thing I've ever seen in science.... It took me several years to absorb how important this was. But in the end, I realized that this one picture contains the clue to what's perhaps the most long-standing mystery in all of science: where, in the end, the complexity of the natural world comes from." In fact, Wolfram was so impressed by Rule 30 that he patented its use as part of a pseudo-random number generator.

In Wolfram's exhaustive survey of all 256 elementary cellular automata, he viewed the behavior over time of each one starting from many different initial configurations. For each elementary cellular automaton and each initial configuration, he applied the cellular automaton rule to the lattice for a number of time steps—until the cellular automaton exhibited a stable type of behavior. He observed the behavior to fall into four classes:

Class 1: Almost all initial configurations settle down to the same uniform final pattern. Rule 8 is an example of this class; for all initial configurations, all cells in the lattice quickly switch to the **off** state and stay that way.

Class 2: Almost all initial configurations settle down to either a uniform final pattern or a cycling between a few final patterns. Here, the specific final pattern or patterns depends on the initial configuration.

Class 3: Most initial configurations produce random-looking behavior, although triangles or other regular structures are present. Rule 30 (figure 10.7) is an example of this class.

Class 4: The most interesting class. As described by Wolfram: "class 4 involves a mixture of order and randomness: localized structures are produced which on their own are fairly simple, but these structures move around and interact with each other in very complicated ways." Rule 110 (figure 10.6) is an example of this class.

Wolfram speculated that, because of this complexity of patterns and interactions, all class 4 rules are capable of universal computation. However, in general it is hard to prove that a particular cellular automaton, Turing machine, or any other device is universal. Turing's proof that there exists a universal Turing machine was a triumph, as was von Neumann's proof that his self-replicating automaton was also a universal computer. Since then several researchers have proved that simple cellular automata (such as the Game of Life) are universal. In the 1990s, Matthew Cook, one of Wolfram's research assistants, finally proved that Rule 110 was indeed universal, and is perhaps the simplest known example of a universal computer.

Wolfram's "New Kind of Science"

I first heard about Cook's result in 1998 when he spoke at a workshop at the Santa Fe Institute. My own reaction, like that of many of my colleagues, was "Very cool! Very ingenious! But not of much practical or scientific significance." Like the Game of Life, Rule 110 is an example of a very simple deterministic system that can create unpredictable complex behavior. But in practice it would be very difficult to design an initial configuration that would result in a desired nontrivial computation. Moreover, Rule 110 would be an even slower computer than the Game of Life.

Wolfram's view of the result is very different. In his 2002 book, *A New Kind of Science*, Wolfram interprets the universality of Rule 110 as strong evidence for "a new law of nature," namely, his Principle of Computational Equivalence. Wolfram's proposed principle consists of four parts:

1. The proper way to think about processes in nature is that they are *computing*.

2. Since even very simple rules (or "programs") such as Rule 110 can support universal computation, the ability to support universal computation is very common in nature.
3. Universal computation is an upper limit on the complexity of computations in nature. That is, no natural system or process can produce behavior that is "noncomputable."
4. The computations done by different processes in nature are almost always equivalent in sophistication.

Got that? I admit, it's kind of hard to figure out what this principle means, and a major purpose of Wolfram's 1,200-page book is to explicate this principle and show how it applies in very diverse areas of science. I read the whole book, and I still don't completely understand what Wolfram is getting at here. However, I'll give my best gloss on it.

What Wolfram means by "processes in nature are computing" is something like what you see in figures 10.6 and 10.7. At any given time a cellular automaton possesses *information*—the configuration of states—and processes information by applying its rule to its current configuration. Wolfram believes that natural systems work much the same way—that they contain information and process that information according to simple rules. In *A New Kind of Science* (or "NKS" as it is called by the cognoscenti), Wolfram discusses scientific topics such as quantum physics, evolutionary and developmental biology, and economics, to name a few, and he attempts to show how each of these can best be described in terms of computation via simple rules. In essence, his "new kind of science" refers to the idea that the universe, and everything in it, can be explained by such simple programs. This is computation writ large, very large.

Now, according to Wolfram, since extremely simple rules such as Rule 110 can support universal computation, then most natural systems—presumably more complicated than Rule 110—can probably support universal computation too. And, Wolfram believes, there is nothing more complex than what can be computed by a universal computer, given the right input. Thus there is an upper limit on the complexity of possible computations in nature.

As I described in chapter 4, Alan Turing showed that universal computers can in principle compute anything that is "computable." However, some computations are simpler than others. Even though they both could run on the same computer, the program "calculate 1 + 1" would result in a less complicated computational process than a program that simulates the earth's climate, correct? But Wolfram's principle in fact asserts that, in nature, the "sophistication" of all computations actually being performed is equivalent.

Does any of this make sense? Wolfram's theories are far from being generally accepted. I'll give you my own evaluation. As for points 1 and 2, I think Wolfram is on the right track in proposing that simple computer models and experiments can lead to much progress in science, as shown in examples throughout this book. As I'll describe in chapter 12, I think that one can interpret the behavior of many natural systems in terms of information processing. I also find it plausible that many such systems can support universal computation, though the significance of this for science hasn't been demonstrated yet.

Regarding point 3, the jury is also still out on whether there are processes in nature that are more powerful than universal computers (i.e., can compute "uncomputable" things). It has been proved that if you could build truly nondigital computers (i.e., that can compute with numbers that have infinitely many decimal places) then you would be able to build such a computer to solve the halting problem, Turing's uncomputable problem that we saw in chapter 4. Some people, including Wolfram, don't believe that numbers with infinitely many decimal places actually exist in nature—that is, they think nature is fundamentally digital. There's no really convincing evidence on either side.

Point 4 makes no sense to me. I find it plausible that my brain can support universal computation (at least as far as my limited memory capacity allows) and that the brain of the worm *C. elegans* is also (approximately) universal, but I don't buy the idea that the actual computations we engage in, respectively, are equivalent in sophistication.

Wolfram goes even further in his speculations, and predicts that there is a single, simple cellular automaton-like rule that is the "definite ultimate model for the universe," the primordial cellular automaton whose computations are the source for everything that exists. How long is this rule? "I'm guessing it's really very short," says Wolfram. But how long? "I don't know. In Mathematica, for example, perhaps three, four lines of code." Computation writ small.

NKS made a big splash when it was published in 2002—it started out as Amazon.com's number one bestseller, and remained on the bestseller list for months. Its publication was followed by a large publicity campaign launched by Wolfram Media, the company Wolfram formed to publish his book. Reactions to the book were bipolar: some readers thought it brilliant and revolutionary, others found it self-aggrandizing, arrogant, and lacking in substance and originality (for example, critics pointed out that physicists Konrad Zuse and Edward Fredkin had both theorized that the universe is a cellular automaton as early as the 1960s). However, whatever its other merits might be, we cellular-automata addicts can all agree that NKS provided a lot of welcome publicity for our obscure passion.

CHAPTER 11 | Computing with Particles

IN 1989 I HAPPENED TO READ AN ARTICLE by the physicist Norman Packard on using genetic algorithms to automatically design cellular automaton rules. I was immediately hooked and wanted to try it myself. Other things got in the way (like finishing my Ph.D. thesis), but working on this idea was always in the back of my mind. A few years later, with thesis finished and a research job at the Santa Fe Institute, I finally had the time to delve into it. A young undergraduate student named Peter Hraber was hanging around the institute at that time, looking for something to work on, so I recruited him to help me with this project. We were soon after joined by a graduate student named Rajarshi Das, who had independently started a similar project.

Like Packard, we used a genetic algorithm to evolve cellular automaton rules to perform a specific task called "majority classification." The task is simple: the cellular automaton must compute whether its initial configuration contains a majority of **on** or **off** states. If **on** states are in the majority, the cellular automaton should signal this fact by turning all the cells **on**. Similarly, if **off** has an initial majority, the cellular automaton should turn all cells **off**. (If the number of initial **on** and **off** states is equal, there is no answer, but this possibility can be avoided by using a cellular automaton with an odd number of cells.) The majority classification task is sort of like having to predict which of two candidates will win an election in your city when all you can see are the political signs on your close neighbors' front lawns.

The majority classification task would be trivial for a von-Neumann-style computer. Simply have the central processing unit count the respective numbers of **on** and **off** states in the initial configuration, storing the count at each step in memory. When done counting, retrieve the two totals from memory, determine which one is larger, and reset the configuration to all **on** or all **off** depending on this comparison. A von-Neumann-style computer can do this easily because it has random access memory in which to store the initial configuration and intermediate sums, and a central processing unit to do the counting, the final comparison, and the resetting of the states.

In contrast, a cellular automaton has no random access memory and no central unit to do any counting. It has only individual cells, each of which has information only about its own state and the states of its neighbors. This situation is an idealized version of many real-world systems. For example, a neuron, with only a limited number of connections to other neurons, must decide whether and at what rate to fire so that the overall firing pattern over large numbers of neurons represents a particular sensory input. Similarly, as I describe in chapter 12, an ant must decide what job it should take on at a given time—in order to benefit the colony as a whole—based only on its interactions with a relatively small number of other ants.

The bottom line is that it is in general difficult to design cellular automata to perform tasks that require collective decision making among all the cells. Peter and I were interested in how a genetic algorithm might solve this design problem.

Using Norman Packard's work as a starting point, we coded up simulations of one-dimensional cellular automata in which each cell is connected to three cells on either side—that is, the size of each cell's neighborhood (including the cell itself) is seven cells.

Think for a minute how you might design a rule for a cellular automaton like this to perform the majority classification task.

One reasonable first guess for a rule might be: "Each cell should update to the state that is currently a majority in its local neighborhood." This would be like basing your election prediction on which candidate is supported by the majority of yard signs among you and your local neighbors. However, this "local majority vote" cellular automaton doesn't work, as illustrated by the space-time diagram in figure 11.1. The initial configuration has a majority of black cells, and each cell updates its state according to the local majority vote rule over 200 time steps. You can see that the lattice quickly settles into a stable pattern of black and white regions. The boundaries between regions are at locations where majority black neighborhoods overlap with majority white neighborhoods. The final configuration, containing both white and black cells, does not give the desired answer to the majority classification task. The problem is that there is no way, using this rule, for one black region, say, to communicate its length to the other black regions, so that collectively they can determine whether or not they are in the majority.

FIGURE 11.1. Space-time behavior of the "local majority vote" cellular automaton starting from a majority black initial configuration. (Figure adapted from Mitchell, M., Crutchfield, J. P., and Das, R., Evolving cellular automata to perform computations: A review of recent work. In *Proceedings of the First International Conference on Evolutionary Computation and Its Applications (EvCA '96)*. Moscow, Russia: Russian Academy of Sciences, 1996.)

It's not immediately obvious how to design a rule for this task, so in the spirit of Robby the Robot from chapter 9 we ran a genetic algorithm in order to see if it could produce a successful rule on its own.

In our genetic algorithm, cellular automaton rules are encoded as strings of 0s and 1s. These bits give the state-update values for each possible neighborhood configuration (figure 11.2).

The genetic algorithm starts out with a population of randomly generated cellular automaton rules. To calculate the fitness of a rule, the GA tests it on many different initial lattice configurations. The rule's fitness is the fraction of times it produces the correct final configuration: all cells **on** for initial majority **on** or all cells **off** for initial majority **off** (figure 11.3). We ran the GA for many generations, and by the end of the run the GA had designed some rules that could do this task fairly well.

As we saw with Robby the Robot, a solution evolved by the genetic algorithm is typically impossible to understand at the level of its "genome." The cellular automata that we evolved for the majority classification task were no different. The genome of one of the highest-performing cellular automata designed by the GA is the following (split into two lines of text):

00000101000001100001010110000111000001110000010000010101010101111
0110010001110111000001010000000101111101111111111011011101111111

Recall that the first bit is the update state for the center cell in the all 0s neighborhood, the second bit is the update state for center cell in the neighborhood 0000001, and so forth. Since there are 128 possible neighborhoods, this genome consists of 128 bits. Looking at this bit string, there is nothing

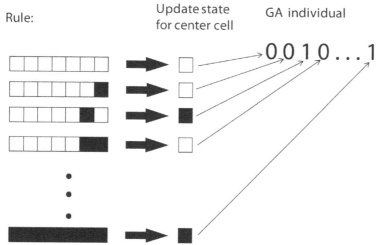

FIGURE 11.2. Illustration of how a cellular automaton rule is encoded as an individual in the genetic algorithm's population. The 128 possible neighborhood configurations are listed in a fixed order. The update state for the center cell of each neighborhood configuration is encoded as a 0 (off) or a 1 (on). An individual in the genetic algorithm's population is a string of 128 bits, encoding the update states in their fixed ordering.

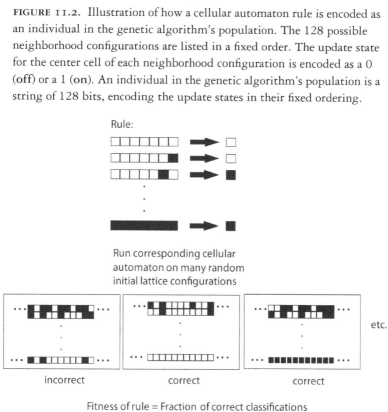

Fitness of rule = Fraction of correct classifications

FIGURE 11.3. To calculate its fitness, each rule is tested on many random initial configurations. The fitness of a rule is the fraction of initial configurations on which the rule produces the correct answer within some number of time steps.

obvious that gives us a hint as to how this rule will behave, or why it obtained high fitness on the majority classification task.

Figure 11.4 gives two space-time diagrams that display the behavior of this rule on two different initial configurations: with (a) a majority of black cells and (b) a majority of white cells. You can see that in both cases the final behavior is correct—all black in (a) and all white in (b). In the time between the initial and final configurations, the cells somehow collectively process information in order to arrive at a correct final configuration. Some interesting patterns form during these intermediate steps, but what do they mean?

It took a lot of staring at pictures like figure 11.4 for us to figure out what was going on. Luckily for us, Jim Crutchfield, a physicist from Berkeley, happened to visit the Santa Fe Institute and became interested in our effort. It turned out that Jim and his colleagues had earlier developed exactly the right conceptual tools to help us understand how these patterns implement the computation.

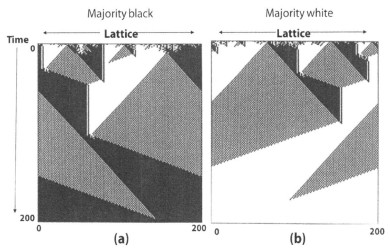

FIGURE 11.4. Space-time behavior of one of the best-performing evolved cellular automaton rules for the majority classification task. In (a), the initial configuration contains a majority of black cells and the cellular automaton iterates to a fixed configuration of all black. In (b), the initial configuration contains a majority of white cells and the cellular automaton iterates to a fixed configuration of all white. (Figure adapted from Mitchell, M., Crutchfield, J. P., and Das, R., Evolving cellular automata to perform computations: A review of recent work. In *Proceedings of the First International Conference on Evolutionary Computation and Its Applications (EvCA '96)*. Moscow, Russia: Russian Academy of Sciences, 1996.)

Three types of patterns can be seen in figure 11.4: all black, all white, and a checkerboard-like region of alternating black and white states (this appears as a grayish region in the low-resolution figure 11.4). It is this checkerboard pattern that transfers information about the density of black or white cells in local regions.

Like the strategy the genetic algorithm evolved for Robby the Robot, the cellular automaton's strategy is quite clever. Take a look at figure 11.5, which is a version of figure 11.4a that I have marked up. Regions in which the initial configuration is either mostly white or mostly black converge in a few time steps to regions that are all white or all black. Notice that whenever a black region on the left meets a white region on the right, there is always a vertical boundary between them. However, whenever a white region on the left meets a black region on the right, a checkerboard triangle forms, composed of alternating black and white cells. You can see the effect of the circular lattice on the triangle as it wraps around the edges.

Sides A and B of the growing checkerboard region travel at the same *velocity* (i.e., distance covered over time). Side A travels southwest until it collides with the vertical boundary. Side B just misses colliding with the vertical boundary on the other side. This means that side A had a shorter distance to travel. That is, the white region bordered by side A is smaller than the black

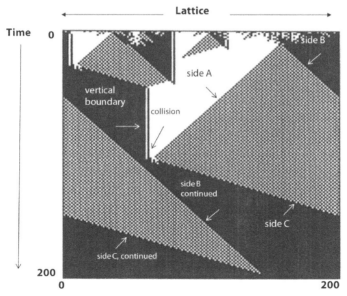

FIGURE 11.5. The space-time diagram of figure 11.4 (a) with important features marked.

region bordered by side B. At the collision point, side A disappears, and the black region is allowed to grow. At the triangle's bottom point, sides B and C disappear, and the entire lattice becomes black, the correct final configuration.

If we try to understand these patterns as carrying out a computation, then the vertical boundary and the checkerboard region can be thought of as *signals*. These signals are created and propagated by local interactions among cells. The signals are what allow the cellular automaton as a whole to determine the relative sizes of the larger-scale adjacent black and white regions, cut off the smaller ones, and allow the larger ones to grow in size.

These signals are the major difference between the local majority voting cellular automaton of figure 11.1 and the cellular automaton of figure 11.5. As I mentioned above, in the former there is no way for separated black or white regions to communicate with one another to figure out which has the majority of cells. In the latter, the signals created by the checkerboard region and the vertical boundary carry out this communication, and the interaction among signals allows the communicated information to be processed so that the answer can be determined.

Jim Crutchfield had earlier invented a technique for detecting what he called "information processing structures" in the behavior of dynamical systems and he suggested that we apply this technique to the cellular automata evolved by the GA. Crutchfield's idea was that the boundaries between simple regions (e.g., sides A, B, C, and the vertical boundary in figure 11.5) are carriers of information and information is processed when these boundaries collide. Figure 11.6 shows the space-time diagram of figure 11.5, but with the black, white, and checkerboard regions filtered out (i.e., colored white), leaving only the boundaries, so we can see them more clearly. The picture looks something like a trace of elementary particles in an old physics device called a bubble chamber. Adopting that metaphor, Jim called these boundaries "particles."

Traditionally in physics particles are denoted with Greek letters, and we have done the same here. This cellular automaton produces six different types of particles: γ (gamma), μ (mu), η (eta), δ (delta), β (beta), and α (alpha, a short-lived particle that decays into γ and μ). Each corresponds to a different kind of boundary—for example, η corresponds to boundaries between black and checkerboard regions. There are five types of particle collisions, three of which ($\beta + \gamma$, $\mu + \beta$, and $\eta + \delta$) create a new particle, and two of which ($\eta + \mu$ and $\gamma + \delta$) are "annihilations," in which both particles disappear.

Casting the behavior of the cellular automaton in terms of particles helps us understand how it is encoding information and performing its computation. For example, the α and β particles encode different types of information about the initial configuration. The α particle decays into γ and μ. The

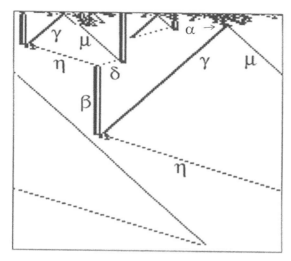

FIGURE 11.6. The space-time diagram of figure 11.4 (a), with regions of "simple patterns" filtered out, leaving the boundaries between these regions ("particles"). (Figure adapted from Mitchell, M., Crutchfield, J. P., and Das, R., Evolving cellular automata to perform computations: A review of recent work. In *Proceedings of the First International Conference on Evolutionary Computation and Its Applications (EvCA '96)*. Moscow, Russia: Russian Academy of Sciences, 1996.)

γ particle carries the information that it borders a white region; similarly, the μ particle carries the information that it borders a black region. When γ collides with β before μ does, the information contained in β and γ is combined to deduce that the large initial white region was smaller than the large initial black region it bordered. This new information is encoded in a newly created particle η, whose job is to catch up with and annihilate the μ (and itself).

We were able to apply this kind of analysis to several different cellular automata evolved to perform the majority classification task as well as other tasks. This analysis allowed us to predict things such as the fitness of a given cellular automaton (without having to run the cellular automaton itself, but using only its "particle" description). It also allowed us to understand why one cellular automaton had higher fitness than another and how to describe the mistakes that were made by different cellular automata in performing computations.

Particles give us something we could not get by looking at the cellular automaton rule or the cellular automaton's space-time behavior alone: they allow us to explain, in information-processing terms, how a cellular automaton performs a computation. Note that particles are a *description* imposed by us (the scientists) rather than anything explicit taking place in a cellular automaton or used by the genetic algorithm to evolve cellular automata. But somehow the genetic algorithm managed to evolve a rule whose behavior can be explained in terms of information-processing particles. Indeed, the language of particles and their interactions form an explanatory vocabulary for decentralized computation in the context of one-dimensional cellular automata. Something like this language may be what Stephen Wolfram was looking for when he posed the last of his "Twenty Problems in the Theory of Cellular Automata": "What higher-level descriptions of information processing in cellular automata can be given?"

All this is relatively recent work and needs to be developed much further. I believe that this approach to understanding computation, albeit unconventional, will turn out to be useful in other contexts in which computation is distributed among simple components with no central control. For example, it is still a mystery how high-level information about sensory data is encoded and processed in the brain. Perhaps the explanation will turn out to be something close to particle-like or, given the brain's three dimensions, wave-like computation, where neurons are the scaffolding for information-carrying waves of activity and their information-processing interactions.

Brain computation is of course a long jump from one-dimensional cellular automata. However, there is one natural system that might be explained by something very much like our particles: the stomata networks of plants. Every leafy plant's leaves are covered with stomata—small apertures that open or close in response to light and humidity. When open, the stomata let in carbon dioxide, which is used in photosynthesis. However, open stomata cause water to evaporate from the plant's fluid stores. Botanist Keith Mott, physicist David Peak, and their colleagues at Utah State University have long been observing the patterns of opening and closing of stomata on leaves, and believe that the stomata constitute a network that is something like a two-dimensional cellular automaton. They also observe that the temporal patterns of opening and closing on the leaves look very much like two-dimensional versions of interacting particles. They hypothesize that plants perform a distributed, decentralized computation via their stomata—namely, how to optimally open and close stomata in order to best balance carbon dioxide gain and water loss—and that the computation may be explainable in terms of such particles.

CHAPTER 12

Information Processing in Living Systems

EVER SINCE SZILARD'S INSIGHT THAT *information* might be the savior of the second law of thermodynamics from the attack of Maxwell's demon, information and its cousin *computation* have increasingly infiltrated science. In many people's minds information has taken on an ontological status equal to that of mass and energy—namely, as a third primitive component of reality. In biology in particular, the description of living systems as *information processing networks* has become commonplace. In fact, the term *information processing* is so widely used that one would think it has a well-understood, agreed-upon meaning, perhaps built on Shannon's formal definition of information. However, like several other central terms of complex systems science, the concept of information processing tends to be ill-defined; it's often hard to glean what is meant by *information processing* or *computation* when they are taken out of the precise formal context defined by Turing machines and von Neumann-style computers. The work described in the previous chapter was an attempt to address this issue in the context of cellular automata.

The purpose of this chapter is to explore the notion of information processing or computation in living systems. I describe three different natural systems in which information processing seems to play a leading role—the immune system, ant colonies, and cellular metabolism—and attempt to illuminate the role of information and computation in each. At the end I attempt to articulate some common qualitative principles of information processing in these and other decentralized systems.

What Is Information Processing?

Let me quote myself from chapter 10: "In what sense do natural systems 'compute'? At a very general level, one might say that *computation* is what a complex system does with *information* in order to succeed or adapt in its environment. But can we make this statement more precise? Where is the information, and what exactly does the complex system do with it?" These questions may seem straightforward, but exploring them will quickly force us to dip our feet into some of the murkiest waters in complex systems science.

When we say a system is *processing information* or *computing* (terms which, for now, I will use synonymously), we need to answer the following:

- What plays the role of "information" in this system?
- How is it communicated and processed?
- How does this information acquire *meaning*? And to whom? (Some will disagree with me that computation requires *meaning* of some sort, but I will go out on a limb and claim it does.)

INFORMATION PROCESSING IN TRADITIONAL COMPUTERS

As we saw in chapter 4, the notion of computation was formalized in the 1930s by Alan Turing as the steps carried out by a Turing machine on a particular input. Ever since then, Turing's formulation has been the basis for designing traditional von-Neumann-style programmable computers. For these computers, questions about information have straightforward answers. We can say that the role of information is played by the tape symbols and possible states of the tape head. Information is communicated and processed by the tape head's actions of reading from and writing to the tape, and changing state. This is all done by following the rules, which constitute the *program*.

We can view all programs written for traditional computers at (at least) two levels of description: a machine-code level and a programming level. The **machine-code** level is the set of specific, step-by-step low-level instructions for the machine (e.g., "move the contents of memory location *n* to CPU register *j*," "perform an *or* logic operation on the contents of CPU registers *j* and *i* and store the result in memory location *m*," and so on). The **programming level** is the set of instructions in a high-level language, such as BASIC or Java, that is more understandable to humans (e.g., "multiply the value of the variable *half_of_total* by 2 and store the result in the variable *total*," etc.). Typically a single high-level instruction corresponds to several low-level instructions, which may be different on different computers. Thus a given high-level program can be implemented in different ways in machine code; it is a more *abstract* description of information processing.

The *meaning* of the input and output information in a Turing machine comes from its interpretation by humans (programmers and users). The meaning of the information created in intermediate steps in the computation also comes from its interpretation (or design) by humans, who understand the steps in terms of commands in a high-level programming language. This higher level of description allows us to understand computations in a human-friendly way that is abstracted from particular details of machine code and hardware.

INFORMATION PROCESSING IN CELLULAR AUTOMATA

For non-von-Neumann-style computers such as cellular automata, the answers are not as straightforward. Consider, for example, the cellular automaton described in the previous chapter that was evolved by a genetic algorithm to perform majority classification. Drawing an analogy with traditional computers, we could say that information in this cellular automaton is located in the configurations of states in the lattice at each time step. The input is the initial configuration, the output is the final configuration, and at each intermediate time step information is communicated and processed within each neighborhood of cells via the actions of the cellular automaton rule. *Meaning* comes from the human knowledge of the task being performed and interpretation of the mapping from the input to the output (e.g., "the final lattice is all white; that means that the initial configuration had a majority of white cells").

Describing information processing at this level, analogous to "machine code," does not give us a human-friendly understanding of *how* the computation is accomplished. As in the case of von-Neumann-style computation, we need a higher-level language to make sense of the intermediate steps in the computation and to abstract from particular details of the underlying cellular automaton.

In the previous chapter I proposed that *particles* and their interactions are one approach toward such a high-level language for describing how information processing is done in cellular automata. Information is communicated via the movement of particles, and information is processed via collisions between particles. In this way, the intermediate steps of information processing acquire "meaning" via the human interpretation of the actions of the particles.

One thing that makes von-Neumann-style computation easier to describe is that there is a clear, unambiguous way to translate from the programming level to the machine code level and vice versa, precisely because these computers were designed to allow such easy translation. Computer science has given us automatic compilers and decompilers that do the translation, allowing us to understand how a particular program is processing information.

For cellular automata, no such compilers or decompilers exist, at least not yet, and there is still no practical and general way to design high-level "programs." Relatively new ideas such as particles as high-level information-processing structures in cellular automata are still far from constituting a theoretical framework for computation in such systems.

The difficulties for understanding information processing in cellular automata arise in spades when we try to understand information processing in actual living systems. My original question, "In what sense do natural systems 'compute'?" is still largely unanswered, and remains a subject of confusion and thorny debate among scientists, engineers, and philosophers. However, it is a tremendously important question for complex systems science, because a high-level description of information processing in living systems would allow us not only to understand in new and more comprehensive ways the mechanisms by which particular systems operate, but also to abstract general principles that transcend the overwhelming details of individual systems. In essence, such a description would provide a "high-level language" for biology.

The rest of this chapter tries to make sense of these ideas by looking at real examples.

The Immune System

I gave a quick description of the immune system way back in chapter 1. Now let's look a bit more in depth at how it processes information in order to protect the body from *pathogens*—viruses, bacteria, parasites, and other unwelcome intruders.

To recap my quick description, the immune system consists of trillions of different cells and molecules circulating in the body, communicating with one another through various kinds of signals.

Of the many many different types of cells in the immune system, the one I focus on here is the *lymphocyte* (a type of white blood cell; see figure 12.1). Lymphocytes are created in the bone marrow. Two of the most important types of lymphocytes are *B cells*, which release *antibodies* to fight viruses and bacteria, and *T cells*, which both kill invaders and regulate the response of other cells.

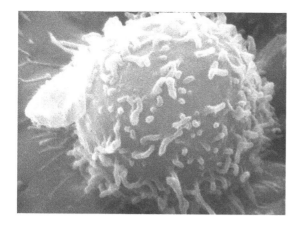

FIGURE 12.1. A human lymphocyte, whose surface is covered with receptors that can bind to certain shapes of molecules that the cell might encounter. (Photograph from National Cancer Institute [http://visualsonline.cancer.gov/details.cfm?imageid=1944].)

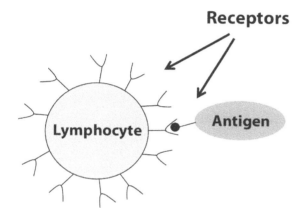

FIGURE 12.2. An illustration of a lymphocyte (here a B cell) receptor binding to an antigen.

Each cell in the body has molecules on its surface called *receptors*. As the name implies, these molecules are a means by which cells receive information. The information is in the form of other molecules that chemically bind to the receptor molecules. Whether or not a receptor binds to a particular molecule depends on whether their physical shapes match sufficiently.

A lymphocyte's surface is covered with identical receptors that bind to a particular range of molecular shapes. If a lymphocyte happens by chance to encounter a special pathogen molecule (called an "antigen") whose shape fits the lymphocyte's receptors, then the molecule binds to one of the lymphocyte's receptors, and the lymphocyte is said to have "recognized" that antigen—the first step in killing off pathogens. The binding can be strong or weak, depending on how closely the molecule's shape fits the receptor. This process is illustrated in figure 12.2.

The main problem facing the immune system is that it doesn't know ahead of time what pathogens will invade the body, so it can't "predesign" a set of lymphocytes with receptors that will bind strongly to just the right shapes. What's more, there are an astronomical number of possible pathogens, so the immune system will never be able to generate enough lymphocytes at any one time to take care of every eventuality. Even with all the many millions of different lymphocytes the body generates per day, the world of pathogens that the system will be likely to encounter is much bigger.

Here's how the immune system solves this problem. In order to "cover" the huge space of possible pathogen shapes in a reasonable way, the population of lymphocytes in the body at any given time is enormously diverse. The immune system employs randomness to allow each individual lymphocyte to recognize a range of shapes that differs from the range recognized by other lymphocytes in the population.

In particular, when a lymphocyte is born, a novel set of identical receptors is created via a complicated random shuffling process in the lymphocyte's DNA. Because of continual turnover of the lymphocyte population (about ten million new lymphocytes are born each day), the body is continually introducing lymphocytes with novel receptor shapes. For any pathogen that enters the body, it will just be a short time before the body produces a lymphocyte that binds to that pathogen's particular marker molecules (i.e., antigens), though the binding might be fairly weak.

Once such a binding event takes place, the immune system has to figure out whether it is indicative of a real threat or is just a nonthreatening situation that can be ignored. Pathogens are harmful, of course, because once they enter the body they start to make copies of themselves in large numbers. However, launching an immune system attack can cause inflammation and other harm to the body, and too strong an attack can be lethal. The immune system as a whole has to determine whether the threat is real and severe enough to warrant the risk of an immune response harming the body. The immune system will go into high-gear attack mode only if it starts picking up *a lot* of sufficiently strong binding events.

The two types of lymphocytes, B and T cells, work together to determine whether an attack is warranted. If the number of strongly bound receptors on a B cell exceeds some threshold, and in the same time frame the B cell gets "go-ahead" signals from T cells with similarly bound receptors, the B cell is *activated*, meaning that it now perceives a threat to the body (figure 12.3). Once activated, the B cell releases antibody molecules into the bloodstream. These antibodies bind to antigens, neutralize them, and mark them for destruction by other immune system cells.

The activated B cell then migrates to a lymph node, where it divides rapidly, producing large numbers of daughter cells, many with mutations

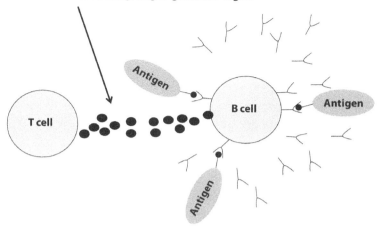

FIGURE 12.3. Illustration of activation of a B cell via binding and "go-ahead" signal from a T cell. This signal prompts the B cell to produce and release antibodies (y-shaped molecules).

that alter the copies' receptor shapes. These copies are then tested on antigens that are captured in the lymph node. The cells that do not bind die after a short time.

The surviving copies are unleashed into the bloodstream, where some of them encounter and bind to antigens, in some cases more strongly than did their mother B cell. These activated daughter B cells also migrate to lymph nodes and create mutated daughters of their own. This massive production of lymphocytes is one reason that your lymph nodes often feel swollen when you are sick.

This cycle continues, with the new B cells that best match antigens themselves producing the most daughter cells. In short, this is a process of natural selection, in which collections of B cells evolve to have receptor shapes that will bind strongly to a target antigen. This results in a growing arsenal of antibodies that have been "designed" via selection to attack this specific antigen. This process of detection and destruction typically takes from a few days to weeks to eradicate the corresponding pathogen from the body.

There are at least two potential problems with this strategy. First, how does the immune system prevent lymphocytes from mistakenly attacking the body's own molecules? Second, how does the immune system stop or tone down its attack if the body is being harmed too much as a result?

Immunologists don't yet have complete answers to these questions, and each is currently an area of active research. It is thought that one major mechanism for avoiding attacking one's own body is a process called *negative selection*. When lymphocytes are born they are not immediately released into the bloodstream. Instead they are tested in the bone marrow and thymus by being exposed to molecules of one's own body. Lymphocytes that bind strongly to "self" molecules tend to be killed off or undergo "editing" in the genes that give rise to receptors. The idea is that the immune system should only use lymphocytes that will not attack the body. This mechanism often fails, sometimes producing autoimmune disorders such as diabetes or rheumatoid arthritis.

A second major mechanism for avoiding autoimmune attacks seems to be the actions of a special subpopulation of T cells called *regulatory T cells*. It's not yet known exactly how these regulatory T cells work, but they do secrete chemicals that suppress the actions of other T cells. A third mechanism has been hypothesized to be the competition among B cells for a limited resource—a particular chemical named BAFF needed for B cells to survive. B cells that slip through the negative selection process and still bind strongly to self-molecules find themselves, due to their continual binding to self-molecules, in need of higher amounts of BAFF than non-self-binding B cells. Competition for this limited resource leads to the increased probability of death for self-binding B cells.

Even if the immune system is attacking foreign pathogens, it needs to balance the virulence of its attack with the obligation to prevent harm to the body as much as possible. The immune system employs a number of (mostly little understood) mechanisms for achieving this balance. Many of these mechanisms rely on a set of signaling molecules called *cytokines*. Harm to the body can result in the secretion of cytokines, which suppress active lymphocytes. Presumably the more harm being done, the higher the concentration of suppression cytokines, which makes it more likely that active cells will encounter them and turn off, thus regulating the immune system without suppressing it altogether.

Ant Colonies

As I described in chapter 1, analogies often have been made between ant colonies and the brain. Both can be thought of as networks of relatively simple elements (neurons, ants) from which emerge larger-scale information-processing behaviors. Two examples of such behavior in ant colonies are the ability to optimally and adaptively forage for food, and the ability to adaptively allocate ants to different tasks as needed by the colony. Both types of behavior are accomplished with no central control, via mechanisms that are surprisingly similar to those described above for the immune system.

In many ant species, foraging for food works roughly as follows. Foraging ants in a colony set out moving randomly in different directions. When an ant encounters a food source, it returns to the nest, leaving a trail made up of a type of signaling chemicals called *pheromones*. When other ants encounter a pheromone trail, they are likely to follow it. The greater the concentration of pheromone, the more likely an ant will be to follow the trail. If an ant encounters the food source, it returns to the nest, reinforcing the trail. In the absence of reinforcement, a pheromone trail will evaporate. In this way, ants collectively build up and communicate information about the locations and quality of different food sources, and this information adapts to changes in these environmental conditions. At any given time, the existing trails and their strengths form a good model of the food environment discovered collectively by the foragers (figure 12.4).

Task allocation is another way in which an ant colony regulates its own behavior in a decentralized way. The ecologist Deborah Gordon has studied task allocation in colonies of Red Harvester ants. Workers in these colonies divide themselves among four types of tasks: foraging, nest-maintenance, patrolling, and refuse-sorting work. The number of workers pursuing each type of task adapts to changes in the environment. Gordon found, for example,

FIGURE 12.4. An ant trail. (Photograph copyright © by Flagstaffotos. Reproduced by permission.)

that if the nest is disturbed in some small way, the number of nest maintenance workers will increase. Likewise, if the food supply in the neighborhood is large and high quality, the number of foragers will increase. How does an individual ant decide which task to adopt in response to nestwide environmental conditions, even though no ant directs the decision of any other ant and each ant interacts only with a small number of other ants?

The answer seems to be that ants decide to switch tasks both as a function of what they encounter in the environment and as a function of the *rate* at which they encounter other ants performing different tasks. For example, an inactive ant—one not currently performing a task—that encounters a foreign object near the nest has increased probability of taking up nest-maintenance work. In addition, an inactive ant that encounters a high rate of nest-maintenance workers entering and leaving the nest will also have an increased probability of adopting the nest-maintenance task; the increased activity in some way signals that there are important nest maintenance tasks to be done. In a similar way, a nest-maintenance worker who encounters a high rate of foragers returning to the nest carrying seeds will have an increased probability of switching to foraging; the increased seed delivery signals in some way that a high-quality food source has been found and needs to be exploited. Ants are apparently able to sense, through direct contact of their antennae with other ants, what task the other ants have been engaged in, by perceiving specific chemical residues associated with each task.

Similar types of mechanisms—based on pheromone signals and direct interaction among individuals—seem to be responsible for other types of collective behavior in ants and other social insects, such as the construction of bridges or shelters formed of ants' bodies described in chapter 1, although many aspects of such behavior are still not very well understood.

Biological Metabolism

Metabolism is the group of chemical processes by which living organisms use the energy they take in from food, air, or sunlight to maintain all the functions needed for life. These chemical processes occur largely inside of cells, via chains of chemical reactions called *metabolic pathways*. In every cell of an organism's body, nutrient molecules are processed to yield energy, and cellular components are built up via parallel metabolic pathways. These components are needed for internal maintenance and repair and for external functions and intercellular communication. At any given time, millions of molecules in the cell drift around randomly in the cytoplasm. The molecules continually encounter one another. Occasionally (on a scale of microseconds), enzymes encounter molecules of matching shape, speeding up the chemical reactions the enzymes control. Sequences of such reactions cause large molecules to be built up gradually.

Just as lymphocytes affect immune system dynamics by releasing cytokines, and as ants affect foraging behavior by releasing pheromones, chemical reactions that occur along a metabolic pathway continually change the speed of and resources given to that particular pathway.

In general, metabolic pathways are complex sequences of chemical reactions, controlled by self-regulating feedback. *Glycolysis* is one example of a metabolic pathway that occurs in all life forms—it is a multistep process in which glucose is transformed into the chemical pryruvate, which is then used by the metabolic pathway called the *citric acid cycle* to produce, among other things, the molecule called ATP (adenosine triphosphate), which is the principal source of usable energy in a cell.

At any given time, hundreds of such pathways are being followed, some independent, some interdependent. The pathways result in new molecules, initiation of other metabolic pathways, and the regulation of themselves or other metabolic pathways.

Similar to the regulation mechanisms I described above for the immune system and ant colonies, metabolic regulation mechanisms are based on feedback. Glycolysis is a great example of this. One of the main purposes of glycolysis is to provide chemicals necessary for the creation of ATP. If there is a large amount of ATP in the cell, this slows down the rate of glycolysis and thus decreases the rate of new ATP production. Conversely, when the cell is lacking in ATP, the rate of glycolysis goes up. In general, the speed of a metabolic pathway is often regulated by the chemicals that are produced by that pathway.

Information Processing in These Systems

Let me now attempt to answer the questions about information processing I posed at the beginning of this chapter:

- What plays the role of "information" in these systems?
- How is it communicated and processed?
- How does this information acquire *meaning*? And to whom?

WHAT PLAYS THE ROLE OF INFORMATION?

As was the case for cellular automata, when I talk about information processing in these systems I am referring not to the actions of individual components such as cells, ants, or enzymes, but to the collective actions of large groups of these components. Framed in this way, information is not, as in a traditional computer, precisely or statically located in any particular place in the system. Instead, it takes the form of statistics and dynamics of patterns over the system's components.

In the immune system the spatial distribution and temporal dynamics of lymphocytes can be interpreted as a dynamic representation of information about the continually changing population of pathogens in the body. Similarly, the spatial distribution and dynamics of cytokine concentrations encode large-scale information about the immune system's success in killing pathogens and avoiding harm to the body.

In ant colonies, information about the colony's food environment is represented, in a dynamic way, by the statistical distribution of ants on various trails. The colony's overall state is represented by the dynamic distribution of ants performing different tasks.

In cellular metabolism information about the current state and needs of the cell are continually reflected in the spatial concentrations and dynamics of different kinds of molecules.

HOW IS INFORMATION COMMUNICATED AND PROCESSED?

Communication via Sampling

One consequence of encoding information as statistical and time-varying patterns of low-level components is that no individual component of the system can perceive or communicate the "big picture" of the state of the system. Instead, information must be communicated via spatial and temporal sampling.

In the immune system, for example, lymphocytes sample their environment via receptors for both antigens and signals from other immune system cells in the form of cytokines. It is the results of the lymphocytes' samples of the spatial and temporal concentration of these molecular signals that cause lymphocytes to become active or stay dormant. Other cells are in turn affected by the samples they take of the concentration and type of active lymphocytes, which can lead pathogen-killer cells to particular areas in the body.

In ant colonies, an individual ant samples pheromone signals via its receptors. It bases its decisions on which way to move on the results of these sampled patterns of concentrations of pheromones in its environment. As I described above, individual ants also use sampling of concentration-based information—via random encounters with other ants—to decide when to adopt a particular task. In cellular metabolism, feedback in metabolic pathways arises from bindings between enzymes and particular molecules as enzymes sample spatial and time-varying concentrations of molecules.

Random Components of Behavior

Given the statistical nature of the information read, the actions of the system need to have random (or at least "unpredictable") components. All three systems described above use randomness and probabilities in essential ways. The receptor shape of each individual lymphocyte has a randomly generated component so as to allow sampling of many possible shapes. The spatial pattern of lymphocytes in the body has a random component due to the distribution of lymphocytes by the bloodstream so as to allow sampling of many possible spatial patterns of antigens. The detailed thresholds for activation of lymphocytes, their actual division rates, and the mutations produced in the offspring all involve random aspects.

Similarly, the movements of ant foragers have random components, and these foragers encounter and are attracted to pheromone trails in a probabilistic way. Ants also task-switch in a probabilistic manner. Biochemist Edward Ziff and science historian Israel Rosenfield describe this reliance on randomness as follows: "Eventually, the ants will have established a detailed map of paths to food sources. An observer might think that the ants are using a map supplied by an intelligent designer of food distribution. However, what appears to be a carefully laid out mapping of pathways to food supplies is really just a consequence of a series of random searches."

Cellular metabolism relies on random diffusion of molecules and on probabilistic encounters between molecules, with probabilities changing as relative concentrations change in response to activity in the system.

It appears that such intrinsic random and probabilistic elements are needed in order for a comparatively small population of simple components (ants, cells, molecules) to explore an enormously larger space of possibilities, particularly when the information to be gained from such explorations is statistical in nature and there is little a priori knowledge about what will be encountered.

However, randomness must be balanced with determinism: self-regulation in complex adaptive systems continually adjusts probabilities of where the components should move, what actions they should take, and, as a result, how deeply to explore particular pathways in these large spaces.

Fine-Grained Exploration

Many, if not all, complex systems in biology have a fine-grained architecture, in that they consist of large numbers of relatively simple elements that work together in a highly parallel fashion.

Several possible advantages are conferred by this type of architecture, including robustness, efficiency, and evolvability. One additional major advantage is that a fine-grained parallel system is able to carry out what Douglas Hofstadter has called a "parallel terraced scan." This refers to a simultaneous exploration of many possibilities or pathways, in which the resources given to each exploration at a given time depend on the perceived success of that exploration at that time. The search is parallel in that many different possibilities are explored simultaneously, but is "terraced" in that not all possibilities are explored at the same speeds or to the same depth. Information is used as it is gained to continually reassess what is important to explore.

For example, at any given time, the immune system must determine which regions of the huge space of possible pathogen shapes should be explored by lymphocytes. Each of the trillions of lymphocytes in the body at any given time can be seen as a particular mini-exploration of a range of shapes. The shape ranges that are most successful (i.e., bind strongly to antigens) are given more exploration resources, in the form of mutated offspring lymphocytes, than the shape ranges that do not pan out (i.e., lymphocytes that do not bind strongly). However, while exploiting the information that has been obtained, the immune system continues at all times to generate new lymphocytes that explore completely novel shape ranges. Thus the system is able to focus on the most promising possibilities seen so far, while never neglecting to explore new possibilities.

Similarly, ant foraging uses a parallel-terraced-scan strategy: many ants initially explore random directions for food. If food is discovered in any of these directions, more of the system's resources (ants) are allocated, via the feedback mechanisms described above, to explore those directions further. At all times, different paths are dynamically allocated exploration resources in proportion to their relative promise (the amount and quality of the food that has been discovered at those locations). However, due to the large number of ants and their intrinsic random elements, unpromising paths continue to be explored as well, though with many fewer resources. After all, who knows—a better source of food might be discovered.

In cellular metabolism such fine-grained explorations are carried out by metabolic pathways, each focused on carrying out a particular task. A pathway can be speeded up or slowed down via feedback from its own results or from other pathways. The feedback itself is in the form of time-varying concentrations of molecules, so the relative speeds of different pathways can continually adapt to the moment-to-moment needs of the cell.

Note that the fine-grained nature of the system not only allows many different paths to be explored, but it also allows the system to continually change its exploration paths, since only relatively simple micro-actions are taken at any time. Employing more coarse-grained actions would involve committing time to a particular exploration that might turn out not to be warranted. In this way, the fine-grained nature of exploration allows the system to fluidly and continuously adapt its exploration as a result of the information it obtains. Moreover, the redundancy inherent in fine-grained systems allows the system to work well even when the individual components are not perfectly reliable and the information available is only statistical in nature. Redundancy allows many independent samples of information to be made, and allows fine-grained actions to be consequential only when taken by large numbers of components.

Interplay of Unfocused and Focused Processes

In all three example systems there is a continual interplay of unfocused, random explorations and focused actions driven by the system's perceived needs.

In the immune system, unfocused explorations are carried out by a continually changing population of lymphocytes with different receptors, collectively prepared to approximately match any antigen. Focused explorations consist of the creation of offspring that are variations of successful lymphocytes, which allow these explorations to zero in on a particular antigen shape.

Likewise, ant foraging consists of unfocused explorations by ants moving at random, looking for food in any direction, and focused explorations in which ants follow existing pheromone trails.

In cellular metabolism, unfocused processes of random exploration by molecules are combined with focused activation or inhibition driven by chemical concentrations and genetic regulation.

As in all adaptive systems, maintaining a correct balance between these two modes of exploring is essential. Indeed, the optimal balance shifts over time. Early explorations, based on little or no information, are largely random and unfocused. As information is obtained and acted on, exploration gradually becomes more deterministic and focused in response to what has been perceived by the system. In short, the system both *explores* to obtain information and *exploits* that information to successfully adapt. This balancing act between unfocused exploration and focused exploitation has been hypothesized to be a general property of adaptive and intelligent systems. John Holland, for example, has cited this balancing act as a way to explain how genetic algorithms work.

HOW DOES INFORMATION ACQUIRE MEANING?

How information takes on *meaning* (some might call it *purpose*) is one of those slippery topics that has filled many a philosophy tome over the eons. I don't think I can add much to what the philosophers have said, but I do claim that in order to understand information processing in living systems we will need to answer this question in some form.

In my view, meaning is intimately tied up with survival and natural selection. Events that happen to an organism *mean* something to that organism if those events affect its well-being or reproductive abilities. In short, the meaning of an event is what tells one how to respond to it. Similarly, events that happen to or within an organism's immune system have meaning in terms of their effects on the fitness of the organism. (I'm using the term *fitness* informally here.) These events mean something *to* the immune system because they tell it how to respond so as to increase the organism's fitness—similarly with ant colonies, cells, and other information-processing systems in living creatures. This focus on fitness is one way I can make sense of the notion of meaning and apply it to biological information-processing systems.

But in a complex system such as those I've described above, in which simple components act without a central controller or leader, *who* or *what* actually perceives the meaning of situations so as to take appropriate actions? This is essentially the question of what constitutes *consciousness* or *self-awareness* in living systems. To me this is among the most profound mysteries in complex systems and in science in general. Although this mystery has been the subject of many books of science and philosophy, it has not yet been completely explained to anyone's satisfaction.

Thinking about living systems as doing computation has had an interesting side effect: it has inspired computer scientists to write programs that mimic such systems in order to accomplish real-world tasks. For example, ideas about information processing in the immune system have inspired so-called artificial immune systems: programs that adaptively protect computers from viruses and other intruders. Similarly ant colonies have inspired what are now called "ant colony optimization algorithms," which use simulated ants, secreting simulated pheromones and switching between simulated jobs, to solve hard problems such as optimal cell-phone communications routing and optimal scheduling of delivery trucks. I don't know of any artificial intelligence programs inspired both by these two systems *and* by cellular metabolism, except for one I myself wrote with my Ph.D. advisor, which I describe in the next chapter.

CHAPTER 13 | How to Make Analogies (if You Are a Computer)

Easy Things Are Hard

The other day I said to my eight-year-old son, "Jake, please put your socks on." He responded by putting them on his head. "See, I put my socks on!" He thought this was hilarious. I, on the other hand, realized that his antics illustrated a deep truth about the difference between humans and computers.

The "socks on head" joke was funny (at least to an eight-year-old) because it violates something we all know is true: even though most statements in human language are, in principle, ambiguous, when you say something to another person, they almost always *know what you mean*. If I say to my husband, "Honey, do you know where my keys are?" and he replies, simply, "yes," I get annoyed—of course I meant "tell me where my keys are." When my best friend says that she is feeling swamped at her job, and I reply "same here," she knows that I don't mean that I am feeling swamped at *her* job, but rather my own. This mutual understanding is what we might call "common sense" or, more formally, "sensitivity to context."

In contrast, we have modern-day computers, which are anything *but* sensitive to context. My computer supposedly has a state-of-the-art spam filter, but sometimes it can't figure out that a message with a "word" such as **V!a&®@** is likely to be spam. As a similar example, a recent *New York Times* article described how print journalists are now learning how to improve the Web accessibility of their stories by tailoring headlines to literal-minded search engines instead of to savvy humans: "About a year ago, the *Sacramento Bee* changed online section titles. 'Real Estate' became 'Homes,' 'Scene' turned into 'Lifestyle,' and dining information found in newsprint under 'Taste,' is online under 'Taste/Food.' "

This is, of course, not to say that computers are dumb about everything. In selected, narrow domains they have become quite intelligent. Computer-controlled vehicles can now drive by themselves across rugged desert terrain. Computer programs can beat human doctors at diagnosing certain diseases, human mathematicians at solving complex equations, and human grand masters at chess. These are only a few examples of a surge of recent successes in artificial intelligence (AI) that have brought a new sense of optimism to the field. Computer scientist Eric Horvitz noted, "At conferences you are hearing the phrase 'human-level AI,' and people are saying that without blushing."

Well, some people, perhaps. There are a few minor "human-level" things computers still can't do, such as understand human language, describe the content of a photograph, and more generally use common sense as in the preceding examples. Marvin Minsky, a founder of the field of artificial intelligence, concisely described this paradox of AI as, "Easy things are hard." Computers can do many things that we humans consider to require high intelligence, but at the same time they are unable to perform tasks that any three-year-old child could do with ease.

Making Analogies

An important missing piece for current-day computers is the ability to make analogies.

The term *analogy* often conjures up people's bad memories of standardized test questions, such as "Shoe is to foot as glove is to _____?" However, what I mean by analogy-making is much broader: analogy-making is the ability to perceive abstract similarity between two things in the face of superficial differences. This ability pervades almost every aspect of what we call intelligence.

Consider the following examples:

> A child learns that dogs in picture books, photographs, and real life are all instances of the same concept.
>
> A person is easily able to recognize the letter A in a vast variety of printed typefaces and handwriting.

Jean says to Simone, "I call my parents once a week." Simone replies "I do that too," meaning, of course, not that she calls Jean's parents once a week, but that she calls her own parents.

A woman says to her male colleague, "I've been working so hard lately, I haven't been able to spend enough time with my husband." He replies, "Same here"—meaning not that he is too busy to spend enough time with the woman's husband, but that he has little time to spend with his girlfriend.

An advertisement describes Perrier as "the Cadillac of bottled waters." A newspaper article describes teaching as "the Beirut of professions." The war in Iraq is called "another Vietnam."

Britain and Argentina go to war over the Falklands (or las Malvinas), a set of small islands located near the coast of Argentina and populated by British settlers. Greece sides with Britain because of its own conflict with Turkey over Cyprus, an island near the coast of Turkey, the majority of whose population is ethnically Greek.

A classical music lover hears an unfamiliar piece on the radio and knows instantly that it is by Bach. An early-music enthusiast hears a piece for baroque orchestra and can easily identify which country the composer was from. A supermarket shopper recognizes the music being piped in as a Muzak version of the Beatles' "Hey Jude."

The physicist Hideki Yukawa explains the nuclear force by using an analogy with the electromagnetic force, on which basis he postulates a mediating particle for the nuclear force with properties analogous to the photon. The particle is subsequently discovered, and its predicted properties are verified. Yukawa wins a Nobel prize.

This list is a small sampling of analogies ranging from the mundane everyday kind to the once-in-a-lifetime-discovery kind. Each of these examples demonstrates, at different levels of impressiveness, how good humans are at perceiving abstract similarity between two entities or situations by letting concepts "slip" from situation to situation in a fluid way. The list taken as a whole illustrates the ubiquity of this ability in human thought. As the nineteenth-century philosopher Henry David Thoreau put it, "All perception of truth is the detection of an analogy."

Perceiving abstract similarities is something computers are notoriously bad at. That's why I can't simply show the computer a picture, say, of a dog swimming in a pool, and ask it to find "other pictures like this" in my online photo collection.

My Own Route to Analogy

In the early 1980s, after I had graduated from college and didn't quite know what to do with my life, I got a job as a high-school math teacher in New York City. The job provided me with very little money, and New York is an expensive city, so I cut down on unnecessary purchases. But one purchase I did make was a relatively new book written by a computer science professor at Indiana University, with the odd title *Gödel, Escher, Bach: an Eternal Golden Braid*. Having majored in math and having visited a lot of museums, I knew who Gödel and Escher were, and being a fan of classical music, I knew very well who Bach was. But putting their names together in a book title didn't make sense to me, and my curiosity was piqued.

Reading the book, written by Douglas Hofstadter, turned out to be one of those life-changing events that one can never anticipate. The title didn't let on that the book was fundamentally about how thinking and consciousness emerge from the brain via the decentralized interactions of large numbers of simple neurons, analogous to the emergent behavior of systems such as cells, ant colonies, and the immune system. In short, the book was my introduction to some of the main ideas of complex systems.

It was clear that Hofstadter's passionate goal was to use similar principles to construct intelligent and "self-aware" computer programs. These ideas

Douglas Hofstadter. (Photograph courtesy of Indiana University.)

quickly became my passion as well, and I decided that I wanted to study artificial intelligence with Hofstadter.

The problem was, I was a young nobody right out of college and Hofstadter was a famous writer of a best-selling book that had won both a Pulitzer Prize and a National Book Award. I wrote him a letter saying I wanted to come work with him as a graduate student. I didn't hear back from him (it turns out he never received my letter). So I settled for biding my time and learning a bit more about AI.

A year later I had moved to Boston with a new job and was taking classes in computer science to prepare for my new career. One day I happened to see a poster advertising a talk by Hofstadter at MIT. Excited, I went to the talk, and afterward mingled among the throng of fans waiting to meet their hero (I wasn't the only one whose life was changed by Hofstadter's book). I finally got to the front of the line, shook Hofstadter's hand, and told him that I wanted to work in AI on ideas like his and that I was interested in applying to Indiana University. I asked if I could visit him sometime at Indiana to talk more. He told me that he was actually living in Boston, visiting the MIT Artificial Intelligence Lab for the year. As the impatient line closed in behind me, Hofstadter handed me off to talk to a former student of his who was hanging around, and quickly went on to the next person in line.

I was disappointed, but not deterred. I managed to find Hofstadter's phone number at the MIT AI Lab, and called several times. Each time the phone was answered by a secretary who told me that Hofstadter was not in, but she would be glad to leave a message. I left several messages but received no response.

Then, one night, I was lying in bed pondering what to do next, when a crazy idea hit me. All my calls to Hofstadter had been in the daytime, and he was never there. If he was never there during the day, then when *was* he there? It must be at night! It was 11:00 P.M., but I got up and dialed the familiar number. Hofstadter answered on the first ring.

He was quite friendly and gracious. We chatted for a while, and he invited me to come by his office the next day to talk about how I could get involved in his group's research. I showed up as requested, and we talked about Hofstadter's current project—writing a computer program that could make analogies.

Sometimes, having the personality of a bulldog can pay off.

Simplifying Analogy

One of Hofstadter's great intellectual gifts is the ability to take a complex problem and simplify it in such a way that it becomes easier to address but still retains its essence, the part that made it interesting in the first place. In this case, Hofstadter took the problem of analogy-making and created a *microworld* that retained many of the problem's most interesting features. The microworld consists of analogies to be made between strings of letters.

For example, consider the following problem: if **abc** changes to **abd**, what is the analogous change to **ijk**? Most people describe the change as something like "Replace the rightmost letter by its alphabetic successor," and answer **ijl**. But clearly there are many other possible answers, among them:

- **ijd** ("Replace the rightmost letter by a d"—similar to Jake putting his socks "on")
- **ijk** ("Replace all c's by d's; there are no c's in **ijk**"), and
- **abd** ("Replace any string by **abd**").

There are, of course, an infinity of other, even less plausible answers, such as **ijxx** ("Replace all c's by d's and each k by two x's"), but almost everyone immediately views **ijl** as the best answer. This being an abstract domain with no practical consequences, I may not be able to convince you that **ijl** is a better answer than, say, **ijd** if you really believe the latter is better. However, it seems that humans have evolved in such a way as to make analogies in the real world that affect their survival and reproduction, and their analogy-making ability seems to carry over into abstract domains as well. This means that almost all of us will, at heart, agree that there is a certain level of abstraction that is "most appropriate," and here it yields the answer **ijl**. Those people who truly believe that **ijd** is a better answer would probably, if alive during the Pleistocene, have been eaten by tigers, which explains why there are not many such people around today.

Here is a second problem: if **abc** changes to **abd**, what is the analogous change to **iijjkk**? The **abc** ⇒ **abd** change can again be described as "Replace the rightmost letter by its alphabetic successor," but if this rule is applied literally to **iijjkk** it yields answer **iijjkl**, which doesn't take into account the double-letter structure of **iijjkk**. Most people will answer **iijjll**, implicitly using the rule "Replace the rightmost *group of letters* by its alphabetic successor," letting the concept *letter* of **abc** slip into the concept *group of letters* for **iijjkk**.

Another kind of conceptual slippage can be seen in the problem

$$\mathbf{abc} \Rightarrow \mathbf{abd}$$
$$\mathbf{kji} \Rightarrow \;?$$

A literal application of the rule "Replace the rightmost letter by its alphabetic successor" yields answer **kjj**, but this ignores the reverse structure of **kji**, in which the increasing alphabetic sequence goes from right to left rather than from left to right. This puts pressure on the concept *rightmost* in **abc** to slip to *leftmost* in **kji**, which makes the new rule "Replace the *leftmost* letter by its alphabetic successor," yielding answer **lji**. This is the answer given by most people. Some people prefer the answer **kjh**, in which the sequence **kji** is seen as going from left to right but decreasing in the alphabet. This entails a slippage from "alphabetic successor" to "alphabetic predecessor," and the new rule is "Replace the rightmost letter by its alphabetic *predecessor*."

Consider

$$\text{abc} \Rightarrow \text{abd}$$
$$\text{mrrjjj} \Rightarrow \ ?$$

You want to make use of the salient fact that **abc** is an alphabetically increasing sequence, but how? This internal "fabric" of **abc** is a very appealing and seemingly central aspect of the string, but at first glance no such fabric seems to weave **mrrjjj** together. So either (like most people) you settle for **mrrkkk** (or possibly **mrrjjk**), or you look more deeply. The interesting thing about this problem is that there happens to be an aspect of **mrrjjj** lurking beneath the surface that, once recognized, yields what many people feel is a more satisfying answer. If you ignore the *letters* in **mrrjjj** and look instead at *group lengths*, the desired successorship fabric is found: the lengths of groups increase as "1-2-3." Once this connection between **abc** and **mrrjjj** is discovered, the rule describing **abc** ⇒ **abd** can be adapted to **mrrjjj** as "Replace the rightmost *group of letters* by its *length* successor," which yields "1-2-4" at the abstract level, or, more concretely, **mrrjjjj**.

Finally, consider

$$\text{abc} \Rightarrow \text{abd}$$
$$\text{xyz} \Rightarrow \ ?$$

At first glance this problem is essentially the same as the problem with target string **ijk** given previously, but there is a snag: *Z* has no successor. Most people answer **xya**, but in Hofstadter's microworld the alphabet is not circular and therefore this answer is excluded. This problem forces an impasse that requires analogy-makers to restructure their initial view, possibly making conceptual slippages that were not initially considered, and thus to discover a different way of understanding the situation.

People give a number of different responses to this problem, including **xy** ("Replace the z by nothing at all"), **xyd** ("Replace the rightmost letter by a d"; given the impasse, this answer seems less rigid and more reasonable than did **ijd** for the first problem above), **xyy** ("If you can't take the z's *successor*, then the next best thing is to take its *predecessor*"), and several other answers. However, there is one particular way of viewing this problem that, to many people, seems like a genuine insight, whether or not they come up with it themselves. The essential idea is that **abc** and **xyz** are "mirror images"—**xyz** is wedged against the end of the alphabet, and **abc** is similarly wedged against the beginning. Thus the z in **xyz** and the a in **abc** can be seen to correspond, and then one naturally feels that the x and the c correspond as well. Underlying these object correspondences is a set of slippages that are conceptually parallel: *alphabetic-first* $\Rightarrow$ *alphabetic-last*, *rightmost* $\Rightarrow$ *leftmost*, and *successor* $\Rightarrow$ *predecessor*. Taken together, these slippages convert the original rule into a rule adapted to the target string **xyz**: "Replace the *leftmost* letter by its *predecessor*." This yields a surprising but strong answer: **wyz**.

It should be clear by now that the key to analogy-making in this microworld (as well as in the real world) is what I am calling *conceptual slippage*. Finding appropriate conceptual slippages given the context at hand is the essence of finding a good analogy.

Being a Copycat

Doug Hofstadter's plan was for me to write a computer program that could make analogies in the letter-string world by employing the same kinds of mechanisms that he believed are responsible for human analogy-making in general. He already had a name for this (as yet nonexistent) program: "Copycat." The idea is that analogy-making is a subtle form of imitation—for example, **ijk** needs to imitate what happened when **abc** changed to **abd**, using concepts relevant in its own context. Thus the program's job was to be a clever and creative copycat.

I began working on this project at MIT in the summer of 1984. That fall, Hofstadter started a new faculty position at the University of Michigan in Ann Arbor. I also moved there and enrolled as a Ph.D. student. It took a total of six years of working closely with Doug for me to construct the program he envisioned—the devil, of course, is in the details. Two results came out of this: a program that could make human-like analogies in its microworld, and (finally) my Ph.D.

How to Do the Right Thing

To be an *intelligent* copycat, you first have to make sense of the object, event, or situation that you are "copycatting." When presented with a situation with many components and potential relations among components, be it a visual scene, a friend's story, or a scientific problem, how does a person (or how might a computer program) mentally explore the typically intractably huge number of possible ways of understanding what is going on and possible similarities to other situations?

The following are two opposite and equally implausible strategies, both to be rejected:

1. Some possibilities are a priori absolutely excluded from being explored. For example, after an initial scan of **mrrjjj**, make a list of candidate concepts to explore (e.g., *letter, group of letters, successor, predecessor, rightmost*) and rigidly stick to it. The problem with this strategy, of course, is that it gives up flexibility. One or more concepts not immediately apparent as relevant to the situation (e.g., *group length*) might emerge later as being central.
2. All possibilities are equally available and easy to explore, so one can do an exhaustive search through all concepts and possible relationships that would ever be relevant in any situation. The problem with this strategy is that in real life there are always too many possibilities, and it's not even clear ahead of time what might constitute a possible concept for a given situation. If you hear a funny clacking noise in your engine and then your car won't start, you might give equal weight to the possibilities that (a) the timing belt has accidentally come off its bearings or (b) the timing belt is old and has broken. If for no special reason you give equal weight to the third possibility that your next-door neighbor has furtively cut your timing belt, you are a bit paranoid. If for no special reason you also give equal weight to the fourth possibility that the atoms making up your timing belt have quantum-tunneled into a parallel universe, you are a bit of a crackpot. If you continue and give equal weight to every other possibility . . . well, you just can't, not with a finite brain. However, there is some chance you might be right about the malicious neighbor, and the quantum-tunneling possibility shouldn't be forever excluded from your cognitive capacities or you risk missing a Nobel prize.

The upshot is that all possibilities have to be potentially available, but they can't all be equally available. Counterintuitive possibilities (e.g., your malicious neighbor; quantum-tunneling) must be potentially available but must require significant pressure to be considered (e.g., you've heard complaints about your neighbor; you've just installed a quantum-tunneling device in your car; every other possibility that you have explored has turned out to be wrong).

The problem of finding an exploration strategy that achieves this goal has been solved many times in nature. For example, we saw this in chapter 12 in the way ant colonies forage for food: the shortest trails leading to the best food sources attain the strongest pheromone scent, and increasing numbers of ants follow these trails. However, at any given time, some ants are still following weaker, less plausible trails, and some ants are still foraging randomly, allowing for the possibility of new food sources to be found.

This is an example of needing to keep a balance between exploration and exploitation, which I mentioned in chapter 12. When promising possibilities are identified, they should be exploited at a rate and intensity related to their estimated promise, which is being continually updated. But at all times exploration for new possibilities should continue. The problem is how to allocate limited resources—be they ants, lymphocytes, enzymes, or thoughts—to different possibilities in a dynamic way that takes new information into account as it is obtained. Ant colonies have solved this problem by having large numbers of ants follow a combination of two strategies: continual random foraging combined with a simple feedback mechanism of preferentially following trails scented with pheromones and laying down additional pheromone while doing so.

The immune system also seems to maintain a near optimal balance between exploration and exploitation. We saw in chapter 12 how the immune system uses randomness to attain the potential for responding to virtually any pathogen it encounters. This potential is realized when an antigen activates a particular B cell and triggers the proliferation of that cell and the production of antibodies with increasing specificity for the antigen in question. Thus the immune system *exploits* the information it encounters in the form of antigens by allocating much of its resources toward targeting those antigens that are actually found to be present. But it always continues to explore additional possibilities that it might encounter by maintaining its huge repertoire of different B cells. Like ant colonies, the immune system combines randomness with highly directed behavior based on feedback.

Hofstadter proposed a scheme for exploring uncertain environments: the "parallel terraced scan," which I referred to in chapter 12. In this scheme many possibilities are explored in parallel, each being allocated resources according to feedback about its current promise, whose estimation is updated continually as new information is obtained. Like in an ant colony or the immune system, all possibilities have the potential to be explored, but at any given time only some are being actively explored, and not with equal resources. When a person (or ant colony or immune system) has little information about the situation facing it, the exploration of possibilities starts out being very random, highly parallel (many possibilities being considered at once) and unfocused: there is no pressure to explore any particular possibility more strongly than any other. As more and more information is obtained, exploration gradually becomes more focused (increasing resources are concentrated on a smaller number of possibilities) and less random: possibilities that have already been identified as promising are exploited. As in ant colonies and the immune system, in Copycat such an exploration strategy emerges from myriad interactions among simple components.

Overview of the Copycat Program

Copycat's task is to use the concepts it possesses to build perceptual structures—descriptions of objects, links between objects in the same string, groupings of objects in a string, and correspondences between objects in different strings—on top of the three "raw," unprocessed letter strings given to it in each problem. The structures the program builds represent its understanding of the problem and allow it to formulate a solution. Since for every problem the program starts out from exactly the same state with exactly the same set of concepts, its concepts have to be adaptable, in terms of their relevance and their associations with one another, to different situations. In a given problem, as the representation of a situation is constructed, associations arise and are considered in a probabilistic fashion according to a parallel terraced scan in which many routes toward understanding the situation are tested in parallel, each at a rate and to a depth reflecting ongoing evaluations of its promise.

Copycat's solution of letter-string analogy problems involves the interaction of the following components:

- The *Slipnet*: A network of concepts, each of which consists of a central node surrounded by potential associations and slippages. A picture of some of the concepts and relationships in the current version of the program is given in figure 13.1. Each node in the Slipnet has a dynamic *activation* value that gives its current perceived relevance to the analogy problem at hand, which therefore changes as the program runs. Activation also spreads from a node to its conceptual neighbors and decays if not reinforced. Each link has a dynamic *resistance* value that gives its current resistance to slippage. This also changes as the program runs. The resistance of a link is inversely proportional to the

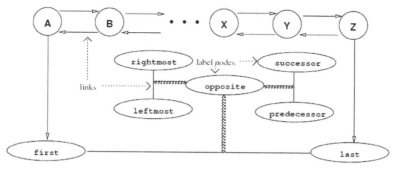

FIGURE 13.1. Part of Copycat's Slipnet. Each node is labeled with the concept it represents (e.g., *A–Z, rightmost, successor*). Some links between nodes (e.g., *rightmost–leftmost*) are connected to a label node giving the link's relationship (e.g., *opposite*). Each node has a dynamic activation value (not shown) and spreads activation to neighboring nodes. Activation decays if not reinforced. Each link has an intrinsic resistance to slippage, which decreases when the label node is activated.

activation of the node naming the link. For example, when *opposite* is highly active, the resistance to slippage between nodes linked by opposite links (e.g., *successor* and *predecessor*) is lowered, and the probability of such slippages is increased.

- The *Workspace*: A working area in which the letters composing the analogy problem reside and in which perceptual structures are built on top of the letters.
- *Codelets*: Agents that continually explore possibilities for perceptual structures to build in the Workspace, and, based on their findings, attempt to instantiate such structures. (The term *codelet* is meant to evoke the notion of a "small piece of code," just as the later term *applet* in Java is meant to evoke the notion of a small application program.) Teams of codelets cooperate and compete to construct perceptual structures defining relationships between objects (e.g., "b is the successor of a in **abc**," or "the two i's in **iijjkk** form a *group*," or "the b in **abc** corresponds to the group of j's in **iijjkk**," or "the c in **abc** corresponds to the k in **kji**"). Each team considers a particular possibility for structuring part of the world, and the resources (codelet time) allocated to each team depends on the promise of the structure it is trying to build, as assessed dynamically as exploration proceeds. In this way, a parallel terraced scan of possibilities emerges as the teams of codelets, via competition and cooperation, gradually build up a

hierarchy of structures that defines the program's "understanding" of the situation with which it is faced.

- *Temperature*, which measures the amount of perceptual organization in the system. As in the physical world, high temperature corresponds to disorganization, and low temperature corresponds to a high degree of organization. In Copycat, temperature both measures organization and feeds back to control the degree of randomness with which codelets make decisions. When the temperature is high, reflecting little perceptual organization and little information on which to base decisions, codelets make their decisions more randomly. As perceptual structures are built and more information is obtained about what concepts are relevant and how to structure the perception of objects and relationships in the world, the temperature decreases, reflecting the presence of more information to guide decisions, and codelets make their decisions more deterministically.

A Run of Copycat

The best way to describe how these different components interact in Copycat is to display graphics from an actual run of the program. These graphics are produced in real-time as the program runs. This section displays snapshots from a run of the program on **abc ⇒ abd, mrrjjj ⇒ ?**

Figure 13.2: The problem is presented. Displayed are: the Workspace (here, the as-yet unstructured letters of the analogy problem); a "thermometer" on the left that gives the current temperature (initially set at 100, its maximum

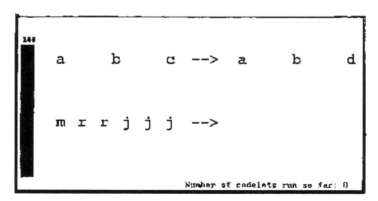

FIGURE 13.2.

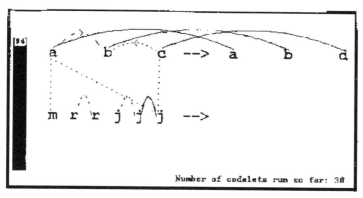

FIGURE 13.3.

value, reflecting the lack of any perceptual structures); and the number of codelets that have run so far (zero).

Figure 13.3: Thirty codelets have run and have investigated a variety of possible structures. Conceptually, codelets can be thought of as antlike agents, each one probabilistically following a path to explore but being guided by the paths laid down by other codelets. In this case the "paths" correspond to candidate perceptual structures. Candidate structures are proposed by codelets looking around at random for plausible descriptions, relationships, and groupings within strings, and correspondences between strings. A proposed structure becomes stronger as more and more codelets consider it and find it worthwhile. After a certain threshold of strength, the structure is considered to be "built" and can then influence subsequent structure building.

In figure 13.3, dotted lines and arcs represent structures in early stages of consideration; dashed lines and arcs represent structures in more serious stages of consideration; finally, solid lines and arcs represent structures that have been built. The speed at which proposed structures are considered depends on codelets' assessments of the promise of the structure. For example, the codelet that proposed the a–m correspondence rated it as highly promising because both objects are *leftmost* in their respective strings: identity relationships such as *leftmost* ⇒ *leftmost* are always strong. The codelet that proposed the a–j correspondence rated it much more weakly, since the mapping it is based on, *leftmost* ⇒ *rightmost*, is much weaker, especially given that *opposite* is not currently active. Thus the a–m correspondence is likely to be investigated more quickly than the less plausible a–j correspondence.

The temperature has gone down from 100 to 94 in response to the single built structure, the "sameness" link between the rightmost two j's in **mrrjjj**. This sameness link activated the node *same* in the Slipnet (not shown), which

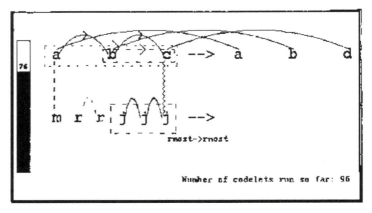

FIGURE 13.4.

creates focused pressure in the form of specifically targeted codelets to look for instances of sameness elsewhere.

Figure 13.4: Ninety-six codelets have run. The successorship fabric of **abc** has been built. Note that the proposed c-to-b predecessor link of figure 13.3 has been out-competed by a successor link. The two successor links in **abc** support each other: each is viewed as stronger due to the presence of the other, making rival predecessor links much less likely to destroy the successor links.

Two rival groups based on successorship links between letters are being considered: bc and abc (a whole-string group). These are represented by dotted or dashed rectangles around the letters in figure 13.4. Although **bc** got off to an early lead (it is dashed while the latter is only dotted), the group **abc** covers more objects in the string. This makes it stronger than **bc**—codelets will likely get around to testing it more quickly and will be more likely to build it than to build **bc**. A strong group, **jjj**, based on sameness is being considered in the bottom string.

Exploration of the crosswise a–j correspondence (dotted line in figure 13.3) has been aborted, since codelets that further investigated it found it too weak to be built. A c–j correspondence has been built (jagged vertical line); the mapping on which it is based (namely, both letters are *rightmost* in their respective strings) is given beneath it.

Since successor and sameness links have been built, along with an identity mapping (*rightmost* ⇒ *rightmost*), these nodes are highly active in the Slipnet and are creating focused pressure in the form of codelets to search explicitly for other instances of these concepts. For example, an identity mapping between the two leftmost letters is being considered.

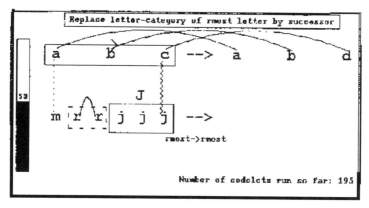

FIGURE 13.5.

In response to the structures that have been built, the temperature has decreased to 76. The lower the temperature, the less random are the decisions made by codelets, so unlikely structures such as the **bc** group are even more unlikely to be built.

Figure 13.5: The **abc** and **jjj** groups have been built, represented by solid rectangles around the letters. For graphical clarity, the links between letters in a group are not displayed. The existence of these groups creates additional pressure to find new successorship and sameness groups, such as the **rr** sameness group that is being strongly considered. Groups, such as the **jjj** sameness group, become new objects in the string and can have their own descriptions, as well as links and correspondences to other objects. The capital J represents the object consisting of the **jjj** group; the **abc** group likewise is a new object but for clarity a single letter representing it is not displayed. Note that the length of a group is not automatically noticed by the program; it has to be noticed by codelets, just like other attributes of an object. Every time a group node (e.g., *successor group*, *sameness group*) is activated in the Slipnet, it spreads some activation to the node *length*. Thus length is now weakly activated and creating codelets to notice lengths, but these codelets are not urgent compared with others and none so far have run and noticed the lengths of groups.

A rule describing the **abc** ⇒ **abd** change has been built: "Replace letter-category of rightmost letter by successor." The current version of Copycat assumes that the example change consists of the replacement of exactly one letter, so rule-building codelets fill in the template "Replace _____ by _____," choosing probabilistically from descriptions that the program has attached to the changed letter and its replacement, with a probabilistic bias

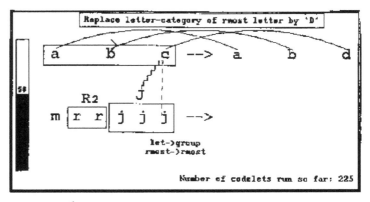

FIGURE 13.6.

toward choosing more abstract descriptions (e.g., usually preferring *rightmost letter* to *C*).

The temperature has fallen to 53, resulting from the increasing perceptual organization reflected in the structures that have been built.

Figure 13.6: Two-hundred twenty-five codelets have run. The letter-to-letter c–j correspondence has been defeated by the letter-to-group c–J correspondence. Reflecting this, the *rightmost* ⇒ *rightmost* mapping has been joined by a *letter* ⇒ *group* mapping underlying the correspondence. The c–J correspondence is stronger than the c–j correspondence because the former covers more objects and because the concept *group* is highly active and thus seen as highly relevant to the problem. However, in spite of its relative weakness, the c–j correspondence is again being considered by a new team of codelets.

Meanwhile, the **rr** group has been built. In addition, its length (represented by the 2 next to the R) has been noticed by a codelet (a probabilistic event). This event activated the node *length*, creating pressure to notice other groups' lengths.

A new rule, "Replace the letter category of the rightmost letter by 'D,'" has replaced the old one at the top of the screen. Although this rule is weaker than the previous one, competitions between rival structures (including rules) are decided probabilistically, and this one simply happened to win. However, its weakness has caused the temperature to increase to 58.

If the program were to stop now (which is quite unlikely, since a key factor in the program's probabilistic decision to stop is the temperature, which is still relatively high), the rule would be adapted for application to the string **mrrjjj** as "Replace the letter category of the rightmost group by 'D,'" obeying the slippage *letter* ⇒ *group* spelled out under the c–J correspondence. This yields

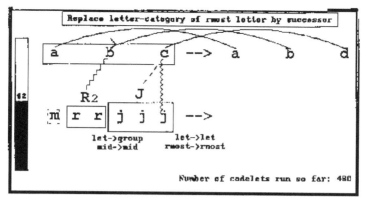

FIGURE 13.7.

answer **mrrddd**, an answer that Copycat does indeed produce, though on very rare occasions.

Codelets that attempt to create an answer run frequently throughout the program (their attempts are not displayed here) but are not likely to succeed unless the temperature is low.

Figure 13.7: Four hundred eighty codelets into the run, the rule "Replace letter-category of rightmost letter by successor" has been restored after it outcompeted the previous weaker rule (a probabilistic event). However, the strong c–J correspondence was broken and replaced by its weaker rival, the c–j correspondence (also a probabilistic event). As a consequence, if the program were to stop at this point, its answer would be **mrrjjk**, since the c in **abc** is mapped to a letter, not to a group. Thus the answer-building codelet would ignore the fact that b has been mapped to a group. However, the (now) candidate correspondence between the c and the group J is again being strongly considered. It will fight again with the c–j correspondence, but will likely be seen as even stronger than before because of the parallel correspondence between the b and the group R.

In the Slipnet the activation of *length* has decayed since the length description given to the R group hasn't so far been found to be useful (i.e., it hasn't yet been connected up with any other structures). In the Workspace, the salience of the group R's length description 2 is correspondingly diminished.

The temperature is still fairly high, since the program is having a hard time making a single, coherent structure out of **mrrjjj**, something that it did easily with **abc**. That continuing difficulty, combined with strong focused pressure from the two sameness groups that have been built inside **mrrjjj**, caused the system to consider the a priori very unlikely idea of making a

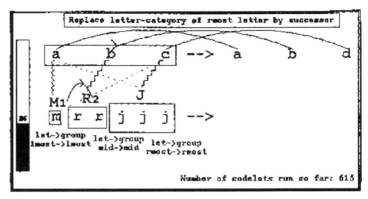

FIGURE 13.8.

single-letter sameness group. This is represented by the dashed rectangle around the letter m.

Figure 13.8: As a result of these combined pressures, the M sameness group was built to parallel the R and J groups in the same string. Its length of 1 has been attached as a description, activating *length*, which makes the program again consider the possibility that group length is relevant for this problem. This activation now more strongly attracts codelets to the objects representing group lengths. Some codelets have already been exploring relations between these objects and, probably due to focused pressures from **abc** to see successorship relationships, have built a successorship link between the 1 and the 2.

A consistent trio of *letter* ⇒ *group* correspondences have been made, and as a result of these promising new structures the temperature has fallen to the relatively low value of 36, which in turn helps to lock in this emerging view.

If the program were to halt at this point, it would produce the answer **mrrkkk**, which is its most frequent answer (see figure 13.12).

Figure 13.9: As a result of *length*'s continued activity, length descriptions have been attached to the remaining two groups in the problem, **jjj** and **abc**, and a successorship link between the 2 and the 3 (for which there is much focused pressure coming from both **abc** and the emerging view of **mrrjjj**) is being considered. Other less likely candidate structures (a **bc** group and a c–j correspondence) continue to be considered, though at considerably less urgency than earlier, now that a coherent perception of the problem is emerging and the temperature is relatively low.

Figure 13.10: The link between the 2 and the 3 was built, which, in conjunction with focused pressure from the **abc** successor group, allowed codelets to propose and build a whole-string group based on successorship links, here

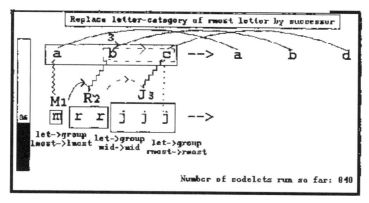

FIGURE 13.9.

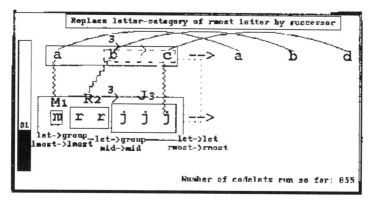

FIGURE 13.10.

between numbers rather than between letters. This group is represented by a large solid rectangle surrounding the three sameness groups. Also, a correspondence (dotted vertical line to the right of the two strings) is being considered between the two whole-string groups **abc** and **mrrjjj**.

Ironically, just as these sophisticated ideas seem to be converging, a small renegade codelet, totally unaware of the global movement, has had some good luck: its bid to knock down the c–J correspondence and replace it with a c–j correspondence was successful. Of course, this is a setback on the global level; while the temperature would have gone down significantly because of the strong **mrrjjj** group that was built, its decrease was offset by the now nonparallel set of correspondences linking together the two strings. If the program were forced to stop at this point, it would answer **mrrjjk**, since at this point, the object that changed, the c, is seen as corresponding to the letter j rather than the group J. However, the two other correspondences will

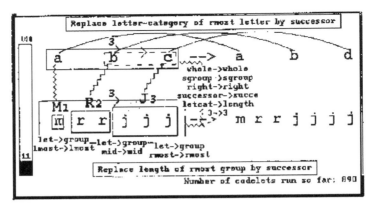

FIGURE 13.11.

continue to put much pressure on the program (in the form of codelets) to go back to the c–J correspondence.

Figure 13.11: Indeed, not much later in the run this happens: the c–j correspondence has been broken and the c–J correspondence has been restored. In addition, the proposed whole-string correspondence between **abc** and **mrrjjj** has been built; underlying it are the mappings *whole* ⇒ *whole*, *successor-group* ⇒ *successor-group*, *right* ⇒ *right* (direction of the links underlying both groups), *successor* ⇒ *successor* (type of links underlying both groups), *letter-category* ⇒ *length*, and $3 \Rightarrow 3$ (size of both groups).

The now very coherent set of perceptual structures built by the program resulted in a very low temperature (11), and (probabilistically) due to this low temperature, a codelet has succeeded in translating the rule according to the slippages present in the Workspace: *letter* ⇒ *group* and *letter-category* ⇒ *length* (all other mappings are identity mappings). The translated rule is "Replace the length of the rightmost group by its successor," and the answer is thus **mrrjjjj**.

It should be clear from the description above that because each run of Copycat is permeated with probabilistic decisions, different answers appear on different runs. Figure 13.12 displays a bar graph showing the different answers Copycat gave over 1,000 runs, each starting from a different random number seed. Each bar's height gives the relative frequency of the answer it corresponds to, and printed above each bar is the actual number of runs producing that answer. The average final temperature for each answer is also given below each bar's label.

The frequency of an answer roughly corresponds to how obvious or immediate it is, given the biases of the program. For example, **mrrkkk**, produced 705 times, is much more immediate to the program than **mrrjjjj**, which was

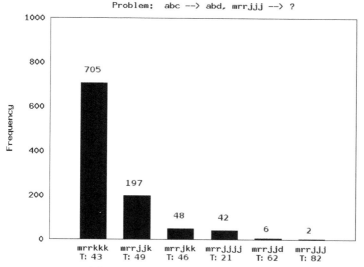

FIGURE 13.12. A bar graph plotting the different answers Copycat gave over 1,000 runs, each starting from a different random number seed.

produced only 42 times. However, the average final temperature on runs producing **mrrjjjj** is much lower than on runs producing **mrrkkk** (21 versus 43), indicating that even though the latter is a more immediate answer, the program judges the former to be a better answer, in terms of the strength and coherence of the structures it built to produce each answer.

Summary

Via the mechanisms illustrated in this run of the program, Copycat avoids the Catch-22 of perception: you can't explore everything, but you don't know which possibilities are worth exploring without first exploring them. You have to be open-minded, but the territory is too vast to explore everything; you need to use probabilities in order for exploration to be fair. In Copycat's biologically inspired strategy, early on there is little information, resulting in high temperature and high degree of randomness, with lots of parallel explorations. As more and more information is obtained and fitting concepts are found, the temperature falls, and exploration becomes more deterministic and more serial as certain concepts come to dominate. The overall result is that the system gradually changes from a mostly random, parallel, bottom-up mode of processing to a deterministic, serial, focused mode in which a coherent perception of the situation at hand is gradually discovered and gradually "frozen in." As I illustrated in chapter 12, this gradual transition between different modes of processing seems to be a feature common to at least some complex adaptive systems.

Analogies such as those between Copycat and biological systems force us to think more broadly about the systems we are building or trying to understand. If one notices, say, that the role of cytokines in immune signaling is similar to that of codelets that call attention to particular sites in an analogy problem, one is thinking at a general *information-processing* level about the function of a biological entity. Similarly, if one sees that temperature-like phenomena in the immune system—fever, inflammation—emerge from the joint actions of many agents, one might get some ideas on how to better model temperature in a system like Copycat.

Finally, there is the ever-thorny issue of meaning. In chapter 12 I said that for traditional computers, information is not meaningful to the computer itself but to its human creators and "end users." However, I would like to think that Copycat, which represents a rather nontraditional mode of computation, perceives a very primitive kind of meaning in the concepts it has and in analogies it makes. For example, the concept *successor group* is embedded in a network in which it is linked to conceptually similar concepts, and Copycat can recognize and use this concept in an appropriate way in a large variety of diverse situations. This is, in my mind, the beginning of meaning. But as I said in chapter 12, meaning is intimately tied up with survival and natural selection, neither of which are relevant to Copycat, except for the very weak "survival" instinct of lowering its temperature. Copycat (and an even more impressive array of successor programs created in Hofstadter's research group) is still quite far from biological systems in this way.

The ultimate goal of AI is to take humans out of the *meaning* loop and have the computer itself perceive meaning. This is AI's hardest problem. The mathematician Gian-Carlo Rota called this problem "the barrier of meaning" and asked whether or when AI would ever "crash" it. I personally don't think it will be anytime soon, but if and when this barrier is unlocked, I suspect that analogy will be the key.

CHAPTER 14 | Prospects of Computer Modeling

BECAUSE COMPLEX SYSTEMS ARE TYPICALLY, as their name implies, hard to understand, the more mathematically oriented sciences such as physics, chemistry, and mathematical biology have traditionally concentrated on studying simple, idealized systems that are more tractable via mathematics. However, more recently, the existence of fast, inexpensive computers has made it possible to construct and experiment with models of systems that are too complex to be understood with mathematics alone. The pioneers of computer science—Alan Turing, John von Neumann, Norbert Wiener, and others—were all motivated by the desire to use computers to simulate systems that develop, think, learn, and evolve. In this fashion a new way of doing science was born. The traditional division of science into theory and experiment has been complemented by an additional category: computer simulation (figure 14.1). In this chapter I discuss what we can learn from computer models of complex systems and what the possible pitfalls are of using such models to do science.

What Is a Model?

A model, in the context of science, is a simplified representation of some "real" phenomenon. Scientists supposedly study nature, but in reality much of what they do is construct and study models of nature.

Think of Newton's law of gravity: the force of gravity between two objects is proportional to the product of their masses divided by the square of the distance between them. This is a mathematical statement of the effects

FIGURE 14.1. The traditional division of science into theory and experiment has been complemented by a new category: computer simulation. (Drawing by David Moser.)

of a particular phenomenon—a *mathematical model*. Another kind of model describes how the phenomenon actually works in terms of simpler concepts—that is, what we call *mechanisms*. In Newton's own time, his law of gravity was attacked because he did not give a mechanism for gravitational force. Literally, he did not show how it could be explained in terms of "size, shape, and motion" of parts of physical objects—the primitive elements that were, according to Descartes, necessary and sufficient components of all models in physics. Newton himself speculated on possible mechanisms of gravity; for example, he "pictured the Earth like a sponge, drinking up the constant stream of fine aethereal matter falling from the heavens, this stream by its impact on bodies above the Earth causing them to descend." Such a conceptualization might be called a *mechanistic model*. Two hundred years later, Einstein proposed a different mechanistic model for gravity, general relativity, in which gravity is conceptualized as being caused by the effects of material bodies on the shape of four-dimensional space-time. At present, some physicists are touting string theory, which proposes that gravity is caused by miniscule, vibrating strings.

Models are ways for our minds to make sense of observed phenomena in terms of concepts that are familiar to us, concepts that we can get our heads around (or in the case of string theory, that only a few very smart people can get their heads around). Models are also a means for predicting the future: for example, Newton's law of gravity is still used to predict planetary orbits, and Einstein's general relativity has been used to successfully predict deviations from those predicted orbits.

Idea Models

For applications such as weather forecasting, the design of automobiles and airplanes, or military operations, computers are often used to run detailed and complicated models that in turn make detailed predictions about the specific phenomena being modeled.

In contrast, a major thrust of complex systems research has been the exploration of *idea models*: relatively simple models meant to gain insights into a general concept without the necessity of making detailed predictions about any specific system. Here are some examples of idea models that I have discussed so far in this book:

- Maxwell's demon: An idea model for exploring the concept of entropy.
- Turing machine: An idea model for formally defining "definite procedure" and for exploring the concept of computation.
- Logistic model and logistic map: Minimal models for predicting population growth; these were later turned into idea models for exploring concepts of dynamics and chaos in general.
- Von Neumann's self-reproducing automaton: An idea model for exploring the "logic" of self-reproduction.
- Genetic algorithm: An idea model for exploring the concept of adaptation. Sometimes used as a minimal model of Darwinian evolution.
- Cellular automaton: An idea model for complex systems in general.
- Koch curve: An idea model for exploring fractal-like structures such as coastlines and snowflakes.
- Copycat: An idea model for human analogy-making.

Idea models are used for various purposes: to explore general mechanisms underlying some complicated phenomenon (e.g., von Neumann's logic of self-reproduction); to show that a proposed mechanism for a phenomenon is plausible or implausible (e.g., dynamics of population growth); to explore the effects of variations on a simple model (e.g., investigating what happens when you change mutation rates in genetic algorithms or the value of the control parameter R in the logistic map); and more generally, to act as what the philosopher Daniel Dennett called "intuition pumps"—thought experiments or computer simulations used to prime one's intuitions about complex phenomena.

Idea models in complex systems also have provided inspiration for new kinds of technology and computing methods. For example, Turing machines inspired programmable computers; von Neumann's self-reproducing automaton inspired cellular automata; minimal models of Darwinian evolution, the immune system, and insect colonies inspired genetic algorithms, computer immune systems, and "swarm intelligence" methods, respectively.

To illustrate the accomplishments and prospects of idea models in science, I now delve into a few examples of particular idea models in the social sciences, starting with the best-known one of all: the Prisoner's Dilemma.

Modeling the Evolution of Cooperation

Many biologists and social scientists have used idea models to explore what conditions can lead to the evolution of *cooperation* in a population of self-interested individuals.

Indeed, living organisms are selfish—their success in evolutionary terms requires living long enough, staying healthy enough, and being attractive enough to potential mates in order to produce offspring. Most living creatures are perfectly willing to fight, trick, kill, or otherwise harm other creatures in the pursuit of these goals. Common sense predicts that evolution will select selfishness and self-preservation as desirable traits that will be passed on through generations and will spread in any population.

In spite of this prediction, there are notable counterexamples to selfishness at all levels of the biological and social realms. Starting from the bottom, sometime in evolutionary history, groups of single-celled organisms cooperated in order to form more complex multicelled organisms. At some point later, social communities such as ant colonies evolved, in which the vast majority of ants not only work for the benefit of the whole colony, but also abnegate their ability to reproduce, allowing the queen ant to be the only source of offspring. Much later, more complex societies emerged in primate populations, involving communal solidarity against outsiders, complicated trading, and eventually human nations, governments, laws, and international treaties.

Biologists, sociologists, economists, and political scientists alike have faced the question of how such cooperation can arise among fundamentally selfish individuals. This is not only a question of science, but also of policy: e.g., is it possible to engender conditions that will allow cooperation to arise and persist among different nations in order to deal with international concerns such as the spread of nuclear weapons, the AIDS epidemic, and global warming?

FIGURE 14.2. Alice and Bob face a "Prisoner's Dilemma." (Drawing by David Moser.)

THE PRISONER'S DILEMMA

In the 1950s, at the height of the Cold War, many people were thinking about how to foster cooperation between enemy nations so as to prevent a nuclear war. Around 1950, two mathematical game theorists, Merrill Flood and Melvin Drescher, invented the Prisoner's Dilemma as a tool for investigating such cooperation dilemmas.

The Prisoner's Dilemma is often framed as follows. Two individuals (call them Alice and Bob) are arrested for committing a crime together and are put into separate rooms to be interrogated by police officers (figure 14.2). Alice and Bob are each separately offered the same deal in return for testifying against the other. If Alice agrees to testify against Bob, then she will go free and Bob will receive a sentence of life in prison. However, if Alice refuses to testify but Bob agrees to testify, he will go free and she will receive the life sentence. If they both testify against the other, they each will go to prison, but for a reduced sentence of ten years. Both Alice and Bob know that if neither testifies against the other they can be convicted only on a lesser charge for which they will go to jail for five years. The police demand a decision from each of them on the spot, and of course don't allow any communication between them.

If you were Alice, what would you do?

You might reason it out this way: Bob is either going to testify against you or keep quiet, and you don't know which. Suppose he plans to testify against you. Then you would come out best by testifying against him (ten years vs. life in prison). Now suppose he plans to keep quiet. Again your best choice is to testify (go free vs. five years in prison). Thus, regardless of what Bob does, your best bet for saving your own skin is to agree to testify.

The problem is that Bob is following the exact same line of reasoning. So the result is, you both agree to testify against the other, giving each of you a worse outcome than if you had both kept quiet.

Let me tell this story in a slightly different context. Imagine you are the U.S. president. You are considering a proposal to build a powerful nuclear weapon, much more powerful than any that you currently have in your stockpile. You suspect, but don't know for sure, that the Russian government is considering the same thing.

Look into the future and suppose the Russians indeed end up building such a weapon. If you also had decided to build the weapon, then the United States and Russia would remain equal in fire power, albeit at significant cost to each nation and making the world a more dangerous place. If you had decided not to build the weapon, then Russia would have a military edge over the United States.

Now suppose Russia does not build the weapon in question. Then if you had decided to build it, the United States would have a military edge over Russia, though at some cost to the nation, and if you had decided not to build, the United States and Russia would remain equal in weaponry.

Just as we saw for Bob and Alice, regardless of what Russia is going to do, the rational thing is for you to approve the proposal, since in each case building the weapon turns out to be the better choice for the United States. Of course the Russians are thinking along similar lines, so both nations end up building the new bomb, producing a worse outcome for both than if neither had built it.

This is the paradox of the Prisoner's Dilemma—in the words of political scientist Robert Axelrod, "The pursuit of self-interest by each leads to a poor outcome for all." This paradox also applies to the all too familiar case of a group of individuals who, by selfishly pursuing their own interests, collectively bring harm to all members of the group (global warming is a quintessential example). The economist Garrett Hardin has famously called such scenarios "the tragedy of the commons."

The Prisoner's Dilemma and variants of it have long been studied as idea models that embody the essence of the cooperation problem, and results from those studies have influenced how scholars, businesspeople, and governments think about real-world policies ranging from weapons control and responses to terrorism to corporate management and regulation.

The Dilemma is typically formulated in terms of a two-person "game" defined by what mathematical game theorists call a *payoff matrix*—an array of all possible outcomes for two players. One possible payoff matrix for the Prisoner's Dilemma is given in figure 14.3. Here, the goal is to get as many

	A cooperates B cooperates	A cooperates B defects	A defects B cooperates	A defects B defects
Player **A** payoff	3	0	5	1
Player **B** payoff	3	5	0	1

FIGURE 14.3. A payoff matrix for the Prisoner's Dilemma game.

points (as opposed to as few years in prison) as possible. A *turn* consists of each player independently making a "cooperate or defect" decision. That is, on each turn, players A and B independently, without communicating, decide whether to *cooperate* with the other player (e.g., refuse to testify; decide not to build the bomb) or to *defect* from the other player (e.g., testify; build the bomb). If both players cooperate, each receives 3 points. If player A cooperates and player B defects, then player A receives zero points and player B gets 5 points, and vice versa if the situation is reversed. If both players defect, each receives 1 point. As I described above, if the game lasts for only one turn, the rational choice for both is to defect. However, if the game is *repeated*, that is, if the two players play several turns in a row, both players' always defecting will lead to a much lower total payoff than the players would receive if they learned to cooperate. How can reciprocal cooperation be induced?

Robert Axelrod, of the University of Michigan, is a political scientist who has extensively studied and written about the Prisoner's Dilemma. His work on the Dilemma has been highly influential in many different disciplines, and has earned him several prestigious research prizes, including a MacArthur foundation "genius" award.

Axelrod began studying the Dilemma during the Cold War as a result of his own concern over escalating arms races. His question was, "Under what conditions will cooperation emerge in a world of egoists without central authority?" Axelrod noted that the most famous historical answer to this question was given by the seventeenth-century philosopher Thomas Hobbes, who concluded that cooperation could develop only under the aegis of a central authority. Three hundred years (and countless wars) later, Albert Einstein similarly proposed that the only way to ensure peace in the nuclear age was to form an effective world government. The League of Nations, and later, the United Nations, were formed expressly for this purpose, but neither has been very successful in either establishing a world government or instilling peace between and within nations.

Robert Axelrod. (Photograph courtesy of the Center for the Study of Complex Systems, University of Michigan.)

Since an effective central government seems out of reach, Axelrod wondered if and how cooperation could come about without one. He believed that exploring strategies for playing the simple repeated Prisoner's Dilemma game could give insight into this question. For Axelrod, "cooperation coming about" meant that cooperative strategies must, over time, receive higher total payoff than noncooperative ones, even in the face of any changes opponents make to their strategies as the repeated game is being played. Furthermore, if the players' strategies are evolving under Darwinian selection, the fraction of cooperative strategies in the population should increase over time.

COMPUTER SIMULATIONS OF THE PRISONER'S DILEMMA

Axlerod's interest in determining what makes for a good strategy led him to organize two Prisoner's Dilemma tournaments. He asked researchers in several disciplines to submit computer programs that implemented particular strategies for playing the Prisoner's Dilemma, and then had the programs play repeated games against one another.

Recall from my discussion of Robby the Robot in chapter 9 that a strategy is a set of rules that gives, for any situation, the action one should take in that situation. For the Prisoner's Dilemma, a strategy consists of a rule for deciding whether to cooperate or defect on the next turn, depending on the opponent's behavior on previous turns.

The first tournament received fourteen submitted programs; the second tournament jumped to sixty-three programs. Each program played with every other program for 200 turns, collecting points according to the payoff matrix in figure 14.3. The programs had some memory—each program could store the results of at least some of its previous turns against each opponent. Some of the strategies submitted were rather complicated, using statistical techniques to characterize other players' "psychology." However, in both tournaments the winner (the strategy with the highest average score over games with all other players) was the simplest of the submitted strategies: TIT FOR TAT. This strategy, submitted by mathematician Anatol Rapoport, cooperates on the first turn and then, on subsequent turns, does whatever the opponent did for its move on the previous turn. That is, TIT FOR TAT offers cooperation and reciprocates it. But if the other player defects, TIT FOR TAT punishes that defection with a defection of its own, and continues the punishment until the other player begins cooperating again.

It is surprising that such a simple strategy could beat out all others, especially in light of the fact that the entrants in the second tournament already knew about TIT FOR TAT so could plan against it. However, out of the dozens of experts who participated, no one was able to design a better strategy.

Axelrod drew some general conclusions from the results of these tournaments. He noted that all of the top scoring strategies have the attribute of being *nice*—that is, they are never the first one to defect. The lowest scoring of all the *nice* programs was the "least forgiving" one: it starts out by cooperating but if its opponent defects even once, it defects against that opponent on every turn from then on. This contrasts with TIT FOR TAT, which will punish an opponent's defection with a defection of its own, but will forgive that opponent by cooperating once the opponent starts to cooperate again.

Axelrod also noted that although the most successful strategies were nice and forgiving, they also were retaliatory—they punished defection soon after it occurred. TIT FOR TAT not only was nice, forgiving, and retaliatory, but it also had another important feature: clarity and predictability. An opponent can easily see what TIT FOR TAT's strategy is and thus predict how it would respond to any of the opponent's actions. Such predictability is important for fostering cooperation.

Interestingly, Axelrod followed up his tournaments by a set of experiments in which he used a genetic algorithm to *evolve* strategies for the Prisoner's Dilemma. The fitness of an evolving strategy is its score after playing many repeated games with the other evolving strategies in the population. The genetic algorithm evolved strategies with the same or similar behavior as TIT FOR TAT.

EXTENSIONS OF THE PRISONER'S DILEMMA

Axelrod's studies of the Prisoner's Dilemma made a big splash starting in the 1980s, particularly in the social sciences. People have studied all kinds of variations on the game—different payoff matrices, different number of players, multiplayer games in which players can decide whom to play with, and so on. Two of the most interesting variations experimented with adding social norms and spatial structure, respectively.

Adding Social Norms

Axelrod experimented with adding *norms* to the Prisoner's Dilemma, where *norms* correspond to social censure (in the form of negative points) for defecting when others catch the defector in the act. In Axelrod's multiplayer game, every time a player defects, there is some probability that some other players will witness that defection. In addition to a strategy for playing a version of the Prisoner's Dilemma, each player also has a strategy for deciding whether to *punish* (subtract points from) a defector if the punisher witnesses the defection.

In particular, each player's strategies consist of two numbers: a probability of defecting (*boldness*) and a probability of punishing a defection that the player witnesses (*vengefulness*). In the initial population of players, these probability values are assigned at random to each individual.

At each generation, the population plays a round of the game: each player in the population plays a single game against all other players, and each time a player defects, there is some probability that the defection is witnessed by other population members. Each witness will punish the defector with a probability defined by the witness's vengefulness value.

At the end of each round, an evolutionary process takes place: a new population of players is created from parent strategies that are selected based on fitness (number of points earned). The parents create offspring that are mutated copies of themselves: each child can have slightly different *boldness* and *vengefulness* numbers than its parent. If the population starts out with most players' *vengefulness* set to zero (e.g., no social norms), then defectors will come to dominate the population. Axelrod initially expected to find that norms would facilitate the evolution of cooperation in the population—that is, *vengefulness* would evolve to counter *boldness*.

However, it turned out that norms alone were not enough for cooperation to emerge reliably. In a second experiment, Axelrod added *metanorms*, in which there were punishers to punish the nonpunishers, if you know what I mean. Sort of like people in the supermarket who give me disapproving looks when I don't discipline my children for chasing each other up the aisles and colliding with innocent shoppers. In my case the metanorm usually works. Axelrod also found that metanorms did the trick—if punishers of nonpunishers were around, the nonpunishers evolved to be more likely to punish, and the punished defectors evolved to be more likely to cooperate. In Axelrod's words, "Meta-norms can promote and sustain cooperation in a population."

Adding Spatial Structure

The second extension that I find particularly interesting is the work done by mathematical biologist Martin Nowak and collaborators on adding spatial structure to the Prisoner's Dilemma. In Axelrod's original simulations, there was no notion of space—it was equally likely for any player to encounter any other player, with no sense of distance between players.

Nowak suspected that placing players on a spatial lattice, on which the notion of *neighbor* is well defined, would have a strong effect on the evolution of cooperation. With his postdoctoral mentor Robert May (whom I mentioned in chapter 2 in the context of the logistic map), Nowak performed computer simulations in which the players were placed in a two-dimensional array, and each player played only with its nearest neighbors. This is illustrated in figure 14.4, which shows a five by five grid with one player at each site (Nowak and May's arrays were considerably larger). Each player has the simplest of strategies—it has no memory of previous turns; it either always cooperates or always defects.

The model runs in discrete time steps. At each time step, each player plays a single Prisoner's Dilemma game against each of its eight nearest neighbors (like a cellular automaton, the grid wraps around at the edges) and its eight resulting scores are summed. This is followed by a *selection* step in which each player is replaced by the highest scoring player in its neighborhood (possibly itself); no mutation is done.

The motivation for this work was biological. As stated by Nowak and May, "We believe that deterministically generated spatial structure within populations may often be crucial for the evolution of cooperation, whether it be among molecules, cells, or organisms."

Nowak and May experimented by running this model with different initial configurations of *cooperate* and *defect* players and by varying the values in the payoff matrix. They found that depending on these settings, the

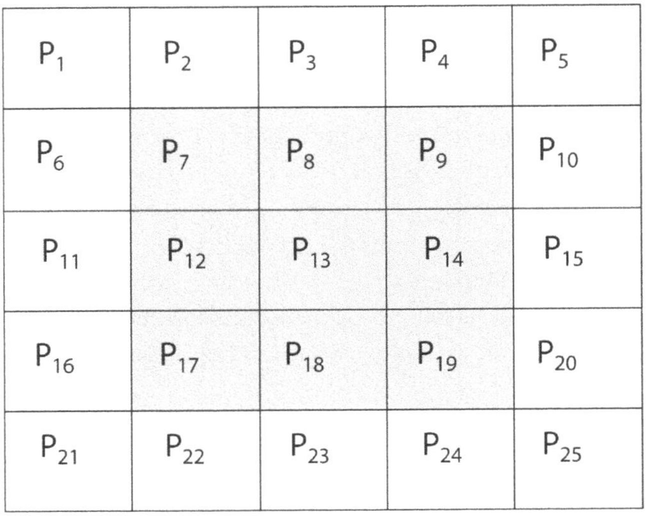

P_1	P_2	P_3	P_4	P_5
P_6	P_7	P_8	P_9	P_{10}
P_{11}	P_{12}	P_{13}	P_{14}	P_{15}
P_{16}	P_{17}	P_{18}	P_{19}	P_{20}
P_{21}	P_{22}	P_{23}	P_{24}	P_{25}

FIGURE 14.4. Illustration of a spatial Prisoner's Dilemma game. Each player interacts only with its nearest neighbors—e.g., player P_{13} plays against, and competes for selection against, only the players in its neighborhood (shaded).

spatial patterns of *cooperate* and *defect* players can either oscillate or be "chaotically changing," in which both cooperators and defectors coexist indefinitely. These results contrast with results from the nonspatial multiplayer Prisoner's Dilemma, in which, in the absence of meta-norms as discussed above, defectors take over the population. In Nowak and May's spatial case, cooperators can persist indefinitely without any special additions to the game, such as norms or metanorms.

Nowak and May believed that their result illustrated a feature of the real world—i.e., the existence of spatial neighborhoods fosters cooperation. In a commentary on this work, biologist Karl Sigmund put it this way: "That territoriality favours cooperation . . . is likely to remain valid for real-life communities."

Prospects of Modeling

Computer simulations of idea models such as the Prisoner's Dilemma, when done well, can be a powerful addition to experimental science and mathematical theory. Such models are sometimes the only available means of investigating complex systems when actual experiments are not feasible and when the math gets too hard, which is the case for almost all of the systems we are most interested in. The most significant contribution of idea models such as the Prisoner's Dilemma is to provide a first hand-hold on a phenomenon—such as the evolution of cooperation—for which we don't yet have precise scientific terminology and well-defined concepts.

The Prisoner's Dilemma models play all the roles I listed above for idea models in science (and analogous contributions could be listed from many other complex-systems modeling efforts as well):

Show that a proposed mechanism for a phenomenon is plausible or implausible. For example, the various Prisoner's Dilemma and related models have shown what Thomas Hobbes might not have believed: that it is indeed possible for cooperation—albeit in an idealized form—to come about in leaderless populations of self-interested (but adaptive) individuals.

Explore the effects of variations on a simple model and prime one's intuitions about a complex phenomenon. The endless list of Prisoner's Dilemma variations that people have studied has revealed much about the conditions under which cooperation can and cannot arise. You might ask, for example, what happens if, on occasion, people who want to cooperate make a mistake that accidentally signals noncooperation—an unfortunate mistranslation into Russian of a U.S. president's comments, for instance? The Prisoner's Dilemma gives an arena in which the effects of miscommunications can be explored. John Holland has likened such models to "flight simulators" for testing one's ideas and for improving one's intuitions.

Inspire new technologies. Results from the Prisoner's Dilemma modeling literature—namely, the conditions needed for cooperation to arise and persist—have been used in proposals for improving peer-to-peer networks and preventing fraud in electronic commerce, to name but two applications.

Lead to mathematical theories. Several people have used the results from Prisoner's Dilemma computer simulations to formulate general mathematical theories about the conditions needed for cooperation. A recent example is work by Martin Nowak, in a paper called "Five Rules for the Evolution of Cooperation."

How should results from idea models such as the Prisoner's Dilemma be used to inform policy decisions, such as the foreign relations strategies of governments or responses to global warming? The potential of idea models in predicting the results of different policies makes such models attractive, and, indeed, the influence of the Prisoner's Dilemma and related models among policy analysts has been considerable.

As one example, New Energy Finance, a consulting firm specializing in solutions for global warming, recently put out a report called "How to Save the Planet: Be Nice, Retaliatory, Forgiving, and Clear." The report argues that the problem of responding to climate change is best seen as a multi-player repeated Prisoner's Dilemma in which countries can either cooperate (mitigate carbon output at some cost to their economies) or defect (do nothing, saving money in the short term). The game is repeated year after year as new agreements and treaties regulating carbon emissions are forged. The report recommends specific policies that countries and global organizations should adopt in order to implement the "nice, retaliatory, forgiving, and clear" characteristics Axelrod cited as requirements for success in the repeated Prisoner's Dilemma.

Similarly, the results of the norms and metanorms models—namely, that not only norms but also metanorms can be important for sustaining cooperation—has had impact on policy-making research regarding government response to terrorism, arms control, and environmental governance policies, among other areas. The results of Nowak and May's spatial Prisoner's Dilemma models have informed people's thinking about the role of space and locality in fostering cooperation in areas ranging from the maintenance of biodiversity to the effectiveness of bacteria in producing new antibiotics. (See the notes for details on these various impacts.)

Computer Modeling Caveats

All models are wrong, but some are useful.
—George Box and Norman Draper

Indeed, the models I described above are highly simplified but have been useful for advancing science and policy in many contexts. They have led to new insights, new ways of thinking about complex systems, better models, and better understanding of how to build useful models. However, some very ambitious claims have been made about the models' results and how they apply in the real world. Therefore, the right thing for scientists to do is to carefully scrutinize the models and ask how general their results actually are. The best way to do that is to try to *replicate* those results.

In an experimental science such as astronomy or chemistry, every important experiment is replicated, meaning that a different group of scientists does the same experiment from scratch to see whether they get the same results as the original group. No experimental result is (or should be) believed if no other group can replicate it in this way. The inability of others to replicate results has been the death knell for uncountable scientific claims.

Computer models also need to be replicated—that is, independent groups need to construct the proposed computer model from scratch and see whether it produces the same results as those originally reported. Axelrod, an outspoken advocate of this idea, writes: "Replication is one of the hallmarks of cumulative science. It is needed to confirm whether the claimed results of a given simulation are reliable in the sense that they can be reproduced by someone starting from scratch. Without this confirmation, it is possible that some published results are simply mistaken due to programming errors, misrepresentation of what was actually simulated, or errors in analyzing or reporting the results. Replication can also be useful for testing the robustness of inferences from models."

Fortunately, many researchers have taken this advice to heart and have attempted to replicate some of the more famous Prisoner's Dilemma simulations. Several interesting and sometimes unexpected results have come out of these attempts.

In 1995, Bernardo Huberman and Natalie Glance re-implemented Nowak and May's spatial Prisoner's Dilemma model. Huberman and Glance ran a simulation with only one change. In the original model, at each time step all games between players in the lattice were played simultaneously, followed by the simultaneous selection in all neighborhoods of the fittest neighborhood player. (This required Nowak and May to simulate parallelism on their nonparallel computer.) Huberman and Glance instead allowed some of the games to be played *asynchronously*—that is, some group of neighboring players would play games and carry out selection, then another group of neighboring players would do the same, and so on. They found that this simple change, arguably making the model more realistic, would typically result in complete replacement of cooperators by defectors over the entire lattice. A similar result was obtained independently by Arijit Mukherji, Vijay Rajan, and James Slagle, who in addition showed that cooperation would die out in the presence of small errors or cheating (e.g., a cooperator accidentally or purposefully defecting). Nowak, May, and their collaborator Sebastian Bonhoeffor replied that these changes did indeed lead to the extinction of all cooperators for some payoff-matrix values, but for others, cooperators were able to stay in the population, at least for long periods.

In 2005 Jose Manuel Galan and Luis Izquierdo published results of their re-implementation of Axelrod's Norms and Metanorms models. Given the increase in computer power over the twenty years that had passed since

Axelrod's work, they were able to run the simulation for a much longer period and do a more thorough investigation of the effects of varying certain model details, such as the payoff matrix values, the probabilities for mutating offspring, and so on. Their results matched well with Axelrod's for some aspects of the simulation, but for others, the re-implementation produced quite different results. For example, they found that whereas metanorms can facilitate the evolution and persistence of cooperation in the short term, if the simulation is run for a long time, defectors end up taking over the population. They also found that the results were quite sensitive to the details of the model, such as the specific payoff values used.

What should we make of all this? I think the message is exactly as Box and Draper put it in the quotation I gave above: all models are wrong in some way, but some are very useful for beginning to address highly complex systems. Independent replication can uncover the hidden unrealistic assumptions and sensitivity to parameters that are part of any idealized model. And of course the replications themselves should be replicated, and so on, as is done in experimental science. Finally, modelers need above all to emphasize the limitations of their models, so that the results of such models are not misinterpreted, taken too literally, or hyped too much. I have used examples of models related to the Prisoner's Dilemma to illustrate all these points, but my previous discussion could be equally applied to nearly all other simplified models of complex systems.

I will give the last word to physicist (and ahead-of-his-time modelbuilding proponent) Phillip Anderson, from his 1977 Nobel Prize acceptance speech:

> The art of model-building is the exclusion of real but irrelevant parts of the problem, and entails hazards for the builder and the reader. The builder may leave out something genuinely relevant; the reader, armed with too sophisticated an experimental probe or too accurate a computation, may take literally a schematized model whose main aim is to be a demonstration of possibility.

PART IV | Network Thinking

In Ersilia, to establish the relationships that sustain the city's life, the inhabitants stretch strings from the corners of the houses, white or black or gray or black-and-white according to whether they mark a relationship of blood, of trade, authority, agency. When the strings become so numerous that you can no longer pass among them, the inhabitants leave: the houses are dismantled; only the strings and their supports remain.

From a mountainside, camping with their household goods, Ersilia's refugees look at the labyrinth of taut strings and poles that rise in the plain. That is the city of Ersilia still, and they are nothing.

They rebuild Ersilia elsewhere. They weave a similar pattern of strings which they would like to be more complex and at the same time more regular than the other. Then they abandon it and take themselves and their houses still farther away.

Thus, when traveling in the territory of Ersilia, you come upon the ruins of abandoned cities, without the walls which do not last, without the bones of the dead which the wind rolls away: spiderwebs of intricate relationships seeking a form.

—Italo Calvino, *Invisible Cities* (Trans. W. Weaver)

CHAPTER 15 | The Science of Networks

Small Worlds

I live in Portland, Oregon, whose metro area is home to over two million people. I teach at Portland State University, which has close to 25,000 students and over 1,200 faculty members. A few years back, my family had recently moved into a new house, somewhat far from campus, and I was chatting with our new next-door neighbor, Dorothy, a lawyer. I mentioned that I taught at Portland State. She said, "I wonder if you know my father. His name is George Lendaris." I was amazed. George Lendaris is one of the three or four faculty members at PSU, including myself, who work on artificial intelligence. Just the day before, I had been in a meeting with him to discuss a grant proposal we were collaborating on. Small world!

Virtually all of us have had this kind of "small world" experience, many much more dramatic than mine. My husband's best friend from high school turns out to be the first cousin of the guy who wrote the artificial intelligence textbook I use in my class. The woman who lived three houses away from mine in Santa Fe turned out to be a good friend of my high-school English teacher in Los Angeles. I'm sure you can think of several experiences of your own like this.

How is it that such unexpected connections seem to happen as often as they do? In the 1950s, a Harvard University psychologist named Stanley Milgram wanted to answer this question by determining, on average, how many links it would take to get from any person to any other person in the United States. He designed an experiment in which ordinary people would attempt to relay a letter to a distant stranger by giving the letter to an acquaintance, having the acquaintance give the letter to one of his or her acquaintances, and so on, until the intended recipient was reached at the end of the chain.

Milgram recruited (from newspaper ads) a group of "starters" in Kansas and Nebraska, and gave each the name, occupation, and home city of a "target," a person unknown to the starter, to whom the letter was addressed. Two examples of Milgram's chosen targets were a stockbroker in Boston and the wife of a divinity student in nearby Cambridge. The starters were instructed to pass on the letter to someone they knew personally, asking that person to continue the chain. Each link in the chain was recorded on the letter; if and when a letter reached the target, Milgram counted the number of links it went through. Milgram wrote of one example:

> Four days after the folders were sent to a group of starting persons in Kansas, an instructor at the Episcopal Theological Seminary approached our target person on the street. "Alice," he said, thrusting a brown folder toward her, "this is for you." At first she thought he was simply returning a folder that had gone astray and had never gotten out of Cambridge, but when we looked at the roster, we found to our pleased surprise that the document had started with a wheat farmer in Kansas. He had passed it on to an Episcopalian minister in his home town, who sent it to the minister who taught in Cambridge, who gave it to the target person. Altogether, the number of intermediate links between starting person and target amounted to two!

In his most famous study, Milgram found that, for the letters that made it to their target, the median number of intermediate acquaintances from starter to target was five. This result was widely quoted and is the source of the popular notion that people are linked by only "six degrees of separation."

Later work by psychologist Judith Kleinfeld has shown that the popular interpretation of Milgram's work was rather skewed—in fact, most of the letters from starters never made it to their targets, and in other studies by Milgram, the median number of intermediates for letters that did reach the targets was higher than five. However, the idea of a *small world* linked by *six degrees of separation* has remained as what may be an urban myth of our culture. As Kleinfeld points out,

> When people experience an unexpected social connection, the event is likely to be vivid and salient in a person's memory.... We have a poor mathematical, as well as a poor intuitive understanding of the nature of coincidence.

Stanley Milgram, 1933–1984. (Photograph by Eric Kroll, reprinted by permission of Mrs. Alexandra Milgram.)

So is it a small world or not? This question has recently received a lot of attention, not only for humans in the social realm, but also for other kinds of networks, ranging from the networks of metabolic and genetic regulation inside living cells to the explosively growing World Wide Web. Over the last decade questions about such networks have sparked a stampede of complex systems researchers to create what has been called the "new science of networks."

The New Science of Networks

You've no doubt seen diagrams of networks like the one in figure 15.1. This one happens to be a map of the domestic flight routes of Continental Airlines. The dots (or *nodes*) represent cities and the lines (or *links*) represent flights between cities.

Airline route maps are an obvious example of the many natural, technological, and cultural phenomena that can usefully be described as networks. The brain is a huge network of neurons linked by synapses. The control of genetic activity in a cell is due to a complex network of genes linked by regulatory proteins. Social communities are networks in which the nodes are people (or

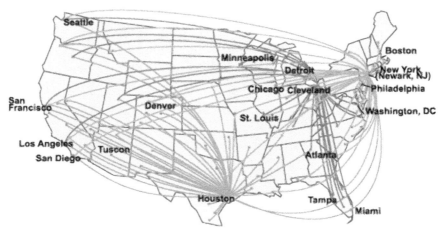

FIGURE 15.1. Slightly simplified route map of Continental Airlines. (From NASA Virtual Skies: [http://virtualskies.arc.nasa.gov/research/tutorial/tutorial2b.html].)

organizations of people) between whom there are many different types of possible relationships. The Internet and the World Wide Web are of course two very prominent networks in today's society. In the realm of national security, much effort has been put into identifying and analyzing possible "terrorist networks."

Until very recently, network science was not seen as a field unto itself. Mathematicians studied abstract network structures in a field called "graph theory." Neuroscientists studied neural networks. Epidemiologists studied the transmission of diseases through networks of interacting people. Sociologists and social psychologists such as Milgram were interested in the structure of human social networks. Economists studied the behavior of economic networks, such as the spread of technological innovation in networks of businesses. Airline executives studied networks like the one in figure 15.1 in order to find a node-link structure that would optimize profits given certain constraints. These different groups worked pretty much independently, generally unaware of one another's research.

However, over the last decade or so, a growing group of applied mathematicians and physicists have become interested in developing a set of unifying principles governing networks of any sort in nature, society, and technology. The seeds of this upsurge of interest in general networks were planted by the publication of two important papers in the late 1990s: "Collective Dynamics of 'Small World Networks'" by Duncan Watts and Steven Strogatz, and "Emergence of Scaling in Random Networks" by Albert-László Barabási and Réka Albert. These papers were published in the world's two top scientific

Duncan Watts (photograph courtesy of Duncan Watts).

journals, *Nature* and *Science*, respectively, and almost immediately got a lot of people *really* excited about this "new" field. Discoveries about networks started coming fast and furiously.

The time and place was right for people to jump on this network-science rushing locomotive. A study of common properties of networks across disciplines is only feasible with computers fast enough to study networks empirically—both in simulation and with massive amounts of real-world data. By the 1990s, such work was possible. Moreover, the rising popularity of using the Internet for social, business, and scientific networking meant that large amounts of data were rapidly becoming available.

In addition, there was a large coterie of very smart physicists who had lost interest in the increasingly abstract nature of modern physics and were looking for something else to do. Networks, with their combination of pristine mathematical properties, complex dynamics, and real-world relevance, were the perfect vehicle. As Duncan Watts (who is an applied mathematician and sociologist) phrased it, "No one descends with such fury and in so great a number as a pack of hungry physicists, adrenalized by the scent of a new problem." All these smart people were trained with just the right mathematical techniques, as well as the ability to simplify complex problems without losing

Steven Strogatz (photograph courtesy of Steven Strogatz).

Albert-László Barabási (photograph courtesy of Albert-László Barabási).

their essential features. Several of these physicists-turned-network-scientists have become major players in this field.

Perhaps most important, there was, among many scientists, a progressive realization that new ideas, new approaches—really, a new way of *thinking*—were direly needed to help make sense of the highly complex, intricately connected systems that increasingly affect human life and well-being. Albert-László Barabási, among others, has labeled the resulting new approaches "network thinking," and proclaimed that "network thinking is poised to invade all domains of human activity and most fields of human inquiry."

What Is Network Thinking?

Network thinking means focusing on relationships between entities rather than the entities themselves. For example, as I described in chapter 7, the fact that humans and mustard plants each have only about 25,000 genes does not seem to jibe with the biological complexity of humans compared with these plants. In fact, in the last few decades, some biologists have proposed that the complexity of an organism largely arises from complexity in the *interactions* among its genes. I say much more about these interactions in chapter 18, but for now it suffices to say that recent results in network thinking are having significant impacts on biology.

Network thinking has recently helped to illuminate additional, seemingly unrelated, scientific and technological mysteries: Why is the typical life span of organisms a simple function of their size? Why do rumors, jokes, and "urban myths" spread so quickly? Why are large, complex networks such as electrical power grids and the Internet so robust in some circumstances, and so susceptible to large-scale failures in others? What types of events can cause a once-stable ecological community to fall apart?

Disparate as these questions are, network researchers believe that the answers reflect commonalities among networks in many different disciplines. The goals of network science are to tease out these commonalities and use them to characterize different networks in a common language. Network scientists also want to understand how networks in nature came to be and how they change over time.

The scientific understanding of networks could have a large impact not only on our understanding of many natural and social systems, but also on our ability to engineer and effectively use complex networks, ranging from better Web search and Internet routing to controlling the spread of diseases, the effectiveness of organized crime, and the ecological damage resulting from human actions.

What Is a 'Network,' Anyway?

In order to investigate networks scientifically, we have to define precisely what we mean by *network*. In simplest terms, a network is a collection of *nodes* connected by *links*. Nodes correspond to the individuals in a network (e.g., neurons, Web sites, people) and links to the connections between them (e.g., synapses, Web hyperlinks, social relationships).

For illustration, figure 15.2 shows part of my own social network—some of my close friends, some of their close friends, et cetera, with a total of 19 nodes. (Of course most "real" networks would be considerably larger.)

At first glance, this network looks like a tangled mess. However, if you look more closely, you will see some structure to this mess. There are some mutually connected clusters—not surprisingly, some of my friends are also friends with one another. For example, David, Greg, Doug, and Bob are all connected to one another, as are Steph, Ginger, and Doyne, with myself as a bridge between the two groups. Even knowing little about my history you might guess that these two "communities" of friends are associated with different interests of mine or with different periods in my life. (Both are true.)

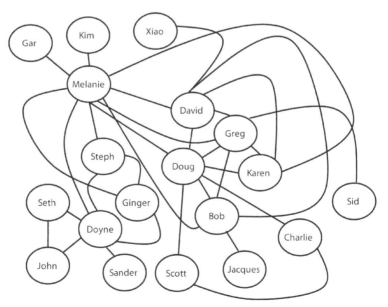

FIGURE 15.2. Part of my own social network.

You also might notice that there are some people with lots of friends (e.g., myself, Doyne, David, Doug, Greg) and some people with only one friend (e.g., Kim, Jacques, Xiao). Here this is due only to the incompleteness of this network, but in most large social networks there will always be some people with many friends and some people with few.

In their efforts to develop a common language, network scientists have coined terminology for these different kinds of network structures. The existence of largely separate tight-knit communities in networks is termed *clustering*. The number of links coming into (or out of) a node is called the *degree* of that node. For example, my degree is 10, and is the highest of all the nodes; Kim's degree is 1 and is tied with five others for the lowest. Using this terminology, we can say that the network has a small number of *high-degree* nodes, and a larger number of *low-degree* ones.

This can be seen clearly in figure 15.3, where I plot the *degree distribution* of this network. For each degree from 1 to 10 the plot gives the number of nodes that have that degree. For example, there are six nodes with degree 1 (first bar) and one node with degree 10 (last bar).

This plot makes it explicit that there are many nodes with low degree and few nodes with high degree. In social networks, this corresponds to the fact that there are a lot of people with a relatively small number of friends, and a much smaller group of very popular people. Similarly, there are a small number of very popular Web sites (i.e., ones to which many other sites link),

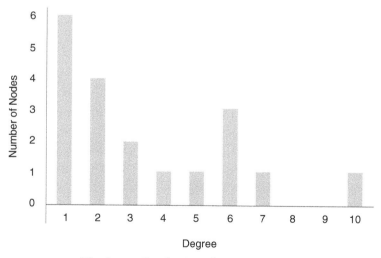

FIGURE 15.3. The degree distribution of the network in figure 15.2. For each degree, a bar is drawn representing the number of nodes with that degree.

such as Google, with more than 75 million incoming links, and a much larger number of Web sites that hardly anyone has heard of—such as my own Web site, with 123 incoming links (many of which are probably from search engines).

High-degree nodes are called *hubs*; they are major conduits for the flow of activity or information in networks. Figure 15.1 illustrates the hub system that most airlines adopted in the 1980s, after deregulation: each airline designated certain cities as hubs, meaning that most of the airline's flights go through those cities. If you've ever flown from the western United States to the East Coast on Continental Airlines, you probably had to change planes in Houston.

A major discovery to date of network science is that high-clustering, skewed degree distributions, and hub structure seem to be characteristic of the vast majority of all the natural, social, and technological networks that network scientists have studied. The presence of these structures is clearly no accident. If I put together a network by randomly sticking in links between nodes, all nodes would have similar degree, so the degree distribution wouldn't be skewed the way it is in figure 15.3. Likewise, there would be no hubs and little clustering.

Why do networks in the real world have these characteristics? This is a major question of network science, and has been addressed largely by developing models of networks. Two classes of models that have been studied in depth are known as *small-world* networks and *scale-free* networks.

Small-World Networks

Although Milgram's experiments may not have established that we actually live in a small world, the world of my social network (figure 15.2) is indeed small. That is, it doesn't take many hops to get from any node to any other node. While they have never met one another (as far as I know), Gar can reach Charlie in only three hops, and John can reach Xiao in only four hops. In fact, in my network people are linked by at most four degrees of separation.

Applied mathematician and sociologist Duncan Watts and applied mathematician Steven Strogatz were the first people to mathematically define the concept of *small-world network* and to investigate what kinds of network structures have this property. (Their work on abstract networks resulted from an unlikely source: research on how crickets synchronize their chirps.) Watts and Strogatz started by looking at the simplest possible "regular" network: a ring of nodes, such as the network of figure 15.4, which has 60 nodes. Each node is

FIGURE 15.4. An example of a *regular* network. This network is a ring of nodes in which each node has a link to its two nearest neighbors.

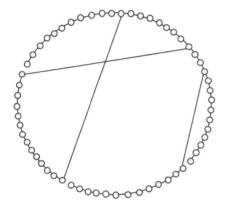

FIGURE 15.5. A random rewiring of three links turns the regular network of figure 15.4 into a small-world network.

linked to its two nearest neighbors in the ring, reminiscent of an elementary cellular automaton. To determine the degree of "small-worldness" in a network, Watts and Strogatz computed the *average path length* in the network. The *path length* between two nodes is simply the number of links on the shortest path between those two nodes. The average path length is simply the average over path lengths between all pairs of nodes in the network. The average path length of the regular network of figure 15.4 turns out to be 15. Thus, as in a children's game of "telephone," on average it would take a long time for a node to communicate with another node on the other side of the ring.

Watts and Strogatz then asked, If we take a regular network like this and rewire it a little bit—that is, change a few of the nearest-neighbor links to long-distance links—how will this affect the average path length? They discovered that the effect is quite dramatic.

As an example, figure 15.5 shows the regular network of figure 15.4 with 5% (here, three) of the links rewired—that is, one end of each of the three links is moved to a randomly chosen node.

This rewired network has the same number of links as the original regular network, but the average path-length has shrunk to about 9. Watts and Strogatz found that as the number of nodes increases, the effect becomes increasingly pronounced. For example, in a regular network with 1,000 nodes, the average path length is 250; in the same network with 5% of the links randomly rewired, the average path length will typically fall to around 20. As Watts puts it, "only a few random links can generate a very large effect ... on average, the first five random rewirings reduce the average path length of the network by one-half, regardless of the size of the network."

These examples illustrate the *small-world property*: a network has this property if it has relatively few long-distance connections but has a small average path-length relative to the total number of nodes. Small-world networks also typically exhibit a high degree of clustering: for any nodes A, B, and C, if node A is connected to nodes B and C, then B and C are also likely to be connected to one another. This is not apparent in figure 15.5, since in this network most nodes are linked to only their two nearest neighbors. However, if the network were more realistic, that is, if each node were initially connected to multiple neighbors, the clustering would be high. An example is my own social network—I'm more likely to be friends with the friends of my friends than with other, random, people.

As part of their work, Watts and Strogatz looked at three examples of real-world networks from completely different domains and showed that they all have the small-world property. The first was a large network of movie actors. In this network, nodes represent individual actors; two nodes are linked if the corresponding actors have appeared in at least one movie together, such as Tom Cruise and Max von Sydow (*Minority Report*), or Cameron Diaz and Julia Roberts (*My Best Friend's Wedding*). This particular social network has received attention via the popular "Kevin Bacon game," in which a player tries to find the shortest path in the network from any given movie actor to the ultra-prolific actor Kevin Bacon. Evidently, if you are in movies and you don't have a short path to Kevin Bacon, you aren't doing so well in your career.

The second example is the electrical power grid of the western United States. Here, nodes represent the major entities in the power grid: electrical generators, transformers, and power substations. Links represent high-voltage transmission lines between these entities. The third example is the brain of the worm *C. elegans*, with nodes being neurons and links being connections between neurons. (Luckily for Watts and Strogatz, neuroscientists had already mapped out every neuron and neural connection in this humble worm's small brain.)

You'd never have suspected that the "high-power" worlds of movie stars and electrical grids (not to mention the low-power world of a worm's brain) would have anything interesting in common, but Watts and Strogatz showed that they are indeed all small-world networks, with low average path lengths and high clustering.

Watts and Strogatz's now famous 1990 paper, "Collective Dynamics of 'Small-World' Networks," helped ignite the spark that set the new science of networks aflame with activity. Scientists are finding more and more examples of small-world networks in the real world, some of which I'll describe in the next chapter. Natural, social, and technological evolution seem to have produced organisms, communities, and artifacts with such structure. Why? It has been hypothesized that at least two conflicting evolutionary selective pressures are responsible: the need for information to travel quickly within the system, and the high cost of creating and maintaining reliable long-distance connections. Small-world networks solve both these problems by having short average path lengths between nodes in spite of having only a relatively small number of long-distance connections.

Further research showed that networks formed by the method proposed by Watts and Strogatz—starting with a regular network and randomly rewiring a small fraction of connections—do not actually have the kinds of degree distributions seen in many real-world networks. Soon, much attention was being paid to a different network model, one which produces *scale-free* networks—a particular kind of small-world network that looks more like networks in the real world.

Scale-Free Networks

I'm sure you have searched the World Wide Web, and you most likely use Google as your search engine. (If you're reading this long after I wrote it in 2008, perhaps a new search engine has become predominant.) Back in the days of the Web before Google, search engines worked by simply looking up the words in your search query in an index that connected each possible English word to a list of Web pages that contained that word. For example, if your search query was the two words "apple records," the search engine would give you a list of all the Web pages that included those words, in order of how many times those words appeared close together on the given page. You might be as likely to get a Web page about the historical price of apples in Washington State, or the fastest times recorded in the Great Apple Race in Tasmania, as you would a page about the famous record label formed in

1968 by the Beatles. It was very frustrating in those days to sort through a plethora of irrelevant pages to find the one with the information you were actually looking for.

In the 1990s Google changed all that with a revolutionary idea for presenting the results of a Web search, called "PageRank." The idea was that the importance (and probable relevance) of a Web page is a function of how many other pages link to it (the number of "in-links"). For example, at the time I write this, the Web page with the *American and Western Fruit Grower* report about Washington State apple prices in 2008 has 39 in-links. The Web page with information about the Great Apple Race of Tasmania has 47 in-links. The Web page www.beatles.com has about 27,000 in-links. This page is among those presented at the top of the list for the "apple records" search. The other two are way down the list of approximately one million pages ("hits") listed for this query. The original PageRank algorithm was a very simple idea, but it produced a tremendously improved search engine whereby the most relevant hits for a given query were usually at the top of the list.

If we look at the Web as a network, with nodes being Web pages and links being hyperlinks from one Web page to another, we can see that PageRank works only because this network has a particular structure: as in typical social networks, there are many pages with low degree (relatively few in-links), and a much smaller number of high-degree pages (i.e., relatively many in-links). Moreover, there is a wide variety in the number of in-links among pages, which allows ranking to mean something—to actually differentiate between pages. In other words, the Web has the skewed degree distribution and hub structure described above. It also turns out to have high clustering—different "communities" of Web pages have many mutual links to one another.

In network science terminology, the Web is a *scale-free* network. This has become one of the most talked-about notions in recent complex systems research, so let's dig into it a bit, by looking more deeply at the Web's degree distribution and what it means to be scale-free.

DEGREE DISTRIBUTION OF THE WEB

How can we figure out what the Web's degree distribution is? There are two kinds of Web links: in-links and out-links. That is, suppose my page has a link to your page but not vice versa: I have an out-link and you have an in-link. One needs to be specific about which kinds of links are counted. The original PageRank algorithm looked only at in-links and ignored out-links—in this discussion I'll do the same. We'll call the number of in-links to a page the *in-degree* of that page.

Now, what is the Web's in-degree distribution? It's hard, if not impossible, to count all the pages and in-links on the Web—there's no complete list stored anywhere and new links are constantly being added and old ones deleted. However, several Web scientists have tried to find approximate values using sampling and clever Web-crawling techniques. Estimates of the total number of Web pages vary considerably; as of 2008, the estimates I have seen range from 100 million to over 10 billion, and clearly the Web is still growing quickly.

Several different research groups have found that the Web's in-degree distribution can be described by a very simple rule: the number of pages with a given in-degree is approximately proportional to 1 divided by the square of that in-degree. Suppose we denote the in-degree by the letter k. Then

Number of Web pages with in-degree k is proportional to $\frac{1}{k^2}$.

(There has been some disagreement in the literature as to the actual exponent on k but it is close to 2—see the notes for details.) It turns out that this rule actually fits the data only for values of in-degree (k) in the thousands or greater.

To demonstrate why the Web is called "scale free," I'll plot the in-degree distribution as defined by this simple rule above, at three different scales. These plots are shown in figure 15.6. The first graph (top) plots the distribution for 9,000 in-degrees, starting at 1,000, which is close to where the rule becomes fairly accurate. Similar to figure 15.3, the in-degree values between 1,000 and 10,000 are shown on the horizontal axis, and their frequency (number of pages with the given in-degree) by the height of the boxes along the vertical axis. Here there are so many boxes that they form a solid black region.

The plots don't give the actual values for frequency since I want to focus on the shape of the graph (not to mention that as far as I know, no one has very good estimates for the actual frequencies). However, you can see that there is a relatively large number of pages with $k = 1,000$ in-links, and this frequency quickly drops as in-degree increases. Somewhere between $k = 5,000$ and $k = 10,000$, the number of pages with k in-links is so small compared with the number of pages with 1,000 in-links that the corresponding boxes have essentially zero height.

What happens if we rescale—that is, zoom in on—this "near-zero-height" region? The second (middle) graph plots the in-degree distribution from $k = 10,000$ to $k = 100,000$. Here I've rescaled the plot so that the $k = 10,000$ box on this graph is at the same height as the $k = 1,000$ box on the previous graph.

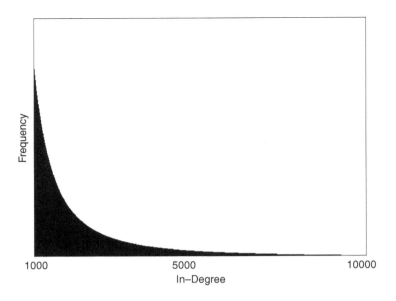

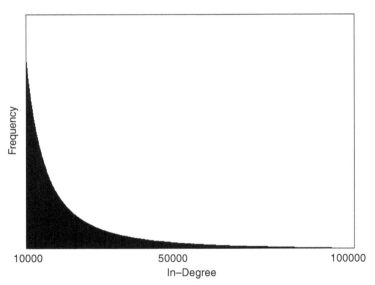

FIGURE 15.6. Approximate shape of the Web's in-degree distribution at three different scales.

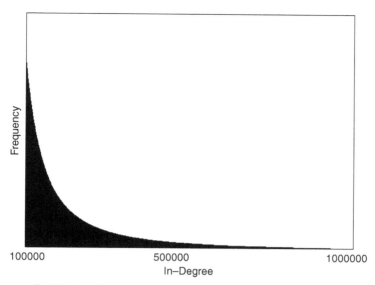

FIGURE 15.6. (*Continued*)

At this scale, there is now a relatively large number of pages with $k = 10,000$ in-links, and now somewhere between $k = 50,000$ and $k = 100,000$ we get "near-zero-height" boxes.

But something else is striking—except for the numbers on the horizontal axis, the second graph is *identical* to the first. This is true even though we are now plotting the distribution over 90,000 values instead of 9,000—what scientists call an *order of magnitude* more.

The third graph (bottom) shows the same phenomenon on an even larger scale. When we plot the distribution over k from 100,000 to 1 million, the shape remains identical.

A distribution like this is called *self-similar*, because it has the same shape at any scale you plot it. In more technical terms, it is "invariant under rescaling." This is what is meant by the term *scale-free*. The term *self-similarity* might be ringing a bell. We saw it back in chapter 7, in the discussion of fractals. There is indeed a connection to fractals here; more on this in chapter 17.

SCALE-FREE DISTRIBUTIONS VERSUS BELL CURVES

Scale-free networks are said to have no "characteristic scale." This is best explained by comparing a scale-free distribution with another well-studied distribution, the so-called bell-curve.

Suppose I plotted the distribution of adult human heights in the world. The smallest (adult) person in the world is a little over 2 feet tall (around

THE SCIENCE OF NETWORKS | 243

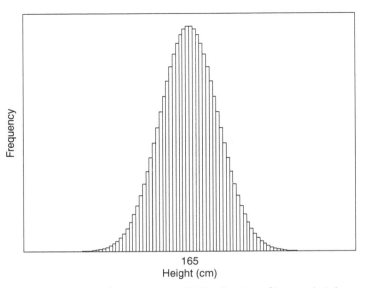

FIGURE 15.7. A bell-curve (normal) distribution of human heights.

70 cm). The tallest person is somewhere close to 9 feet tall (around 270 cm). The average adult height is about 5 feet 5 inches (165 cm), and the vast majority of all adults have height somewhere between 5 and 7 feet.

The distribution of human heights looks something like figure 15.7. The plot's bell-like shape is why it is often called a bell curve. Lots of things have approximate bell-curve distributions—height, weight, test scores on some exams, winning scores in basketball games, abundance of different species, and so on. In fact, because so many quantities in the natural world have this distribution, the bell curve is also called the *normal distribution*.

Normal distributions are characterized by having a particular *scale*—e.g., 70–270 cm for height, 0–100 for test scores. In a bell-curve distribution, the value with highest frequency is the average—e.g., 165 cm for height. Most other values don't vary much from the average—the distribution is fairly homogeneous. If in-degrees in the Web were normally distributed, then PageRank wouldn't work as well, since nearly all Web pages would have close to the average number of in-links. The Web page www.beatles.com would have more or less the same number of in-links as all other pages containing the phrase "apple records"; thus "number of in-links" could not be used as a way to rank such pages in order of probable relevance.

Fortunately for us (and even more so for Google stockholders), the Web degree distribution has a scale-free rather than bell-curve structure. Scale-free networks have three notable properties: (1) a relatively small number of very high-degree nodes (hubs); (2) nodes with degrees over a very large range of different values (i.e., *heterogeneity of degree values*); (3) self-similarity.

In more scientific terms, a scale-free network always has a *power law* degree distribution. Recall that the approximate in-degree distribution for the Web is

$$\text{Number of Web pages with in-degree } k \text{ is proportional to } \frac{1}{k^2}.$$

Perhaps you will remember from high school math that $\frac{1}{k^2}$ also can be written as k^{-2}. This is a "power law with exponent -2." Similarly, $\frac{1}{k}$ (or, equivalently, k^{-1}) is a power law with exponent -1." In general, a power-law distribution has the form of x^d, where x is a quantity such as in-degree. The key number describing the distribution is the exponent d; different exponents cause very different-looking distributions.

I will have more to say about power laws in chapter 17. The important thing to remember for now is *scale-free network = power-law degree distribution*.

Network Resilience

A very important property of scale-free networks is their resilience to the deletion of nodes. This means that if a set of random nodes (along with their links) are deleted from a large scale-free network, the network's basic properties do not change: it still will have a heterogeneous degree distribution, short average path length, and strong clustering. This is true even if the number of deleted nodes is fairly large. The reason for this is simple: if nodes are deleted at random, they are overwhelmingly likely to be low-degree nodes, since these constitute nearly all nodes in the network. Deleting such nodes will have little effect on the overall degree distribution and path lengths. We can see many examples of this in the Internet and the Web. Many individual computers on the Internet fail or are removed all the time, but this doesn't have any obvious effect on the operation of the Internet or on its average path length. Similarly, although individual Web pages and their links are deleted all the time, Web surfing is largely unaffected.

However, this resilience comes at a price: if one or more of the *hubs* is deleted, the network will be likely to lose all its scale-free properties and cease to function properly. For example, a blizzard in Chicago (a big airline hub) will probably cause flight delays or cancellations all over the country. A failure in Google will wreak havoc throughout the Web.

In short, scale-free networks are resilient when it comes to random deletion of nodes but highly vulnerable if hubs go down or can be targeted for attack.

In the next chapter I discuss several examples of real-world networks that have been found to have small-world or scale-free properties and describe some theories of how they got that way.

CHAPTER 16 | Applying Network Science to Real-World Networks

NETWORK THINKING IS EVIDENTLY ON a lot of people's minds. According to my search on the Google Scholar Web site, at the time of this writing over 14,000 academic papers on small-world or scale-free networks have been published in the last five years (since 2003), nearly 3,000 in the last year alone. I did a scan of the first 100 or so titles in the list and found that 11 different disciplines are represented, ranging from physics and computer science to geology and neuroscience. I'm sure that the range of disciplines I found would grow substantially if I did a more comprehensive scan.

In this chapter I survey some diverse examples of real-world networks and discuss how advances in network science are influencing the way scientists think about networks in many disciplines.

Examples of Real-World Networks

THE BRAIN

Several groups have found evidence that the brain has small-world properties. The brain can be viewed as a network at several different levels of description; for example, with neurons as nodes and synapses as links, or with entire *functional areas* as nodes and larger-scale connections between them (i.e., groups of neural connections) as links.

As I mentioned in the previous chapter, the neurons and neural connections of the brain of the worm *C. elegans* have been completely mapped by neuroscientists and have been shown to form a small-world network. More recently, neuroscientists have mapped the connectivity structure in certain higher-level functional brain areas in animals such as cats, macaque monkeys, and even humans and have found the small-world property in those structures as well.

Why would evolution favor brain networks with the small-world property? Resilience might be one major reason: we know that individual neurons die all the time, but, happily, the brain continues to function as normal. The *hubs* of the brain are a different story: if a stroke or some other mishap or disease affects, say, the hippocampus (which is a hub for networks encoding short-term memory), the failure can be quite devastating.

In addition, researchers have hypothesized that a scale-free degree distribution allows an optimal compromise between two modes of brain behavior: processing in local, segregated areas such as parts of the visual cortex or language areas versus global processing of information, for example when information from the visual cortex is communicated to areas doing language processing, and vice versa.

If every neuron were connected to every other neuron, or all different functional areas were fully connected to one another, then the brain would use up a mammoth amount of energy in sending signals over the huge number of connections. Evolution presumably selected more energy-efficient structures. In addition, the brain would probably have to be much larger to fit all those connections. At the other extreme, if there were no long-distance links in the brain, it would take too long for the different areas to communicate with one another. The human brain size—and corresponding skull size—seems to be exquisitely balanced between being large enough for efficient complex cognition and small enough for mothers to give birth. It has been proposed that the small-world property is exactly what allows this balance.

It has also been widely speculated that *synchronization*, in which groups of neurons repeatedly fire simultaneously, is a major mechanism by which information in the brain is communicated efficiently, and it turns out that a small-world connectivity structure greatly facilitates such synchronization.

GENETIC REGULATORY NETWORKS

As I mentioned in chapter 7, humans have about 25,000 genes, roughly the same number as the mustard plant *arabidopsis*. What seems to generate the complexity of humans as compared to, say, plants is not how many genes we have but how those genes are organized into networks.

There are many genes whose function is to *regulate* other genes—that is, control whether or not the regulated genes are expressed. A well-known simple example of gene regulation is the control of lactose metabolism in *E. coli* bacteria. These bacteria usually live off of glucose, but they can also metabolize lactose. The ability to metabolize lactose requires the cell to contain three particular protein enzymes, each encoded by a separate gene. Let's call these genes A, B, and C. There is a fourth gene that encodes a protein, called a *lactose repressor*, which binds to genes A, B, and C, in effect, turning off these genes. If there is no lactose in the bacterium's local environment, lactose repressors are continually formed, and no lactose metabolism takes place. However, if the bacterium suddenly finds itself in a glucose-free but lactose-rich environment, then lactose molecules bind to the lactose repressor and detach it from genes A, B, and C, which then proceed to produce the enzymes that allow lactose metabolism.

Regulatory interactions like this, some much more intricate, are the heart and soul of complexity in genetics. Network thinking played a role in understanding these interactions as early as the 1960s, with the work of Stuart Kauffman (more on this in chapter 18). More recently, network scientists teaming up with geneticists have demonstrated evidence that at least some networks of these interactions are approximately scale-free. Here, the nodes are individual genes, and each node links to all other genes it regulates (if any).

Resilience is mandatory for genetic regulatory networks. The processes of gene transcription and gene regulation are far from perfect; they are inherently error-ridden and often affected by pathogens such as viruses. Having a scale-free structure helps the system to be mostly impervious to such errors.

Metabolic Networks

As I described in chapter 12, cells in most organisms have hundreds of different metabolic pathways, many interconnecting, forming networks of metabolic reactions. Albert-László Barabási and colleagues looked in detail at the structure of metabolic networks in forty-three different organisms and found that they all were "well fitted" by a power-law distribution—i.e., are scale free. Here the nodes in the network are chemical *substrates*—the fodder and product of chemical reactions. One substrate is considered to be linked to another if the first participates in a reaction that produces the second. For example, in the second step of the pathway called glycolysis, the substrate *glucose-6-phosphate* produces the substrate *fructose-6-phosphate*, so there would be a link in the network from the first substrate to the second.

Since metabolic networks are scale-free, they have a small number of hubs that are the products of a large number of reactions involving many different substrates. These hubs turn out to be largely the same chemicals in all the diverse organisms studied—the chemicals that are known to be most essential for life. It has been hypothesized that metabolic networks evolved to be scale-free so as to ensure robustness of metabolism and to optimize "communication" among different substrates.

Epidemiology

In the early 1980s, in the early stages of the worldwide AIDS epidemic, epidemiologists at the Centers for Disease Control in Atlanta identified a Canadian flight attendant, Gaetan Dugas, as part of a cluster of men with AIDS who were responsible for infecting large numbers of other gay men in many different cities around the world. Dugas was later vilified in the media as "patient zero," the first North American with AIDS, who was responsible for introducing and widely spreading the AIDS virus in the United States and elsewhere. Although later studies debunked the theory that Dugas was the source of the North American epidemic, there is no question that Dugas, who claimed to have had hundreds of different sexual partners each year, infected many people. In network terms, Dugas was a hub in the network of sexual contacts.

Epidemiologists studying sexually transmitted diseases often look at networks of sexual contacts, in which nodes are people and links represent sexual partnerships between two people. Recently, a group consisting of sociologists and physicists analyzed data from a Swedish survey of sexual behavior and found that the resulting network has a scale-free structure; similar results have been found in studies of other sexual networks.

In this case, the vulnerability of such networks to the removal of hubs can work in our favor. It has been suggested that safe-sex campaigns, vaccinations, and other kinds of interventions should mainly be targeted at such hubs.

How can these hubs be identified without having to map out huge networks of people, for which data on sexual partners may not be available?

A clever yet simple method was proposed by another group of network scientists: choose a set of random people from the at-risk population and ask each to name a partner. Then vaccinate that partner. People with many partners will be more likely to be named, and thus vaccinated, under this scheme.

This strategy, of course, can be exported to other situations in which "hub-targeting" is desired, such as fighting computer viruses transmitted by e-mail: in this case, one should target anti-virus methods to the computers of people with large address books, rather than depending on all computer users to perform virus detection.

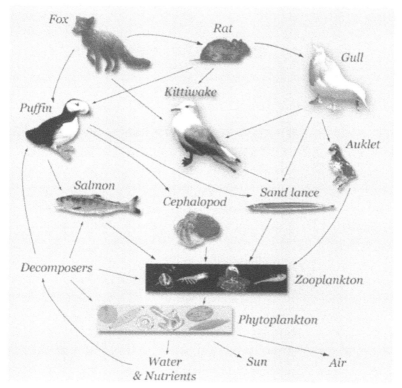

FIGURE 16.1. Example of a food web. (Illustration from USGS Alaska Science Center, [http://www.absc.usgs.gov/research/seabird_foragefish/marinehabitat/home.html].)

Ecologies and Food Webs

In the science of ecology, the common notion of *food chain* has been extended to *food web*, a network in which a node represents a species or group of species; if species B is part of the diet of species A, then there is a link from node A to node B. Figure 16.1 shows a simple example of a food web.

Mapping the food webs of various ecosystems has been an important part of ecological science for some time. Recently, researchers have been applying network science to the analysis of these webs in order to understand biodiversity and the implications of different types of disruptions to that biodiversity in ecosystems.

Several ecologists have claimed that (at least some) food webs possess the small-world property, and that some of these have scale-free degree distributions, which evolved presumably to give food webs resilience to the random deletion of species. Others ecologists have disagreed that food webs have scale-free structure, and the ecology research community has recently seen a lot of debate on this issue, mainly due to the difficulty of interpreting real-world data.

Significance of Network Thinking

The examples above are only a small sampling of the ways in which network thinking is affecting various areas of science and technology. Scale-free degree distributions, clustering, and the existence of hubs are the common themes; these features give rise to networks with small-world communication capabilities and resilience to deletion of random nodes. Each of these properties is significant for understanding complex systems, both in science and in technology.

In science, network thinking is providing a novel language for expressing commonalities across complex systems in nature, thus allowing insights from one area to influence other, disparate areas. In a self-referential way, network science itself plays the role of a *hub*—the common connection among otherwise far-flung scientific disciplines.

In technology, network thinking is providing novel ways to think about difficult problems such as how to do efficient search on the Web, how to control epidemics, how to manage large organizations, how to preserve ecosystems, how to target diseases that affect complex networks in the body, how to target modern criminal and terrorist organizations, and, more generally, what kind of resilience and vulnerabilities are intrinsic to natural, social, and technological networks, and how to exploit and protect such systems.

Where Do Scale-Free Networks Come From?

No one purposely designed the Web to be scale-free. The Web's degree distribution, like that of the other networks I've mentioned above, is an emergent outcome of the way in which the network was formed, and how it grows.

In 1999 physicists Albert-László Barabási and Réka Albert proposed that a particular growing process for networks, which they called *preferential attachment*, is the explanation for the existence of most (if not all) scale-free networks in the real world. The idea is that networks grow in such a way that nodes with higher degree receive more new links than nodes with lower degree. Intuitively this makes sense. People with many friends tend to meet more new people and thus make more new friends than people with few friends. Web pages with many incoming links are easier to find than those with few incoming links, so more new Web pages link to the high-degree ones. In other words, the rich get richer, or perhaps the linked get more linked. Barabási and Albert showed that growth by preferential attachment leads to scale-free degree distributions. (Unbeknownst to them at the time, this process and its power-law outcome had been discovered independently at least three times before.)

The growth of so-called scientific citation networks is one example of the effects of preferential attachment. Here the nodes are papers in the scientific literature; each paper receives a link from all other papers that cite it. Thus the more citations others have given to your paper, the higher its degree in the network. One might assume that a large number of citations is an indicator of good work; for example, in academia, this measure is routinely used in making decisions about tenure, pay increases, and other rewards. However, it seems that preferential attachment often plays a large role. Suppose you and Joe Scientist have independently written excellent articles about the same topic. If I happen to cite your article but not Joe's in my latest opus, then others who read only my paper will be more likely to cite yours (usually without reading it). Other people will read *their* papers, and also be more likely to cite you than to cite Joe. The situation for Joe gets worse and worse as your situation gets better and better, even though your paper and Joe's were both of the same quality. Preferential attachment is one mechanism for getting to what the writer Malcolm Gladwell called *tipping points*—points at which some process, such as citation, spread of fads, and so on, starts increasing dramatically in a positive-feedback cycle. Alternatively, tipping points can refer to failures in a system that induce an accelerating systemwide spread of additional failures, which I discuss below.

Power Laws and Their Skeptics

So far I have implied that scale-free networks are ubiquitous in nature due to the adaptive properties of robustness and fast communication associated with power-law degree distributions, and that the mechanism by which they form is growth by preferential attachment. These notions have given scientists new ways of thinking about many different scientific problems.

However compelling all this may seem, scientists are supposed to be skeptical by nature, especially of new, relatively untested ideas, and even more particularly of ideas that claim generality over many disciplines. Such skepticism is not only healthy, it is also essential for the progress of science. Thus, fortunately, not everyone has jumped on the network-science bandwagon, and even many who have are skeptical concerning some of the most optimistic statements about the significance of network science for complex systems research. This skepticism is founded on the following arguments.

1. **Too many phenomena are being described as power-law or scale-free.** It's typically rather difficult to obtain good data about real-world network degree distributions. For example, the data used by Barabási and colleagues for analyzing metabolic networks came from a Web-based database to which biologists from all over the world contributed information. Such biological databases, while invaluable to research, are invariably incomplete and error-ridden. Barabási and colleagues had to rely on statistics and curve-fitting to determine the degree distributions in various metabolic networks—an imperfect method, yet the one that is most often used in analyzing real-world data. A number of networks previously identified to be "scale-free" using such techniques have in fact later been shown to have non-scale-free distributions.

   As noted by philosopher and historian of biology Evelyn Fox Keller, "Current assessments of the commonality of power laws are probably overestimates." Physicist and network scientist Cosma Shalizi had a less polite phrasing of the same sentiments: "Our tendency to hallucinate power laws is a disgrace." As I write this, there are still considerable controversies over which real-world networks are indeed scale-free.

2. **Even for networks that are actually scale-free, there are many possible causes for power law degree distributions in networks; preferential attachment is not necessarily the one that actually occurs in nature.** As Cosma Shalizi succinctly said: "there turn out to be nine and sixty ways of constructing power laws, and *every single one of them is right*." When I was at the Santa Fe Institute, it seemed that there was a lecture every other day on a new hypothesized mechanism that resulted in power law distributions. Some are similar to preferential attachment, some work quite differently. It's not obvious how to decide which ones are the mechanisms that are actually causing the power laws observed in the real world.

3. **The claimed significance of network science relies on models that are overly simplified and based on unrealistic assumptions.** The small-world and scale-free network models are just that—models—which means that they make simplifying assumptions that might not be true of real-world networks. The hope in creating such simplified models is that they will capture at least some aspects of the phenomenon they are designed to represent. As we have seen,

these two network models, in particular the scale-free model, indeed seem to capture something about degree-distributions, clustering, and resilience in a large number of real-world systems (though point 1 above suggests that the number might not be as large as some think).

However, simplified models of networks, in and of themselves, cannot explain everything about their real-world counterparts. In both the small-world and scale-free models, all nodes are assumed to be identical except for their degree; and all links are the same type and have the same strength. This is not the case in real-world networks. For example, in the real version of my social network (whose simplified model was shown in figure 14.2), some friendship links are stronger than others. Kim and Gar are both friends of mine but I know Kim much better, so I might be more likely to tell her about important personal events in my life. Furthermore, Kim is a woman and Gar is a man, which might increase my likelihood of confiding in her but not in Gar. Similarly, my friend Greg knows and cares a lot more about math than Kim, so if I wanted to share some neat mathematical fact I learned, I'd be much more likely to tell Greg about it than Kim. Such differences in link and node types as well as link strength can have very significant effects on how information spreads in a network, effects that are not captured by the simplified network models.

Information Spreading and Cascading Failure in Networks

In fact, understanding the ways in which information spreads in networks is one of the most important open problems in network science. The results I have described in this and the previous chapter are all about the *structure* of networks—e.g., their static degree distributions—rather than *dynamics* of spreading information in a network.

What do I mean by "spreading information in a network"? Here I'm using the term *information* to capture any kind of communication among nodes. Some examples of information spreading are the spread of rumors, gossip, fads, opinions, epidemics (in which the communication between people is via germs), electrical currents, Internet packets, neurotransmitters, calories (in the case of food webs), vote counts, and a more general network-spreading phenomenon called "cascading failure."

The phenomenon of cascading failure emphasizes the need to understand information spreading and how it is affected by network structure. Cascading failure in a network happens as follows: Suppose each node in the network is responsible for performing some task (e.g., transmitting electrical power).

If a node fails, its task gets passed on to other nodes. This can result in the other nodes getting overloaded and failing, passing on their task to still other nodes, and so forth. The result is an accelerating domino effect of failures that can bring down the entire network.

Examples of cascading failure are all too common in our networked world. Here are two fairly recent examples that made the national news:

- August 2003: A massive power outage hit the Midwestern and Northeastern United States, caused by cascading failure due to a shutdown at one generating plant in Ohio. The reported cause of the shutdown was that electrical lines, overloaded by high demand on a very hot day, sagged too far down and came into contact with overgrown trees, triggering an automatic shutdown of the lines, whose load had to be shifted to other parts of the electrical network, which themselves became overloaded and shut down. This pattern of overloading and subsequent shutdown spread rapidly, eventually resulting in about 50 million customers in the Eastern United States and Canada losing electricity, some for more than three days.
- August 2007: The computer system of the U.S. Customs and Border Protection Agency went down for nearly ten hours, resulting in more than 17,000 passengers being stuck in planes sitting on the tarmac at Los Angeles International Airport. The cause turned out to be a malfunction in a single network card on a desktop computer. Its failure quickly caused a cascading failure of other network cards, and within about an hour of the original failure, the entire system shut down. The Customs agency could not process arriving international passengers, some of whom had to wait on airplanes for more than five hours.

A third example shows that cascading failures can also happen when network nodes are not electronic devices but rather corporations.

- August–September 1998: Long-Term Capital Management (LTCM), a private financial hedge fund with credit from several large financial firms, lost nearly all of its equity value due to risky investments. The U.S. Federal Reserve feared that this loss would trigger a cascading failure in worldwide financial markets because, in order to cover its debts, LTCM would have to sell off much of its investments, causing prices of stocks and other securities to drop, which would force other companies to sell off *their* investments, causing a further drop in prices, et cetera. At the end of September 1998, the Federal Reserve acted to prevent such a cascading failure by brokering a bailout of LTCM by its major creditors.

The network resilience I talked about earlier—the ability of networks to maintain short average path lengths in spite of the failure of random nodes—doesn't take into account the cascading failure scenario in which the failure of one node *causes* the failure of other nodes. Cascading failures provide another example of "tipping points," in which small events can trigger accelerating feedback, causing a minor problem to balloon into a major disruption. Although many people worry about malicious threats to our world's networked infrastructure from hackers or "cyber-terrorists," it may be that cascading failures pose a much greater risk. Such failures are becoming increasingly common and dangerous as our society becomes more dependent on computer networks, networked voting machines, missile defense systems, electronic banking, and the like. As Andreas Antonopoulos, a scientist who studies such systems, has pointed out, "The threat is complexity itself."

Indeed, a general understanding of cascading failures and strategies for their prevention are some of the most active current research areas in network science. Two current approaches are theories called Self-Organized Criticality (SOC) and Highly Optimized Tolerance (HOT). SOC and HOT are examples of the many theories that propose mechanisms different from preferential attachment for how scale-free networks arise. SOC and HOT each propose a general set of mechanisms for cascading failures in both evolved and engineered systems.

The simplified models of small-world networks and scale-free networks described in the previous chapter have been extraordinarily useful, as they have opened up the idea of network thinking to many different disciplines and established network science as a field in its own right. The next step is understanding the dynamics of information and other quantities in networks. To understand the dynamics of information in networks such as the immune system, ant colonies, and cellular metabolism (cf. chapter 12), network science will have to characterize networks in which the nodes and links continually change in both time and space. This will be a major challenge, to say the least. As Duncan Watts eloquently writes: "Next to the mysteries of dynamics on a network—whether it be epidemics of disease, cascading failures in power systems, or the outbreak of revolutions—the problems of networks that we have encountered up to now are just pebbles on the seashore."

CHAPTER 17 | The Mystery of Scaling

THE PREVIOUS TWO CHAPTERS SHOWED how network thinking is having profound effects on many areas of science, particularly biology. Quite recently, a kind of network thinking has led to a proposed solution for one of biology's most puzzling mysteries: the way in which properties of living organisms scale with size.

Scaling in Biology

Scaling describes how one property of a system will change if a related property changes. The scaling mystery in biology concerns the question of how the average energy used by an organism while resting—the *basal metabolic rate*—scales with the organism's body mass. Since metabolism, the conversion by cells of food, water, air, and light to usable energy, is the key process underlying all living systems, this relation is enormously important for understanding how life works.

It has long been known that the metabolism of smaller animals runs faster relative to their body size than that of larger animals. In 1883, German physiologist Max Rubner tried to determine the precise scaling relationship by using arguments from thermodynamics and geometry. Recall from chapter 3 that processes such as metabolism, that convert energy from one form to another, always give off heat. An organism's metabolic rate can be defined as the rate at which its cells convert nutrients to energy, which is used for all the cell's functions and for building new cells. The organism gives off heat at this same rate as a by-product. An organism's metabolic rate can thus be inferred by measuring this heat production.

If you hadn't already known that smaller animals have faster metabolisms relative to body size than large ones, a naïve guess might be that metabolic rate scales linearly with body mass—for example, that a hamster with eight times the body mass of a mouse would have eight times that mouse's metabolic rate, or even more extreme, that a hippopotamus with 125,000 times the body mass of a mouse would have a metabolic rate 125,000 times higher.

The problem is that the hamster, say, would generate eight times the amount of heat as the mouse. However, the total surface area of the hamster's body—from which the heat must radiate—would be only about four times the total surface of the mouse. This is because as an animal gets larger, its surface area grows more slowly than its mass (or equivalent, its volume).

This is illustrated in figure 17.1, in which a mouse, hamster, and hippo are represented by spheres. You might recall from elementary geometry that the formula for the volume of a sphere is four-thirds *pi* times the radius cubed, where $pi \approx 3.14159$. Similarly, the formula for the surface area of a sphere is four times *pi* times the radius squared. We can say that "volume scales as the cube of the radius" whereas "surface area scales as the square of the radius." Here "scales as" just means "is proportional to"—that is, ignore the

FIGURE 17.1. Scaling properties of animals (represented as spheres). (Drawing by David Moser.)

constants $4/3 \times pi$ and $4 \times pi$. As illustrated in figure 17.1, the hamster sphere has twice the radius of the mouse sphere, and it has four times the surface area and eight times the volume of the mouse sphere. The radius of the hippo sphere (not drawn to scale) is fifty times the mouse sphere's radius; the hippo sphere thus has 2,500 times the surface area and 125,000 times the volume of the mouse sphere. You can see that as the radius is increased, the surface area grows (or "scales") much more slowly than the volume. Since the surface area scales as the radius squared and the volume scales as the radius cubed, we can say that "the surface area scales as the volume raised to the two-thirds power." (See the notes for the derivation of this.)

Raising volume to the two-thirds power is shorthand for saying "square the volume, and then take its cube root."

Generating eight times the heat with only four times the surface area to radiate it would result in one very hot hamster. Similarly, the hippo would generate 125,000 times the heat of the mouse but that heat would radiate over a surface area of only 2,500 times the mouse's. Ouch! That hippo is seriously burning.

Nature has been very kind to animals by *not* using that naïve solution: our metabolisms thankfully **do not** scale linearly with our body mass. Max Rubner reasoned that nature had figured out that in order to safely radiate the heat we generate, our metabolic rate should scale with body mass in the same way as surface area. Namely, he proposed that metabolic rate scales with body mass to the two-thirds power. This was called the "surface hypothesis," and it was accepted for the next fifty years. The only problem was that the actual data did not obey this rule.

This was discovered in the 1930s by a Swiss animal scientist, Max Kleiber, who performed a set of careful measures of metabolism rate of different animals. His data showed that metabolic rate scales with body mass to the three-fourths power: that is, metabolic rate is proportional to *bodymass*$^{3/4}$. You'll no doubt recognize this as a power law with exponent $3/4$. This result was surprising and counterintuitive. Having an exponent of $3/4$ rather than $2/3$ means that animals, particularly large ones, are able to maintain a higher metabolic rate than one would expect, given their surface area. This means that animals are more efficient than simple geometry predicts.

Figure 17.2 illustrates such scaling for a number of different animals. The horizontal axis gives the body mass in kilograms and the vertical axis gives the average basal metabolic rate measured in watts. The labeled dots are the actual measurements for different animals, and the straight line is a plot of metabolic rate scaling with body mass to exactly the three-fourths power.

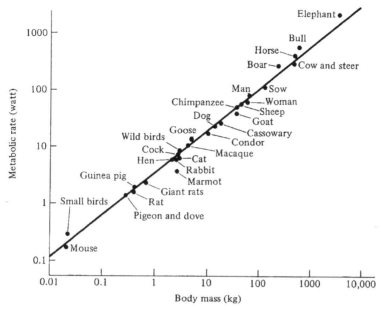

FIGURE 17.2. Metabolic rate of various animals as a function of their body mass. (From K. Schmidt-Nielsen, *Scaling: Why Is Animal Size So Important?* Copyright © 1984 by Cambridge University Press. Reprinted with permission of Cambridge University Press.)

The data do not exactly fit this line, but they are pretty close. Figure 17.2 is a special kind of plot—technically called a *double logarithmic* (or log-log) plot—in which the numbers on both axes increase by a power of ten with each tic on the axis. If you plot a power law on a double logarithmic plot, it will look like a straight line, and the slope of that line will be equal to the power law's exponent. (See the notes for an explanation of this.)

This power law relation is now called Kleiber's law. Such 3/4-power scaling has more recently been claimed to hold not only for mammals and birds, but also for the metabolic rates of many other living beings, such as fish, plants, and even single-celled organisms.

Kleiber's law is based only on observation of metabolic rates and body masses; Kleiber offered no explanation for *why* his law was true. In fact, Kleiber's law was baffling to biologists for over fifty years. The mass of living systems has a huge range: from bacteria, which weigh less than one one-trillionth of a gram, to whales, which can weigh over 100 million grams. Not only does the law defy simple geometric reasoning; it is also surprising that such a law seems to hold so well for organisms over such a vast variety of sizes, species types, and habitat types. What common aspect of nearly all organisms could give rise to this simple, elegant law?

Several other related scaling relationships had also long puzzled biologists. For example, the larger a mammal is, the longer its life span. The life span for a mouse is typically two years or so; for a pig it is more like ten years, and for an elephant it is over fifty years. There are some exceptions to this general rule, notably humans, but it holds for most mammalian species. It turns out that if you plot average life span versus body mass for many different species, the relationship is a power law with exponent $1/4$. If you plot average heart rate versus body mass, you get a power law with exponent $-1/4$ (the larger an animal, the slower its heart rate). In fact, biologists have identified a large collection of such power law relationships, all having fractional exponents with a 4 in the denominator. For that reason, all such relationships have been called *quarter-power scaling laws*. Many people suspected that these quarter-power scaling laws were a signature of something very important and common in all these organisms. But no one knew what that important and common property was.

An Interdisciplinary Collaboration

By the mid-1990s, James Brown, an ecologist and professor at the University of New Mexico, had been thinking about the quarter-power scaling problem for many years. He had long realized that solving this problem—understanding the reason for these ubiquitous scaling laws—would be a key step in developing any general theory of biology. A biology graduate student named Brian Enquist, also deeply interested in scaling issues, came to work with Brown, and they attempted to solve the problem together.

Brown and Enquist suspected that the answer lay somewhere in the structure of the systems in organisms that transport nutrients to cells. Blood constantly circulates in blood vessels, which form a branching network that carries nutrient chemicals to all cells in the body. Similarly, the branching structures in the lungs, called bronchi, carry oxygen from the lungs to the blood vessels that feed it into the blood (figure 17.3). Brown and Enquist believed that it is the universality of such branching structures in animals that give rise to the quarter-power laws. In order to understand how such structures might give rise to quarter-power laws, they needed to figure out how to describe these structures mathematically and to show that the math leads directly to the observed scaling laws.

Most biologists, Brown and Enquist included, do not have the math background necessary to construct such a complex geometric and topological analysis. So Brown and Enquist went in search of a "math buddy"—a

Left to right: Geoffrey West, Brian Enquist, and James Brown. (Photograph copyright © by Santa Fe Institute. Reprinted with permission.)

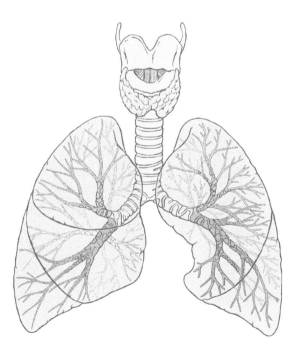

FIGURE 17.3. Illustration of bronchi, branching structures in the lungs. (Illustration by Patrick Lynch, licensed under Creative Commons [http://creativecommons.org/licenses/by/3.0/].)

mathematician or theoretical physicist who could help them out with this problem but not simplify it so much that the biology would get lost in the process.

Enter Geoffrey West, who fit the bill perfectly. West, a theoretical physicist then working at Los Alamos National Laboratory, had the ideal mathematical skills to address the scaling problem. Not only had he already worked on the topic of scaling, albeit in the domain of quantum physics, but he himself had been mulling over the biological scaling problem as well, without knowing very much about biology. Brown and Enquist encountered West at the Santa Fe Institute in the mid-1990s, and the three began to meet weekly at the institute to forge a collaboration. I remember seeing them there once a week, in a glass-walled conference room, talking intently while someone (usually Geoffrey) was scrawling reams of complex equations on the white board. (Brian Enquist later described the group's math results as "pyrotechnics.") I knew only vaguely what they were up to. But later, when I first heard Geoffrey West give a lecture on their theory, I was awed by its elegance and scope. It seemed to me that this work was at the apex of what the field of complex systems had accomplished.

Brown, Enquist, and West had developed a theory that not only explained Kleiber's law and other observed biological scaling relationships but also predicted a number of new scaling relationships in living systems. Many of these have since been supported by data. The theory, called *metabolic scaling theory* (or simply *metabolic theory*), combines biology and physics in equal parts, and has ignited both fields with equal parts excitement and controversy.

Power Laws and Fractals

Metabolic scaling theory answers two questions: (1) why metabolic scaling follows a power law at all; and (2) why it follows the particular power law with exponent 3/4. Before I describe how it answers these questions, I need to take a brief diversion to describe the relationship between power laws and fractals.

Remember the Koch curve and our discussion of fractals from chapter 7? If so, you might recall the notion of "fractal dimension." We saw that in the Koch curve, at each level the line segments were one-third the length of the previous level, and the structure at each level was made up of four copies of the structure at the previous level. In analogy with the traditional definition of dimension, we defined the fractal dimension of the Koch curve this way: $3^{dimension} = 4$, which yields $dimension = 1.26$. More generally, if each level is scaled by a factor of x from the previous level and is made up of N copies of the previous level, then $x^{dimension} = N$. Now, after having read chapter 15, you can recognize that this is a power law, with *dimension* as the exponent. This illustrates the intimate relationship between power laws and fractals. Power law distributions, as we saw in chapter 15, figure 15.6, *are* fractals—they are self-similar at all scales of magnification, and a power-law's exponent gives the dimension of the corresponding fractal (cf. chapter 7), where the dimension quantifies precisely how the distribution's self-similarity scales with level of magnification. Thus one could say, for example, that the degree distributions of the Web has a fractal structure, since it is self-similar. Similarly one could say that a fractal like the Koch curve gives rise to a power-law—the one that describes precisely how the curve's self-similarity scales with level of magnification.

The take-home message is that fractal structure is one way to generate a power-law distribution; and if you happen to see that some quantity (such as metabolic rate) follows a power-law distribution, then you can hypothesize that there is something about the underlying system that is self-similar or "fractal-like."

Metabolic Scaling Theory

Since metabolic rate is the rate at which the body's cells turn fuel into energy, Brown, Enquist, and West reasoned that metabolic rate must be largely determined by how efficiently that fuel is delivered to cells. It is the job of the organism's circulatory system to deliver this fuel.

Brown, Enquist, and West realized that the circulatory system is not just characterized in terms of its mass or length, but rather in terms of its network structure. As West pointed out, "You really have to think in terms of two separate scales—the length of the superficial you and the real you, which is made up of networks."

In developing their theory, Brown, Enquist, and West assumed that evolution has produced circulatory and other fuel-transport networks that are maximally "space filling" in the body—that is, that can transport fuel to cells in every part of the body. They also assumed that evolution has designed these networks to minimize the energy and time that is required to distribute this fuel to cells. Finally, they assume that the "terminal units" of the network, the sites where fuel is provided to body tissue, do not scale with body mass, but rather are approximately the same size in small and large organisms. This property has been observed, for example, with capillaries in the circulatory system, which are the same size in most animals. Big animals just have more of them. One reason for this is that cells themselves do not scale with body size: individual mouse and hippo cells are roughly the same size. The hippo just has more cells so needs more capillaries to fuel them.

The maximally space-filling geometric objects are indeed fractal branching structures—the self-similarity at all scales means that space is equally filled at all scales. What Brown, Enquist, and West were doing in the glass-walled conference room all those many weeks and months was developing an intricate mathematical model of the circulatory system as a space-filling fractal. They adopted the energy-and-time-minimization and constant-terminal-unit-size assumptions given above, and asked, What happens in the model when body mass is scaled up? Lo and behold, their calculations showed that in the model, the rate at which fuel is delivered to cells, which determines metabolic rate, scales with body mass to the 3/4 power.

The mathematical details of the model that lead to the 3/4 exponent are rather complicated. However, it is worth commenting on the group's interpretation of the 3/4 exponent. Recall my discussion above of Rubner's surface hypothesis—that metabolic rate must scale with body mass the same way in which volume scales with surface area, namely, to the 2/3 power. One way to look at the 3/4 exponent is that it would be the result of the surface hypothesis applied to four-dimensional creatures! We can see this via a simple dimensional analogy. A two-dimensional object such as a circle has a circumference and an area. In three dimensions, these correspond to surface area and volume, respectively. In four dimensions, surface area and volume correspond, respectively, to "surface" volume and what we might call *hypervolume*—a quantity that is hard to imagine since our brains are wired to think in three, not four dimensions. Using arguments that are analogous to the discussion of how surface area scales with volume to the 2/3 power, one can show that in four dimensions surface volume scales with hypervolume to the 3/4 power.

In short, what Brown, Enquist, and West are saying is that evolution structured our circulatory systems as fractal networks to approximate a "fourth dimension" so as to make our metabolisms more efficient. As West, Brown, and Enquist put it, "Although living things occupy a three-dimensional space, their internal physiology and anatomy operate as if they were four-dimensional ... Fractal geometry has literally given life an added dimension."

Scope of the Theory

In its original form, metabolic scaling theory was applied to explain metabolic scaling in many animal species, such as those plotted in figure 17.2. However,

Brown, Enquist, West, and their increasing cadre of new collaborators did not stop there. Every few weeks, it seems, a new class of organisms or phenomena is added to the list covered by the theory. The group has claimed that their theory can also be used to explain other quarter-power scaling laws such as those governing heart rate, life span, gestation time, and time spent sleeping.

The group also believes that the theory explains metabolic scaling in plants, many of which use fractal-like *vascular* networks to transport water and other nutrients. They further claim that the theory explains the quarter-power scaling laws for tree trunk circumference, plant growth rates, and several other aspects of animal and plant organisms alike. A more general form of the metabolic scaling theory that includes body temperature was proposed to explain metabolic rates in reptiles and fish.

Moving to the microscopic realm, the group has postulated that their theory applies at the cellular level, asserting that 3/4 power metabolic scaling predicts the metabolic rate of single-celled organisms as well as of metabolic-like, molecule-sized distribution processes inside the cell itself, and even to metabolic-like processes inside *components* of cells such as mitochondria. The group also proposed that the theory explains the rate of DNA changes in organisms, and thus is highly relevant to both genetics and evolutionary biology. Others have reported that the theory explains the scaling of mass versus growth rate in cancerous tumors.

In the realm of the very large, metabolic scaling theory and its extensions have been applied to entire ecosystems. Brown, Enquist, and West believe that their theory explains the observed $-3/4$ scaling of species population density with body size in certain ecosystems.

In fact, because metabolism is so central to all aspects of life, it's hard to find an area of biology that this theory doesn't touch on. As you can imagine, this has got many scientists very excited and looking for new places to apply the theory. Metabolic scaling theory has been said to have "the potential to unify all of biology" and to be "as potentially important to biology as Newton's contributions are to physics." In one of their papers, the group themselves commented, "We see the prospects for the emergence of a general theory of metabolism that will play a role in biology similar to the theory of genetics."

Controversy

As to be expected for a relatively new, high-profile theory that claims to explain so much, while some scientists are bursting with enthusiasm for metabolic scaling theory, others are roiling with criticism. Here are the two main criticisms that are currently being published in some of the top scientific journals:

- **Quarter-power scaling laws are not as universal as the theory claims.** As a rule, given any proposed general property of living systems, biology exhibits exceptions to the rule. (And maybe even exceptions to this rule itself.) Metabolic scaling theory is no exception, so to speak. Although most biologists agree that a large number of species seem to follow the various quarter-power scaling laws, there are also many exceptions, and sometimes there is considerable variation in metabolic rate even within a single species. One familiar example is dogs, in which smaller breeds tend to live at least as long as larger breeds. It has been argued that, while Kleiber's law represents a statistical *average*, the variations from this average can be quite large, and metabolic theory does not explain this because it takes into account only body mass and temperature. Others have argued that there are laws predicted by the theory that real-world data strongly contradict. Still others argue that Kleiber was wrong all along, and the best fit to the data is actually a power law with exponent 2/3, as proposed over one hundred years ago by Rubner in his surface hypothesis. In most cases, this is an argument about how to correctly interpret data on metabolic scaling and about what constitutes a "fit" to the theory. The metabolic scaling group stands by its theory, and has diligently replied to many of these arguments, which become increasingly technical and obscure as the authors discuss the intricacies of advanced statistics and biological functions.
- **The Kleiber scaling law is valid but the metabolic scaling theory is wrong.** Others have argued that metabolic scaling theory is oversimplified, that life is too complex and varied to be covered by one overreaching theory, and that positing fractal structure is by no means the only way to explain the observed power-law distributions. One ecologist put it this way: "The more detail that one knows about the particular physiology involved, the less plausible these explanations become." Another warned, "It's nice when things are simple, but the real world isn't always so." Finally, there have been arguments that the mathematics in metabolic scaling theory is incorrect. The authors of metabolic scaling theory have vehemently disagreed with these critiques and in some cases have pointed out what they believed to be fundamental mistakes in the critic's mathematics.

The authors of metabolic scaling theory have strongly stood by their work and expressed frustration about criticisms of details. As West said, "Part of me doesn't want to be cowered by these little dogs nipping at our heels." However, the group also recognizes that a deluge of such criticisms is a good sign—whatever they end up believing, a very large number of people have sat up and taken notice of metabolic scaling theory. And of course, as I have mentioned, skepticism is one of the most important jobs of scientists, and the more prominent the theory and the more ambitious its claims are, the more skepticism is warranted.

The arguments will not end soon; after all, Newton's theory of gravity was not widely accepted for more than sixty years after it first appeared, and many other of the most important scientific advances have faced similar fates. The main conclusion we can reach is that metabolic scaling theory is an exceptionally interesting idea with a huge scope and some experimental support. As ecologist Helene Müller-Landau predicts: "I suspect that West, Enquist et al. will continue repeating their central arguments and others will continue repeating the same central critiques, for years to come, until the weight of evidence finally leads one way or the other to win out."

The Unresolved Mystery of Power Laws

We have seen a lot of power laws in this and the previous chapters. In addition to these, power-law distributions have been identified for the size of cities, people's incomes, earthquakes, variability in heart rate, forest fires, and stock-market volatility, to name just a few phenomena.

As I described in chapter 15, scientists typically assume that most natural phenomena are distributed according to the bell curve or *normal* distribution. However, power laws are being discovered in such a great number and variety of phenomena that some scientists are calling them "more normal than 'normal.'" In the words of mathematician Walter Willinger and his colleagues: "The presence of [power-law] distributions in data obtained from complex natural or engineered systems should be considered the norm rather than the exception."

Scientists have a pretty good handle on what gives rise to bell curve distributions in nature, but power laws are something of a mystery. As we have seen, there are many different explanations for the power laws observed in nature (e.g., preferential attachment, fractal structure, self-organized criticality, highly optimized tolerance, among others), and little agreement on which observed power laws are caused by which mechanisms.

In the early 1930s, a Harvard professor of linguistics, George Kingsley Zipf, published a book in which he included an interesting property of language. First take any large text such as a novel or a newspaper, and list each word in the order of how many times it appears. For example, here is a partial list of words and frequencies from Shakespeare's "To be or not to be" monologue from the play *Hamlet*:

Word	Frequency	Rank
the	22	1
to	15	2
of	15	3
and	12	4
that	7	5
a	5	6
sleep	5	7
we	4	8
be	3	9
us	3	10
bear	3	11
with	3	12
is	3	13
'tis	2	14
death	2	15
die	2	16
in	2	17
have	2	18
make	2	19
end	2	20

Putting this list in order of decreasing frequencies, we can assign a rank of 1 to the most frequent word (here, "the"), a rank of 2 to the second most frequent word, and so on. Some words are tied for frequency (e.g., "a" and "sleep" both have five occurrences). Here, I have broken ties for ranking at random.

In figure 17.4, I have plotted the to-be-or-not-to-be word frequency as a function of rank. The shape of the plot indeed looks like a power law. If the text I had chosen had been larger, the graph would have looked even more power-law-ish.

Zipf analyzed large amounts of text in this way (without the help of computers!) and found that, given a large text, the frequency of a word is

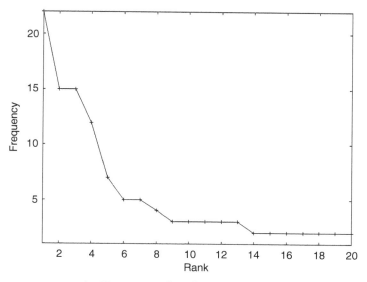

FIGURE 17.4. An illustration of Zipf's law using Shakespeare's "To be or not to be" monologue.

approximately proportional to the inverse of its rank (i.e., $1/rank$). This is a power law, with exponent -1. The second highest ranked word will appear about half as often as the first, the third about one-third as often, and so forth. This relation is now called *Zipf's law*, and is perhaps the most famous of known power laws.

There have been many different explanations proposed for Zipf's law. Zipf himself proposed that, on the one hand, people in general operate by a "Principle of Least Effort": once a word has been used, it takes less effort to use it again for similar meanings than to come up with a different word. On the other hand, people want language to be unambiguous, which they can accomplish by using different words for similar but nonidentical meanings. Zipf showed mathematically that these two pressures working together could produce the observed power-law distribution.

In the 1950s, Benoit Mandelbrot, of fractal fame, had a somewhat different explanation, in terms of information content. Following Claude Shannon's formulation of information theory (cf. chapter 3), Mandelbrot considered a word as a "message" being sent from a "source" who wants to maximize the amount of information while minimizing the cost of sending that information. For example, the words *feline* and *cat* mean the same thing, but the latter, being shorter, costs less (or takes less energy) to transmit. Mandelbrot showed that if the information content and transmission costs are simultaneously optimized, the result is Zipf's law.

At about the same time, Herbert Simon proposed yet another explanation, presaging the notion of preferential attachment. Simon envisioned a person adding words one at a time to a text. He proposed that at any time, the probability of that person reusing a word is proportional to that word's current frequency in the text. All words that have not yet appeared have the same, nonzero probability of being added. Simon showed that this process results in text that follows Zipf's law.

Evidently Mandelbrot and Simon had a rather heated argument (via dueling letters to the journal *Information and Control*) about whose explanation was correct.

Finally, also around the same time, to everyone's amusement or chagrin, the psychologist George Miller showed, using simple probability theory, that the text generated by monkeys typing randomly on a keyboard, ending a word every time they (randomly) hit the space bar, will follow Zipf's law as well.

The many explanations of Zipf's law proposed in the 1930s through the 1950s epitomize the arguments going on at present concerning the physical or informational mechanisms giving rise to power laws in nature. Understanding power-law distributions, their origins, their significance, and their commonalities across disciplines is currently a very important open problem in many areas of complex systems research. It is an issue I'm sure you will hear more about as the science behind these laws becomes clearer.

CHAPTER 18 | Evolution, Complexified

IN CHAPTER 1 I ASKED, "HOW DID evolution produce creatures with such an enormous contrast between their individual simplicity and their collective sophistication?" Indeed, as illustrated by the examples we've seen in this book, the closer one looks at living systems, the more astonishing it seems that such intricate complexity could have been formed by the gradual accumulation of favorable mutations or the whims of historical accident. This very argument has been used from Darwin's time to the present by believers in divine creation or other supernatural means of "intelligent design."

The questions of how, why, and even *if* evolution creates complexity, and how complexity in biology might be characterized and measured, are still very much open. One of the most important contributions of complex systems research over the last few decades has been to demonstrate new ways to approach these age-old questions. In this chapter I describe some of the recent discoveries in genetics and the dynamics of genetic regulation that are giving us surprising new insights into the evolution of complex systems.

Genetics, Complexified

Often in science new technologies can open a floodgate of discoveries that change scientists' views of a previously established field of study. We saw an example of this back in chapter 2—it was the invention of the electronic computer, and its capacity for modeling complex systems such as weather, that allowed for the demonstration of the existence of chaos. More recently, extremely powerful land and space-based telescopes have led to a flurry of discoveries in astronomy concerning so-called dark matter and dark energy, which seem to call into question much of what was previously accepted in cosmology.

No new set of technologies has had a more profound impact on an established field than the so-called molecular revolution in genetics over the last four decades. Technologies for rapidly copying, sequencing, synthesizing, and engineering DNA, for imaging molecular-level structures that had never been seen before, and for viewing expression patterns of thousands of different genes simultaneously; these are only a few examples of the feats of biotechnology in the late twentieth and early twenty-first centuries. And it seems that with each new technology allowing biologists to peer closer into the cell, more unexpected complexities appear.

At the time Watson and Crick discovered its structure, DNA was basically thought of as a string of genes, each of which coded for a particular protein that carried out some function in the cell. This string of genes was viewed essentially as the "computer program" of the cell, whose commands were translated and enacted by RNA, ribosomes, and the like, in order to synthesize the proteins that the genes stood for. Small random changes to the genome occurred when copying errors were made during the DNA duplication process; the long-term accumulation of those small random changes that happened to be favorable were the ultimate cause of adaptive change in biology and the origin of new species.

This conventional view has undergone monumental changes in the last 40 years. The term *molecular revolution* refers not only to the revolutionary new techniques in genetics, but also to the revolutionary new view of DNA, genes, and the nature of evolution that these techniques have provided.

What Is a Gene?

One casualty of the molecular revolution is the straightforward concept of *gene*. The mechanics of DNA that I sketched in chapter 6 still holds true—chromosomes contain stretches of DNA that are transcribed and translated to create proteins—but it turns out to be only part of the story. The following are a few examples that give the flavor of the many phenomena that have been and are being discovered; these phenomena are confounding the straightforward view of how genes and inheritance work.

- Genes are not like "beads on a string." When I took high-school biology, genes and chromosomes were explained using the beads-on-a-string metaphor (and I think we even got to put together a model using pop-together plastic beads). However, it turns out that genes are not so discretely separated from one another. There are genes that overlap with other genes—i.e., they each code for a different protein, but they share DNA nucleotides. There are genes that are wholly contained inside other genes.

- Genes move around on their chromosome and between chromosomes. You may have heard of "jumping genes." Indeed, genes can move around, rearranging the makeup of chromosomes. This can happen in any cell, including sperm and egg cells, meaning that the effects can be inherited. The result can be a much higher rate of mutation than comes from errors in DNA replication. Some scientists have proposed that these "mobile genetic elements" might be responsible for the differences observed between close relatives, and even between identical twins. The phenomenon of jumping genes has even been proposed as one of the mechanisms responsible for the diversity of life.

- A single gene can code for more than one protein. It had long been thought that there was a one-to-one correspondence between genes and proteins. A problem for this assumption arose when the human genome was sequenced, and it was discovered that while the number of different types of proteins encoded by genes may exceed 100,000, the human genome contains only about 25,000 genes. The recently discovered phenomena of *alternative splicing* and *RNA editing* help explain this discrepancy. These processes can alter messenger RNA in various ways after it has transcribed DNA but before it is translated into amino acids. This means that different transcription events of the same gene can produce different final proteins.

- In light of all these complications, even professional biologists don't always agree on the definition of "gene." Recently a group of science philosophers and biologists performed a survey in which 500 biologists were independently given certain unusual but real DNA sequences and asked whether each sequence qualified as a "gene," and how confident they were of their answer. It turned out that for many of the sequences, opinion was split, with about 60% confident of one answer and 40% confident of the other answer. As stated in an article in *Nature* reporting on this work, "The more expert scientists become in molecular genetics, the less easy it is to be sure about what, if anything, a gene actually is."

- The complexity of living systems is largely due to networks of genes rather than the sum of independent effects of individual genes. As I described in chapter 16, genetic regulatory networks are currently a major focus of the field of genetics. In the old genes-as-beads-on-a-string view, as in Mendel's laws, genes are *linear*—each gene independently contributes to the entire phenotype. The new, generally accepted view, is that genes in a cell operate in *nonlinear* information-processing networks, in which some genes control the actions of other genes in response to changes in the cell's state—that is, genes do not operate independently.

- There are heritable changes in the function of genes that can occur without any modification of the gene's DNA sequence. Such changes are studied in the growing field of *epigenetics*. One example is so-called *DNA methylation*, in which an enzyme in a cell attaches particular molecules to some parts of a DNA sequence, effectively "turning off" those parts. When this occurs in a cell, all descendents of that cell will have the same DNA methylation. Thus if DNA methylation occurs in a sperm or egg cell, it will be inherited.

  On the one hand, this kind of epigenetic effect happens all the time in our cells, and is essential for life in many respects, turning off genes that are no longer needed (e.g., once we reach adulthood, we no longer need to grow and develop like a child; thus genes controlling juvenile development are methylated). On the other hand, incorrect or absent methylation is the cause of some genetic disorders and diseases. In fact, the absence of necessary methylation during embryo development is thought by some to be the reason so many cloned embryos do not survive to birth, or why so many cloned animals that do survive have serious, often fatal disorders.

- It has recently been discovered that in most organisms a large proportion of the DNA that is transcribed by RNA is *not* subsequently translated into proteins. This so-called noncoding RNA can have many regulatory effects on genes, as well as functional roles in cells, both of which jobs were previously thought to be the sole purview of proteins. The significance of noncoding RNAs is currently a very active research topic in genetics.

Genetics has become very complicated indeed. And the implications of all these complications for biology are enormous. In 2003 the Human Genome Project published the entire human genome—that is, the complete sequence of human DNA. Although a tremendous amount was learned from this project, it was less than some had hoped. Some had believed that a complete mapping of human genes would provide a nearly complete understanding of how genetics worked, which genes were responsible for which traits, and that this would guide the way for revolutionary medical discoveries and targeted gene therapies. Although there have been several discoveries of certain genes that are implicated in particular diseases, it has turned out that simply knowing the sequence of DNA is not nearly enough to understand a person's (or any complex organism's) unique collection of traits and defects.

One sector that pinned high hopes on the sequencing of genes is the international biotechnology industry. A recent *New York Times* article reported on the effects that all this newly discovered genetic complexity was having on biotech: "The presumption that genes operate independently has been institutionalized since 1976, when the first biotech company was founded. In fact, it is the economic and regulatory foundation on which the entire biotechnology industry is built."

The problem is not just that the science underlying genetics is being rapidly revised. A major issue lurking for biotech is the status of gene patents. For decades biotech companies have been patenting particular sequences of human DNA that were believed to "encode a specific functional product." But as we have seen above, many, if not most, complex traits are not determined by the exact DNA sequence of a particular gene. So are these patents defensible? What if the "functional product" is the result of epigenetic processes acting on the gene or its regulators? Or what if the product requires not only the patented gene but also the genes that regulate it, and the genes that regulate those genes, and so on? And what if those regulatory genes are patented by someone else? Once we leave the world of linear genes and encounter essential nonlinearity, the meaning of these patents becomes very murky and may guarantee the employment of patent lawyers and judges for a long time to come. And patents aren't the only problem. As the *New York Times* pointed out, "Evidence of a networked genome shatters the scientific basis for virtually every official risk assessment of today's commercial biotech products, from genetically engineered crops to pharmaceuticals."

Not only genetics, but evolutionary theory as a whole has been profoundly challenged by these new genetic discoveries. A prominent example of this is the field of "Evo-Devo."

Evo-Devo

Evo-Devo is the nickname for "evolutionary developmental biology." Many people are very excited about this field and its recent discoveries, which are claimed to explain at least three big mysteries of genetics and evolution: (1) Humans have only about 25,000 genes. What is responsible for our complexity? (2) Genetically, humans are very similar to many other species. For example, more than 90% of our DNA is shared with mice and more than 95%

with chimps. Why are our bodies so different from those of other animals? (3) Supposing that Stephen Jay Gould and others are correct about punctuated equilibria in evolution, how could big changes in body morphology happen in short periods of evolutionary time?

It has recently been proposed that the answer to these questions lies, at least in part, in the discovery of *genetic switches*.

The fields of developmental biology and embryology study the processes by which a fertilized single egg cell becomes a viable multibillion-celled living organism. However, the Modern Synthesis's concern was with genes; in the words of developmental biologist Sean Carroll, it treated developmental biology and embryology as a " 'black box' that somehow transformed genetic information into three-dimensional, functional animals." This was in part due to the view that the huge diversity of animal morphology would eventually be explained by large differences in the number of and DNA makeup of genes.

In the 1980s and 1990s, this view became widely challenged. As I noted above, DNA sequencing had revealed the extensive similarities in DNA among many different species. Advances in genetics also produced a detailed understanding of the mechanisms of gene expression in cells during embryonic and fetal development. These mechanisms turned out to be quite different from what was generally expected. Embryologists discovered that, in all complex animals under study, there is a small set of "master genes" that regulate the formation and morphology of many of the animal's body parts. Even more surprising, these master genes were found to share many of *the same sequences of DNA* across many species with extreme morphological differences, ranging from fruit flies to humans.

Given that their developmental processes are governed by the same genes, how is it that these different animals develop such different body parts? Proponents of Evo-Devo propose that morphological diversity among species is, for the most part, not due to differences in genes but in genetic switches that are used to turn genes on and off. These switches are sequences of DNA—often several hundred base pairs in length—that do not code for any protein. Rather they are part of what used to be called "junk DNA," but now have been found to be used in gene regulation.

Figure 18.1 illustrates how switches work. A switch is a sequence of noncoding DNA that resides nearby a particular gene. This sequence of molecules typically contains on the order of a dozen *signature* subsequences, each of which chemically binds with a particular protein, that is, the protein attaches to the DNA string. Whether or not the nearby gene gets transcribed, and how quickly, depends on the combination of proteins attached to these subsequences. Proteins that allow transcription create strong binding sites for RNA

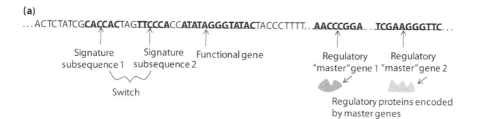

FIGURE 18.1. Illustration of genetic "switches." (a) A DNA sequence, containing a switch with two signature subsequences, a functional gene turned on by that switch, and two regulatory master genes. The regulatory master genes give rise to regulatory proteins. (b) The regulatory proteins bind to the signature subsequences, switching on the functional gene—that is, allowing it to be transcribed.

molecules that will do the transcribing; proteins that prevent transcription block these same RNA molecules from binding to the DNA. Some of these proteins can negate the effects of others.

Where do these special regulator proteins come from? Like all proteins, they come from genes, in this case *regulatory* genes that encode such proteins in order to turn other genes on or off, depending on the current state of the cell. How do these regulatory genes determine the current state of the cell? By the presence or absence of proteins that signal the state of the cell by binding to the regulatory genes' own switches. Such proteins are often encoded by other regulatory genes, and so forth.

In summary, genetic regulatory networks are made up of several different kinds of entities, including *functional* genes that encode proteins (and sometimes noncoding RNA) for cellular maintenance or building, and *regulatory* genes that encode proteins (and sometimes noncoding RNA) that turn other genes on or off by binding to DNA "switches" near to the gene in question.

I can now give Evo-Devo's answers to the three questions posed at the beginning of this section. Humans (and other animals) can be more complex than their number of genes would suggest for many reasons, some listed above in the "What Is a Gene" section. But a primary reason is that genetic regulatory networks allow a huge number of possibilities for gene expression patterns, since there are so many possible ways in which proteins can be attached to switches.

The reason we humans can share so many genes with other creatures quite different from us is that, although the genes might be the same, the sequences making up switches have often evolved to be different. Small changes in switches can produce very different patterns of genes turning on and off during development. Thus, according to Evo-Devo, the diversity of organisms is largely due to evolutionary modifications of switches, rather than genes. This is also the reason that large changes in morphology—possibly including speciation—can happen swiftly in evolutionary time: the master genes remain the same, but switches are modified. According to Evo-Devo, such modifications—in the parts of DNA long thought of as "junk"—are the major force in evolution, rather than the appearance of new genes. Biologist John Mattick goes so far as to say, "The irony . . . is that what was dismissed as junk [DNA] because it wasn't understood will turn out to hold the secret of human complexity."

One striking instance of Evo-Devo in action is the famous example of the evolution of finches' beaks. As I described in chapter 5, Darwin observed large variations in beak size and shape among finches native to the Galápagos Islands. Until recently, most evolutionary biologists would have assumed that such variations resulted from a gradual process in which chance mutations of several different genes accumulated. But recently, a gene called BMP4 was discovered that helps control beak size and shape by regulating other genes that produce bones. The more strongly BMP4 is expressed during the birds' development, the larger and stronger their beaks. A second gene, called *calmodulin*, was discovered to be associated with long, thin beaks. As Carol Kaesuk Yoon reported in the *New York Times*, "To verify that the BMP4 gene itself could indeed trigger the growth of grander, bigger, nut-crushing beaks, researchers artificially cranked up the production of BMP4 in the developing beaks of chicken embryos. The chicks began growing wider, taller, more robust beaks similar to those of a nut-cracking finch. . . . As with BMP4, the more that calmodulin was expressed, the longer the beak became. When scientists artificially increased calmodulin in chicken embryos, the chicks began growing extended beaks, just like a cactus driller. . . . So, with just these two genes, not tens or hundreds, the scientists found the potential to re-create beaks, massive or stubby or elongated." The conclusion is that large changes in the morphology of beaks (and other traits) can take place rapidly without the necessity of waiting for many chance mutations over a long period of time.

Another example where Evo-Devo is challenging long-held views about evolution concerns the notion of *convergent evolution*. In my high school biology class, we learned that the octopus eye and the human eye—greatly different in morphology—were examples of convergent evolution: eyes in these two species evolved completely independently of one another as a consequence of natural selection acting in two different environments in which eyes were a useful adaptation.

However, recent evidence has indicated that the evolution of these two eyes was not as independent as previously thought. Humans, octopi, flies, and many other species have a common gene called PAX6, which helps direct the development of eyes. In a strange but revealing experiment, the Swiss biologist Walter Gehring took PAX6 genes from mice and inserted them into the genomes of fruit flies. In particular, in different studies, PAX6 was inserted in three different parts of the genome: those that direct the development of legs, wings, and antennae, respectively. The researchers got eerie results: eye-like structures formed on flies' legs, wings, and antennae. Moreover, the structures were like fly eyes, not mouse eyes. Gehring's conclusion: the eye evolved not many times independently, but only once, in a common ancestor with the PAX6 gene. This conclusion is still quite controversial among evolutionary biologists.

Although genetic regulatory networks directed by master genes can produce great diversity, they also enforce certain constraints on evolution. Evo-Devo scientists claim that the types of body morphology (called *body plans*) any organism can have are highly constrained by the master genes, and that is why only a few basic body plans are seen in nature. It's possible that genomes vastly different from ours could result in new types of body plans, but in practice, evolution can't get us there because we are so reliant on the existing regulatory genes. Our possibilities for evolution are constrained. According to Evo-Devo, the notion that "every trait can vary indefinitely" is wrong.

Genetic Regulation and Kauffman's "Origins of Order"

Stuart Kauffman is a theoretical biologist who has been thinking about genetic regulatory networks and their role in constraining evolution for over forty years, long before the ascendency of Evo-Devo. He has also thought about the implications for evolution of the "order" we see emerging from such complex networks.

Kauffman is a legendary figure in complex systems. My first encounter with him was at a conference I attended during my last year of graduate school. His talk was the very first one at the conference, and I must say that, for me at the time, it was the most inspiring talk I had ever heard. I don't remember the exact topic; I just remember the feeling I had while listening that what he was saying was profound, the questions he was addressing were the most important ones, and I wanted to work on this stuff too.

Kauffman started his career with a short stint as a physician but soon moved to genetics research. His work was original and influential; it earned

Stuart Kauffman (Photograph by Daryl Black, reprinted with permission.)

him many academic accolades, including a MacArthur "genius" award, as well as a faculty position at the Santa Fe Institute. At SFI seminars, Kauffman would sometimes chime in from the audience with, "I know I'm just a simple country doctor, but . . . " and would spend a good five minutes or more fluently and eloquently giving his extemporaneous opinion on some highly technical topic that he had never thought about before. One science journalist called him a "world-class intellectual riffer," which is an apt description that I interpret as wholly complimentary.

Stuart's "simple country doctor" humble affect belies his personality. Kauffman is one of Complex Systems' big thinkers, a visionary, and not what you would call a "modest" or "humble" person. A joke at SFI was that Stuart had "patented Darwinian evolution," and indeed, he holds a patent on techniques for evolving protein sequences in the laboratory for the purpose of discovering new useful drugs.

RANDOM BOOLEAN NETWORKS

Kauffman was perhaps the first person to invent and study simplified computer models of genetic regulatory networks. His model was a structure called

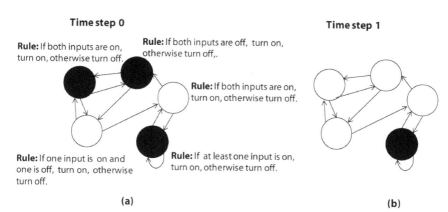

FIGURE 18.2. (a) A random Boolean network with five nodes. The in-degree (K) of each node is equal to 2. At time step 0, each node is in a random initial state: on (black) or off (white). (b) Time step 1 shows the network after each node has updated its state.

a *Random Boolean Network* (RBN), which is an extension of cellular automata. Like any network, an RBN consists of a set of nodes and links between the nodes. Like a cellular automaton, an RBN updates its nodes' states in discrete time steps. At each time step each node can be in either state **on** or state **off**.

The property that **on** and **off** are the only allowed states is where the term *Boolean* comes in: a Boolean rule (or *function*) is one that gets some number of inputs, each equal to either 0 or 1, and from those inputs it produces an output of a 0 or 1. Such rules are named after the mathematician George Boole, who did extensive mathematical research on them.

In an RBN, links are directional: if node A links to node B, node B does not necessarily (but can possibly) link to node A. The in-degree of each node (the number of links from other nodes to that node) is the same for each node—let's call that number K.

Here is how to build an RBN: for each node, create in-links to that node from K other randomly chosen nodes (including, possibly, a self-link), and give that node a Boolean rule, chosen randomly, that inputs K **on** or **off** states and outputs a single **on** or **off** state (figure 18.2a).

To *run* the RBN, give each node an initial state of **on** or **off** chosen at random. Then at each time step, each node transmits its state to the nodes it links to, and receives as input the states from the nodes that link to it. Each node then applies its rule to its input to determine its state at the next time step. All this is illustrated in figure 18.2, which shows the action of an RBN of five nodes, each with two inputs, over one time step.

RBNs are similar to cellular automata, but with two major differences: nodes are connected not to spatially neighboring nodes but at random, and rather than all nodes having an identical rule, each node has its own rule.

In Kauffman's work, the RBN as a whole is an idealized model of a genetic regulatory network, in which "genes" are represented by nodes, and "gene A regulates gene B" is represented by node A linking to node B. The model is of course vastly simpler than real genetic networks. Using such idealized models in biology is now becoming common, but when Kauffman started this work in the 1960s, it was less well accepted.

LIFE AT THE EDGE OF CHAOS

Kauffman and his students and collaborators have done a raft of simulations of RBNs with different values of the in-degree K for each node. Starting from a random initial state, and iterated over a series of time steps, the nodes in the RBN change state in random ways for a while, and finally settle down to either a fixed point (all nodes' states remain fixed) or an oscillation (the state of the whole network oscillates with some small period), or do not settle down at all, with random-looking behavior continuing over a large number of iterations. Such behavior is chaotic, in that the precise trajectory of states of the network have sensitive dependence on the initial state of the network.

Kauffman found that the typical final behavior is determined by both the number of nodes in the network and each node's in-degree K. As K is increased from 1 (i.e., each node has exactly one input) all the way up to the total number of nodes (i.e., each node gets input from all other nodes, including itself), the typical behavior of the RBNs moves through the three different "regimes" of behavior (fixed-point, oscillating, chaotic). You might notice that this parallels the behavior of the logistic map as R is increased (cf. chapter 2). At $K = 2$ Kauffman found an "interesting" regime—neither fixed point, oscillating, or completely chaotic. In analogy with the term "onset of chaos" used with the logistic map, he called this regime the "edge of chaos."

Assuming the behavior of his RBNs reflected the behavior of real genetic networks, and making an analogy with the phases of water as temperature changes, he concluded that "the genomic networks that control development from zygote to adult can exist in three major regimes: a frozen ordered regime, a gaseous chaotic regime, and a kind of liquid regime located in the region between order and chaos."

Kauffman reasoned that, for an organism to be both alive and stable, the genetic networks his RBNs modeled had to be in the interesting "liquid" regime—not too rigid or "frozen," and not too chaotic or "gaseous." In his own words, "life exists at the edge of chaos."

Kauffman used the vocabulary of dynamical systems theory—attractors, bifurcations, chaos—to describe his findings. Suppose we call a possible configuration of the nodes' states a *global state* of the network. Since RBNs have a finite number of nodes, there are only a finite number of possible global states, so if the network is iterated for long enough it will repeat one of the global states it has already been in, and hence cycle through the next series of states until it repeats that global state again. Kauffman called this cycle an "attractor" of the network. By performing many simulations of RBNs, he estimated that the average number of different attractors produced in different networks with $K = 2$ was approximately equal to the square root of the number of nodes.

Next came a big leap in Kauffman's interpretation of this model. Every cell in the body has more or less identical DNA. However, the body has different types of cells: skin cells, liver cells, and so forth. Kauffman asserted that what determines a particular cell type is the pattern of gene expression in the cell over time—I have described above how gene expression patterns can be quite different in different cells. In the RBN model, an attractor, as defined above, is a pattern over time of "gene expression." Thus Kauffman proposed that an attractor in his network represents a cell type in an organism's body. Kauffman's model thus predicted that for an organism with 100,000 genes, the number of different cell types would be approximately the square root of 100,000, or 316. This is not too far from the actual number of cell types identified in humans—somewhere around 256.

At the time Kauffman was doing these calculations, it was generally believed that the human genome contained about 100,000 genes (since the human body uses about 100,000 types of proteins). Kauffman was thrilled that his model had come close to correctly predicting the number of cell types in humans. Now we know that the human genome contains only about 25,000 genes, so Kauffman's model would predict about 158 cell types.

THE ORIGIN OF ORDER

The model wasn't perfect, but Kauffman believed it illustrated his most important general point about living systems: that natural selection is in principle not necessary to create a complex creature. Many RBNs with $K = 2$ exhibited what he termed "complex" behavior, and no natural selection or evolutionary algorithm was involved. His view was that once a network structure becomes sufficiently complex—that is, has a large number of nodes controlling other nodes—complex and "self-organized" behavior will emerge. He says,

> Most biologists, heritors of the Darwinian tradition, suppose that the order of ontogeny is due to the grinding away of a molecular Rube Goldberg machine, slapped together piece by piece by evolution. I present a countering thesis: most of the beautiful order seen in ontogeny is spontaneous, a natural expression of the stunning self-organization that abounds in very complex regulatory networks. We appear to have been profoundly wrong. Order, vast and generative, arises naturally.

Kauffman was deeply influenced by the framework of statistical mechanics, which I described in chapter 3. Recall that statistical mechanics explains how properties such as temperature arise from the statistics of huge numbers of molecules. That is, one can predict the behavior of a system's temperature without having to follow the Newtonian trajectory of every molecule. Kauffman similarly proposed that he had found a statistical mechanics law governing the emergence of complexity from huge numbers of interconnected, mutually regulating components. He termed this law a "candidate fourth law of thermodynamics." Just as the second law states that the universe has an innate tendency toward increasing entropy, Kauffman's "fourth law" proposes that life has an innate tendency to become more complex, which is independent of any tendency of natural selection. This idea is discussed at length in Kauffman's book, *The Origins of Order*. In Kauffman's view, the evolution of complex organisms is due in part to this self-organization and in part to natural selection, and perhaps self-organization is really what predominates, severely limiting the possibilities for selection to act on.

Reactions to Kauffman's Work

Given that Kauffman's work implies "a fundamental reinterpretation of the place of selection in evolutionary theory," you can imagine that people react rather strongly to it. There are a lot of huge fans of this work ("His approach opens up new vistas"; it is "the first serious attempt to model a complete biology"). On the other side, many people are highly skeptical of both his results and his broad interpretations of them. One reviewer called Kauffman's writing style "dangerously seductive" and said of *The Origins of Order*, "There are times when the bracing walk through hyperspace seems unfazed by the nagging demands of reality."

Indeed, the experimental evidence concerning Kauffman's claims is not all on his side. Kauffman himself admits that regarding RBNs as models of genetic regulatory networks requires many unrealistic assumptions: each node can be in only one of two states (whereas gene expression has different degrees of strength), each has an identical number of nodes that regulate it, and all nodes are updated in synchrony at discrete time steps. These simplifications may ignore important details of genetic activity.

Most troublesome for his theory are the effects of "noise"—errors and other sources of nondeterministic behavior—that are inevitable in real-world complex systems, including genetic regulation. Biological genetic networks make errors all the time, yet they are resilient—most often our health is not affected by these errors. However, simulations have shown that noise has a significant effect on the behavior of RBNs, and sometimes will prevent RBNs from reaching a stable attractor. Even some of the claims Kauffman made specifically about his RBN results are not holding up to further scrutiny. For example, recall Kauffman's claim that the number of attractors that occur in a typical network is close to the square root of the number of nodes, and his interpretation of this fact in terms of cell-types. Additional simulations have shown that the number of attractors is actually not well approximated by the square root of the number of nodes. Of course this doesn't necessarily mean that Kauffman is wrong in his broader claims; it just shows that there is considerably more work to be done on developing more accurate models. Developing accurate models of genetic regulatory networks is currently a very active research area in biology.

Summary

Evolutionary biology is still working on answering its most important question: How does complexity in living systems come about through evolution? As we have seen in this chapter, the degree of complexity in biology is only beginning to be fully appreciated. We also have seen that many major steps are being taken toward understanding the evolution of complexity. One step has been the development of what some have called an "extended Synthesis," in which natural selection still plays an important role, but other forces—historical accidents, developmental constraints, and self-organization—are joining natural selection as explanatory tools. Evolutionists, particularly in the United States, have been under attack from religious extremists and are often on the defensive, reluctant to admit that natural selection may not be the entire story. As biologists Guy Hoelzer, John Pepper, and Eric Smith have written about this predicament: "It has essentially become a matter of social responsibility for evolutionary biologists to join the battle in defense of Darwinism, but there is a scientific cost associated with this cultural norm. Alternative ways of describing evolutionary processes, complementary to natural selection, can elicit the same defensive posture without critical analysis."

Evolutionary biologist Dan McShea has given me a useful way to think about these various issues. He classifies evolutionists into three categories: adaptationists, who believe that natural selection is primary; historicists, who give credit to historical accident for many evolutionary changes; and structuralists, such as Kauffman, who focus on how organized structure comes about even in the absence of natural selection. Evolutionary theory will be unified only when these three groups are able to show how their favored forces work as an integrated whole.

Dan also gave me an optimistic perspective on this prospect: "Evolutionary biology is in a state of intellectual chaos. But it's an intellectual chaos of a very productive kind."

PART V | Conclusion

I will put Chaos into fourteen lines
And keep him there; and let him thence escape
If he be lucky; let him twist, and ape
Flood, fire, and demon—his adroit designs
Will strain to nothing in the strict confines
Of this sweet order, where, in pious rape,
I hold his essence and amorphous shape,
Till he with Order mingles and combines.
Past are the hours, the years of our duress,
His arrogance, our awful servitude:
I have him. He is nothing more nor less
Than something simple not yet understood;
I shall not even force him to confess;
Or answer. I will only make him good.

— Edna St. Vincent Millay, *Mine the Harvest: A Collection of New Poems*

CHAPTER 19 | The Past and Future of the Sciences of Complexity

IN 1995, THE SCIENCE JOURNALIST John Horgan published an article in *Scientific American*, arguably the world's leading popular science magazine, attacking the field of complex systems in general and the Santa Fe Institute in particular. His article was advertised on the magazine's cover under the label "Is Complexity a Sham?" (figure 19.1).

The article contained two main criticisms. First, in Horgan's view, it was unlikely that the field of complex systems would uncover any useful general principles, and second, he believed that the predominance of computer modeling made complexity a "fact-free science." In addition, the article made several minor jabs, calling complexity "pop science" and its researchers "complexologists." Horgan speculated that the term "complexity" has little meaning but that we keep it for its "public-relations value."

To add insult to injury, Horgan quoted me as saying, "At some level you can say all complex systems are aspects of the same underlying principles, but I don't think that will be very useful." Did I really say this? I wondered. What was the context? Do I believe it? Horgan had interviewed me on the phone for an hour or more and I had said a lot of things; he chose the single most negative comment to use in his article. I hadn't had very much experience with science journalists at that point and I felt really burned.

I wrote an angry, heartfelt letter to the editor at *Scientific American*, listing all the things I thought were wrong and unfair in Horgan's article. Of course a dozen or more of my colleagues did the same; the magazine published only one of these letters and it wasn't mine.

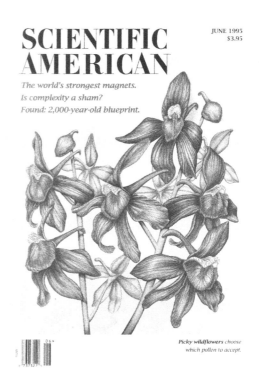

FIGURE 19.1. Complexity is "dissed" on the cover of *Scientific American*. (Cover art by Rosemary Volpe, reprinted by permission.)

The whole incident taught me some lessons. Mostly, be careful what you say to journalists. But it did force me to think harder and more carefully about the notion of "general principles" and what this notion might mean.

Horgan's article grew into an equally cantankerous book, called *The End of Science*, in which he proposed that all of the really important discoveries of science have already been made, and that humanity would make no more. His *Scientific American* article on complexity was expanded into a chapter, and included the following pessimistic prediction: "The fields of chaos, complexity, and artificial life will continue.... But they will not achieve any great insights into nature—certainly none comparable to Darwin's theory of evolution or quantum mechanics."

Is Horgan right in any sense? Is it futile to aim for the discovery of general principles or a "unified theory" covering all complex systems?

On Unified Theories and General Principles

The term *unified theory* (or Grand Unified Theory, quaintly abbreviated as GUT) usually refers to a goal of physics: to have a single theory that unifies the basic forces in the universe. String theory is one attempt at a GUT, but there is no consensus in physics that string theory works or even if a GUT exists.

Imagine that string theory turns out to be correct—physics' long sought-after GUT. That would be an enormously important achievement, but it would not be the end of science, and in particular it would be far from the end of complex systems science. The behaviors of complex systems that interest us are not understandable at the level of elementary particles or ten-dimensional strings. Even if these elements make up all of reality, they are the wrong vocabulary for explaining complexity. It would be like answering the question, "Why is the logistic map chaotic?" with the answer, "Because $x_{t+1} = R\,x_t(1-x_t)$." The explanation for chaos is, in some sense, embedded in the equation, just as the behavior of the immune system would in some sense be embedded in a Grand Unified Theory of physics. But not in the sense that constitutes human understanding—the ultimate goal of science. Physicists Jim Crutchfield, Doyne Farmer, Norman Packard, and Robert Shaw voiced this view very well: "[T]he hope that physics could be complete with an increasingly detailed understanding of fundamental physical forces and constituents is unfounded. The interaction of components on one scale can lead to complex global behavior on a larger scale that in general cannot be deduced from knowledge of the individual components." Or, as Albert Einstein supposedly quipped, "Gravitation is not responsible for people falling in love."

So if fundamendal physics is not to be a unified theory for complex systems, what, if anything, is? Most complex systems researchers would probably say that a unified theory of complexity is not a meaningful goal at this point. The science of physics, being over two thousand years old, is conceptually way ahead in that it has identified two main kinds of "stuff"—mass and energy—which Einstein unified with $E = mc^2$. It also has identified the four basic forces in nature, and has unified at least three of them. Mass, energy, and force, and the elementary particles that give rise to them, are the building blocks of theories in physics.

As for complex systems, we don't even know what corresponds to the elemental "stuff" or to a basic "force"; a unified theory doesn't mean much until you figure out what the conceptual components or building blocks of that theory should be.

Deborah Gordon, the ecologist and entomologist voiced this opinion:

Recently, ideas about complexity, self-organization, and emergence—when the whole is greater than the sum of its parts—have come into fashion as alternatives for metaphors of control. But such explanations offer only smoke and mirrors, functioning merely to provide names for what we can't explain; they elicit for me the same dissatisfaction I feel when a physicist says that a particle's behavior is caused by the equivalence of two terms in an equation. Perhaps there can be a general theory of complex systems, but it is clear we don't have one yet. A better route to understanding the dynamics of apparently self-organizing systems is to focus on the details of specific systems. This will reveal whether there are general laws.... The hope that general principles will explain the regulation of all the diverse complex dynamical systems that we find in nature can lead to ignoring anything that doesn't fit a pre-existing model. When we learn more about the specifics of such systems, we will see where analogies between them are useful and where they break down.

Of course there are many general principles that are not very useful, for example, "all complex systems exhibit emergent properties," because, as Gordon says, they "provide names for what we can't explain." This is, I think, what I was trying to say in the statement of mine that Horgan quoted. I think Gordon is correct in her implication that no single set of useful principles is going to apply to all complex systems.

It might be better to scale back and talk of *common* rather than *general* principles: those that provide new insight into—or new conceptualizations of—the workings of a set of systems or phenomena that would be very difficult to glean by studying these systems or phenomena separately, and trying to make analogies after the fact.

The discovery of common principles might be part of a feedback cycle in complexity research: knowledge about specific complex systems is synthesized into common principles, which then provide new ideas for understanding the specific systems. The specific details and common principles inform, constrain, and enrich one another.

This all sounds well and good, but where are examples of such principles? Of course proposals for common or universal principles abound in the literature, and we have seen several such proposals in this book: the universal properties of chaotic systems; John von Neumann's principles of self-reproduction; John Holland's principle of balancing exploitation and exploration; Robert Axelrod's general conditions for the evolution of cooperation; Stephen Wolfram's principle of computational equivalence; Albert-László Barabási and Réka Albert's proposal that preferential attachment is a general mechanism for the development of real-world networks; West, Brown, and Enquist's proposal that fractal circulation networks explain scaling relationships; et cetera. There are also many proposals that I have had to leave out because of limited time and space.

I stuck my neck out in chapter 12 by proposing a number of common principles of adaptive information processing in decentralized systems. I'm not sure Gordon would agree, but I believe those principles might actually be useful for people who study specific complex systems such as the ones I covered—the principles might give them new ideas about how to understand the systems they study. As one example, I proposed that "randomness and probabilities are essential." When I gave a lecture recently and outlined those principles, a neuroscientist in the audience responded by speculating where randomness might come from in the brain and what its uses might be. Some people in the room had never thought about the brain in these terms, and this idea changed their view a bit, and perhaps gave them some new concepts to use in their own research.

On the other hand, feedback must come from the specific to the general. At that same lecture, several people pointed out examples of complex adaptive systems that they believed did not follow all of my principles. This forced me to rethink what I was saying and to question the generality of my assertions. As Gordon so rightly points out, we should be careful to not ignore "anything that doesn't fit a pre-existing model." Of course what are thought to be facts about nature are sometimes found to be wrong as well, and perhaps some common principles will help in directing our skepticism. Albert Einstein, a theorist *par excellence*, supposedly said, "If the facts don't fit the theory, change the facts." Of course this depends on the theory and the facts. The more established the theory or principles, the more skeptical you have to be of any contradicting facts, and conversely the more convincing the contradicting facts are, the more skeptical you need to be of your supposed principles. This is the nature of science—an endless cycle of proud proposing and disdainful doubting.

Roots of Complex Systems Research

The search for common principles governing complex systems has a long history, particularly in physics, but the quest for such principles became most prominent in the years after the invention of computers. As early as the 1940s, some scientists proposed that there are strong analogies between computers and living organisms.

In the 1940s, the Josiah Macy, Jr. Foundation sponsored a series of interdisciplinary scientific meetings with intriguing titles, including "Feedback Mechanisms and Circular Causal Systems in Biological and Social Systems,"

Norbert Wiener, 1894–1964 (AIP Emilio Segre Visual Archives)

"Teleological Mechanisms in Society," and "Teleological Mechanisms and Circular Causal Systems." These meetings were organized by a small group of scientists and mathematicians who were exploring common principles of widely varying complex systems. A prime mover of this group was the mathematician Norbert Wiener, whose work on the control of anti-aircraft guns during World War II had convinced him that the science underlying complex systems in both biology and engineering should focus not on the *mass*, *energy*, and *force* concepts of physics, but rather on the concepts of feedback, control, information, communication, and purpose (or "teleology").

In addition to Norbert Wiener, the series of Macy Foundation conferences included several scientific luminaries of the time, such as John von Neumann, Warren McCulloch, Margaret Mead, Gregory Bateson, Claude Shannon, W. Ross Ashby, among others. The meetings led Wiener to christen a new discipline of *cybernetics*, from the Greek word for "steersman"—that is, one who controls a ship. Wiener summed up cybernetics as "the entire field of control and communication theory, whether in the machine or in the animal."

The discussions and writings of this loose-knit cybernetics group focused on many of the issues that have come up in this book. They asked: What are information and computation? How are they manifested in living organisms? What analogies can be made between living systems and machines? What is the role of feedback in complex behavior? How do *meaning* and *purpose* arise from information processing?

There is no question that much important work on analogies between living systems and machines came out of the cybernetics group. This work includes von Neumann's self-reproducing automaton, which linked notions of information and reproduction; H. Ross Ashby's "Design for a Brain," an influential proposal for how the ideas of dynamics, information, and feedback should inform neuroscience and psychology; Warren McCulloch and Walter Pitts' model of neurons as logic devices, which was the impetus for the later field of neural networks; Margaret Mead and Gregory Bateson's application of cybernetic ideas in psychology and anthropology; and Norbert Wiener's books *Cybernetics* and *The Human Use of Human Beings*, which attempted to provide a unified overview of the field and its relevance in many disciplines. These are only a few examples of works that are still influential today.

In its own time, the research program of cybernetics elicited both enthusiasm and disparagement. Proponents saw it as the beginning of a new era of science. Critics argued that it was too broad, too vague, and too lacking in rigorous theoretical foundations to be useful. The anthropologist Gregory Bateson adopted the first view, writing, "The two most important historical events in my life were the Treaty of Versailles and the discovery of Cybernetics." On the other side, the biophysicist and Nobel prize–winner Max Delbrück characterized the cybernetics meeting he attended as "vacuous in the extreme and positively inane." Less harshly, the decision theorist Leonard Savage described one of the later Macy Foundation meetings as "bull sessions with a very elite group."

In time, the enthusiasm of the cyberneticists for attending meetings faded, along with the prospects of the field itself. William Aspray, a historian of science who has studied the cybernetics movement writes that "in the end Wiener's hope for a unified science of control and communication was not fulfilled. As one participant in these events explained, cybernetics had 'more extent than content.' It ranged over too disparate an array of subjects, and its theoretical apparatus was too meager and cumbersome to achieve the unification Wiener desired."

A similar effort toward finding common principles, under the name of General System Theory, was launched in the 1950s by the biologist Ludwig von Bertalanffy, who characterized the effort as "the formulation and deduction of those principles which are valid for 'systems' in general." A *system* is defined in a very general sense: a collection of interacting elements that together produce, by virtue of their interactions, some form of system-wide behavior. This, of course, can describe just about anything. The general system theorists were particularly interested in general properties of living systems. System theorist Anatol Rapoport characterized the main themes of general system theory (as applied to living systems, social systems, and other complex systems) as *preservation of identity* amid changes, *organized complexity*, and *goal-directedness*. Biologists Humberto Maturana and Francisco Varela attempted to make sense of the first two themes in terms of their notion of *autopoiesis*, or "self-construction"—a self-maintaining process by which a system (e.g., a biological cell) functions as a whole to continually produce the components (e.g., parts of a cell) which themselves make up the system that produces them. To Maturana, Varela, and their many followers, autopoiesis was a key, if not *the* key feature of life.

Like the research program of the cyberneticists, these ideas are very appealing, but attempts to construct a rigorous mathematical framework—one that explains and predicts the important common properties of such systems—were not generally successful. However, the central scientific questions posed by these efforts formed the roots of several modern areas of science and engineering. Artificial intelligence, artificial life, systems ecology, systems biology, neural networks, systems analysis, control theory, and the sciences of complexity have all emerged from seeds sown by the cyberneticists and general system theorists. Cybernetics and general system theory are still active areas of research in some quarters of the scientific community, but have been largely overshadowed by these offspring disciplines.

Several more recent approaches to general theories of complex systems have come from the physics community. For example, Hermann Haken's *Synergetics* and Ilya Prigogine's theories of *dissipative structures* and *nonequilibrium systems* both have attempted to integrate ideas from thermodynamics, dynamical systems theory, and the theory of "critical phenomena" to explain self-organization in physical systems such as turbulent fluids and complex chemical reactions, as well as in biological systems. In particular, Prigogine's goal was to determine a "vocabulary of complexity": in the words of Prigogine and his colleague, Grégoire Nicolis, "a number of concepts that deal with mechanisms that are encountered repeatedly throughout the different phenomena; they are nonequilibrium, stability, bifurcation and symmetry breaking, and long-range order ... they become the basic elements of what we believe to be a new scientific vocabulary." Work continues along these lines, but to date these efforts have not yet produced the coherent and general vocabulary of complexity envisioned by Prigogine, much less a general theory that unifies these disparate concepts in a way that explains complexity in nature.

Five Questions

As you can glean from the wide variety of topics I have covered in this book, what we might call modern complex systems science is, like its forebears, still not a unified whole but rather a collection of disparate parts with some overlapping concepts. What currently unifies different efforts under this rubric are common questions, methods, and the desire to make rigorous mathematical and experimental contributions that go beyond the less rigorous analogies characteristic of these earlier fields. There has been much debate about what, if anything, modern complex systems science is contributing that was lacking in previous efforts. To what extent is it succeeding?

There is a wide spectrum of opinions on this question. Recently, a researcher named Carlos Gershenson sent out a list of questions on complex systems to a set of his colleagues (including myself) and plans to publish the responses in a book called *Complexity: 5 Questions*. The questions are

1. Why did you begin working with complex systems?
2. How would you define complexity?
3. What is your favorite aspect/concept of complexity?
4. In your opinion, what is the most problematic aspect/concept of complexity?
5. How do you see the future of complexity?

I have so far seen fourteen of the responses. Although the views expressed are quite diverse, some common opinions emerge. Most of the respondents dismiss the possibility of "universal laws" of complexity as being too ambitious or too vague. Moreover, most respondents believe that defining complexity is one of the most problematic aspects of the field and is likely to be the wrong goal altogether. Many think the word *complexity* is not meaningful; some even avoid using it. Most do not believe that there is yet a "science of complexity," at least not in the usual sense of the word *science*—complex systems often seems to be a fragmented subject rather than a unified whole.

Finally a few of the respondents worry that the field of complex systems will share the fate of cybernetics and related earlier efforts—that is, it will pinpoint intriguing analogies among different systems without producing a coherent and rigorous mathematical theory that explains and predicts their behavior.

However, in spite of these pessimistic views of the limitations of current complex systems research, most of the respondents are actually highly enthusiastic about the field and the contributions it has and probably will make to science. In the life sciences, brain science, and social sciences, the more carefully scientists look, the more complex the phenomena are. New technologies have enabled these discoveries, and what is being discovered is in dire need of new concepts and theories about how such complexity comes about and operates. Such discoveries will *require* science to change so as to grapple with the questions being asked in complex systems research. Indeed, as we have seen in examples in previous chapters, in recent years the themes and results of complexity science have touched almost every scientific field, and some areas of study, such as biology and social sciences, are being profoundly transformed by these ideas. Going further, several of the survey participants voiced opinions similar to that stated by one respondent: "I see some form of complexity science taking over the whole of scientific thinking."

Apart from important individual discoveries such as Brown, Enquist, and West's work on metabolic scaling or Axelrod's work on the evolution of cooperation (among many other examples), perhaps the most significant contributions of complex systems research to date have been the questioning of many long-held scientific assumptions and the development of novel ways of conceptualizing complex problems. Chaos has shown us that intrinsic randomness is not necessary for a system's behavior to look random; new discoveries in genetics have challenged the role of gene change in evolution; increasing appreciation of the role of chance and self-organization has challenged the centrality of natural selection as an evolutionary force. The importance of thinking in terms of nonlinearity, decentralized control, networks, hierarchies, distributed feedback, statistical representations of information, and essential randomness is gradually being realized in both the scientific community and the general population.

New conceptual frameworks often require the broadening of existing concepts. Throughout this book we have seen how the concepts of information and computation are being extended to encompass living systems and even complex social systems; how the notions of adaptation and evolution have been extended beyond the biological realm; and how the notions of life and intelligence are being expanded, perhaps even to include self-replicating machines and analogy-making computer programs.

This way of thinking is progressively moving into mainstream science. I could see this clearly when I interacted with young graduate students and postdocs at the SFI summer schools. In the early 1990s, the students were extremely excited about the new ideas and novel scientific worldview presented at the school. But by the early 2000s, largely as a result of the educational efforts of SFI and similar institutes, these ideas and worldview had already permeated the culture of many disciplines, and the students were much more blasé, and, in some cases, disappointed that complex systems science seemed so "mainstream." This should be counted as a success, I suppose.

Finally, complex systems research has emphasized above all interdisciplinary collaboration, which is seen as essential for progress on the most important scientific problems of our day.

The Future of Complexity, or Waiting for Carnot

In my view complex systems science is branching off in two separate directions. Along one branch, ideas and tools from complexity research will be refined and applied in an increasingly wide variety of specific areas. In this book we've seen ways in which similar ideas and tools are being used in fields as disparate as physics, biology, epidemiology, sociology, political science, and computer science, among others. Some areas I didn't cover in which these ideas are gaining increasing prominence include neuroscience, economics, ecology, climatology, and medicine—the seeds of complexity and interdisciplinary science are being widely sowed.

The second branch, more controversial, is to view all these fields from a higher level, so as to pursue explanatory and predictive mathematical theories that make commonalities among complex systems more rigorous, and that can describe and predict emergent phenomena.

At one complexity meeting I attended, a heated discussion took place about what direction the field should take. One of the participants, in a moment of frustration, said, " 'Complexity' was once an exciting thing and is now a cliché. We should start over."

What should we call it? It is probably clear by now that this is the crux of the problem—we don't have the right vocabulary to precisely describe what we're studying. We use words such as *complexity, self-organization,* and *emergence* to represent phenomena common to the systems in which we're interested but we can't yet characterize the commonalities in a more rigorous way. We need a new vocabulary that not only captures the conceptual building blocks of self-organization and emergence but that can also describe how these come to encompass what we call *functionality, purpose,* or *meaning* (cf. chapter 12). These ill-defined terms need to be replaced by new, better-defined terms that reflect increased understanding of the phenomena in question. As I have illustrated in this book, much work in complex systems involves the integration of concepts from dynamics, information, computation, and evolution. A new conceptual vocabulary and a new kind of mathematics will have to be forged from this integration. The mathematician Steven Strogatz puts it this way: "I think we may be missing the conceptual equivalent of calculus, a way of seeing the consequences of myriad interactions that define a complex system. It could be that this ultracalculus, if it were handed to us, would be forever beyond human comprehension. We just don't know."

Having the right conceptual vocabulary and the right mathematics is essential for being able to understand, predict, and in some cases, direct or control self-organizing systems with emergent properties. Developing such

Sadi Carnot, 1796–1832 (Boilly lithograph, Photographische Gesellschaft, Berlin, courtesy AIP Emilio Segre Visual Archives, Harvard University Collection.)

concepts and mathematical tools has been, and remains, the greatest challenge facing the sciences of complex systems.

An in-joke in our field is that we're "waiting for Carnot." Sadi Carnot was a physicist of the early nineteenth century who originated some of the key concepts of thermodynamics. Similarly, we are waiting for the right concepts and mathematics to be formulated to describe the many forms of complexity we see in nature.

Accomplishing all of this will require something more like a modern Isaac Newton than a modern Carnot. Before the invention of calculus, Newton faced a conceptual problem similar to what we face today. In his biography of Newton, the science writer James Gleick describes it thus: "He was hampered by the chaos of language—words still vaguely defined and words not quite existing. . . . Newton believed he could marshal a complete science of motion, if only he could find the appropriate lexicon. . . ." By inventing calculus, Newton finally created this lexicon. Calculus provides a mathematical language to rigorously describe change and motion, in terms of such notions as *infinitesimal, derivative, integral*, and *limit*. These concepts already existed in mathematics but in a fragmented way; Newton was able to see how they are related and to construct a coherent edifice that unified them and made them completely general. This edifice is what allowed Newton to create the science of dynamics.

Can we similarly invent the calculus of complexity—a mathematical language that captures the origins and dynamics of self-organization, emergent behavior, and adaptation in complex systems? There are some people who have embarked on this monumental task. For example, as I described in chapter 10, Stephen Wolfram is using the building blocks of dynamics and computation in cellular automata to create what he thinks is a new, fundamental theory of nature. As I noted above, Ilya Prigogine and his followers have attempted to identify the building blocks and build a theory of complexity in terms of a small list of physical concepts. The physicist Per Bak introduced the notion of *self-organized criticality*, based on concepts from dynamical systems theory and phase transitions, which he presented as a general theory of self-organization and emergence. The physicist Jim Crutchfield has proposed a theory of *computational mechanics*, which integrates ideas from dynamical systems, computation theory, and the theory of statistical inference to explain the emergence and structure of complex and adaptive behavior.

While each of these approaches, along with several others I don't describe here, is still far from being a comprehensive explanatory theory for complex systems, each contains important new ideas and all are still areas of active research. Of course it's still unclear if there even exists such a theory; it may be that complexity arises and operates by very different processes in different systems. In this book I've presented some of the likely pieces of a complex systems theory, if one exists, in the domains of information, computation, dynamics, and evolution. What's needed is the ability to see their deep relationships and how they fit into a coherent whole—what might be referred to as "the simplicity on the other side of complexity."

While much of the science I've described in this book is still in its early stages, to me, the prospect of fulfilling such ambitious goals is part of what makes complex systems a truly exciting area to work in. One thing is clear: pursuing these goals will require, as great science always does, an adventurous intellectual spirit and a willingness to risk failure and reproach by going beyond mainstream science into ill-defined and uncharted territory. In the words of the writer and adventurer André Gide, "One doesn't discover new lands without consenting to lose sight of the shore." Readers, I look forward to the day when we can together tour those new territories of complexity.

NOTES

Preface

ix. "REDUCTIONISM is": Hofstadter, D. R., *Gödel, Escher, Bach: an Eternal Golden Braid*. New York: Basic Books, 1979, p. 312.

ix. "to divide all the difficulties under examination": Descartes, R., *A Discourse on the Method*. Translated by Ian Maclean. Oxford: Oxford University Press, 1637/2006, p. 17.

ix. "it seems probable that most of the grand underlying principles": Quoted in Horgan, J., *The End of Science: Facing the Limits of Knowledge in the Twilight of the Scientific Age*. Reading, MA: Addison-Wesley, 1996, p. 19.

x. "emerging syntheses in science": The proceedings of this meeting were published as a book: Pines, D., *Emerging Syntheses in Science*. Reading, MA: Addison-Wesley, 1988.

x. "pursue research on a large number of highly complex and interactive systems"; "promote a unity of knowledge": G. Cowan, Plans for the future. In Pines, D., *Emerging Syntheses in Science*. Reading, MA: Addison-Wesley, 1988, pp. 235, 237.

xi. "a conference ... on the subject of 'emergent computation' ": The proceedings of this meeting were published as a book: Forrest, S., *Emergent Computation*. Cambridge, MA: MIT Press, 1991.

Part I

1. "Science has explored the microcosmos and the macrocosmos": Pagels, H., *The Dreams of Reason*. New York: Simon & Schuster, 1988, p. 12.

Chapter 1

3. "Ideas thus made up": Locke, J., *An Essay Concerning Human Understanding*. Edited by P. H. Nidditch. Oxford: Clarendon Press, 1690/1975, p. 2.12.1.

3. "Half a million army ants": This description of army ant behavior was gleaned from the following sources: Franks, N. R., Army ants: A collective intelligence. *American Scientist*, 77(2), 1989, pp. 138–145; and Hölldobler, B. and Wilson, E. O., *The Ants*. Cambridge, MA: Belknap Press, 1990.

3. "The solitary army ant": Franks, N. R., Army ants: A collective intelligence. *American Scientist*, 77(2), 1989, pp. 138–145.

3. "what some have called a 'superorganism' ": E.g., Hölldobler, B. and Wilson, E. O., *The Ants*. Cambridge, MA: Belknap Press, 1990, p. 107.

4. "I have studied *E. burchelli*": Franks, N. R., Army ants: A collective intelligence. *American Scientist*, 77(2), 1989, p. 140.

5. "Douglas Hofstadter, in his book *Gödel, Escher, Bach*": Hofstadter, D. R., Ant fugue. In *Gödel, Escher, Bach: an Eternal Golden Braid*. New York: Basic Books, 1979.

Chapter 2

15. "It makes me so happy": Stoppard, T., *Arcadia*. New York: Faber & Faber, 1993, pp. 47–48.

19. "nature is exceedingly simple": Quoted in Westfall, R. S., *Never at Rest: A Biography of Isaac Newton*. Cambridge: Cambridge University Press, 1983, p. 389.

19. "it was possible, in principle, to predict everything for all time": Laplace, P. S., *Essai Philosophique Sur Les Probabilites*. Paris: Courcier, 1814.

20. "influences whose physical magnitude": Liu, Huajie, A brief history of the concept of chaos, 1999 [http://members.tripod.com/~huajie/Paper/chaos.htm].

21. "If we knew exactly the laws of nature": Poincaré, H., *Science and Method*. Translated by Francis Maitland. London: Nelson and Sons, 1914.

22. "Edward Lorenz found": Lorenz, E. N., Deterministic nonperiodic flow. *Journal of Atmospheric Science*, 357, 1963, pp. 130–141.

24. "This is a linear system": One could argue that this is not actually a linear system, since the population increases exponentially over time: $n_t = 2^t n_0$. However, it is the map from n_t to n_{t+1} that is the linear system being discussed here.

25. "an equation called the logistic model": From [http://mathworld.wolfram.com/LogisticEquation.html]: "The logistic equation (sometimes called the Verhulst model, logistic map, or logistic growth curve) is a model of population growth first published by Pierre Verhulst (1845). The model is continuous in time, but a modification of the continuous equation to a discrete quadratic recurrence equation is also known as the logistic equation." The logistic *map* is the name given to one particularly useful way of expressing the logistic model.

25. "I won't give the actual equation ": Here is the logistic model:

$$n_{t+1} = (birthrate - deathrate)[kn_t - n_t^2]/k,$$

where n_t is the population at the current generation and k is the carrying capacity. To derive the logistic map from this model, let $x_t = n_t/k$, and $R = (birthrate - deathrate)$. Note that x_t is the "fraction of carrying capacity": the ratio of the current population to the maximum possible population. Then

$$x_{t+1} = Rx_t(1 - x_t).$$

Because the population size n_t is always between 0 and k, x_t is always between 0 and 1.

27. "the *logistic map*": The following references provide technical discussions of the logistic map, aimed at the general scientifically educated reader: Feigenbaum, M. J., Universal behavior in nonlinear systems. *Los Alamos Science*, 1 (1), 1980, pp. 4–27; Hofstadter, D. R., Mathematical chaos and strange attractors. In *Metamagical Themas*. New York: Basic Books, 1985; Kadanoff, Leo P., Chaos, A view of complexity in the physical sciences. In *From Order to Chaos: Essays: Critical, Chaotic, and Otherwise*. Singapore: World Scientific, 1993.

28. "a 1971 article by the mathematical biologist Robert May": May, R. M., Simple mathematical models with very complicated dynamics. *Nature*, 261, pp. 459–467, 1976.

28. "Stanislaw Ulam, John von Neumann, Nicholas Metropolis, Paul Stein, and Myron Stein": Ulam, S. M., and von Neumann, J., *Bulletin of the American Mathematical Society*, 53, 1947, p. 1120. Metropolis, N., Stein, M. L., & Stein, P. R., On finite limit

sets for transformations on the unit interval. *Journal of Combinatorial Theory*, 15(A), 1973, pp. 25–44.

31. "The values of x... become chaotic": How do we know that the system will not settle down to a regular oscillation after a large number of iterations? This can be shown mathematically; e.g., see Strogtaz, S., *Nonlinear Dynamics and Chaos*. Reading, MA: Addison-Wesley, 1994, pp. 368–369.

33. "a basis for constructing pseudo-random numbers": A pseudo-random number generator is a deterministic function or algorithm that outputs values whose distribution and lack of correlation satisfies certain statistical tests of randomness. Such algorithms are used in modern computing for many different applications. Using the logistic map as a basis for pseudo-random number generators was first suggested in Ulam, S. M. and von Neumann, J., On combination of stochastic and deterministic processes (abstract). *Bulletin of the American Mathematical Society*, 53, 1947, p. 1120. This has been further investigated by many others, for example, Wagner, N. R., The logistic equation in random number generation. In *Proceedings of the Thirtieth Annual Allerton Conference on Communications, Control, and Computing*, University of Illinois at Urbana-Champaign, 1993, pp. 922–931.

33. "The fact that the simple and deterministic equation": May, R. M., Simple mathematical models with very complicated dynamics. *Nature*, 261, 1976, pp. 459–467.

34. "The term *chaos* ... T. Y. Li and James York": Li, T. Y. and Yorke, J. A., Period three implies chaos. *American Mathematical Monthly* 82, 1975, p. 985.

35. "The period-doubling route to chaos has a rich history": For an interesting history of the study of chaos, see Aubin, D. and Dalmedico, A. D., Writing the history of dynamical systems and chaos: *Longue Durée* and revolution, disciplines, and cultures. *Historia Mathematica* 29, 2002, pp. 273–339.

36. "Feigenbaum adapted it for dynamical systems theory": For an accessible explanation of Feigenbaum's theory, see Hofstadter, D. R., Mathematical chaos and strange attractors. In *Metamagical Themas*. New York: Basic Books, 1985.

37. "the best thing that can happen to a scientist": Quoted in Gleick, J., *Chaos: Making a New Science*. New York: Viking, 1987, p. 189.

37. "certain computer models of weather": see, e.g., Selvam, A. M.. The dynamics of deterministic chaos in numerical weather prediction models. *Proceedings of the American Meteorological Society, 8th Conference on Numerical Weather Prediction*, Baltimore, MD, 1988; and Lee, B. and Ajjarapu, V., Period-doubling route to chaos in an electrical power system. *IEE Proceedings, Part C*, 140, 1993, pp. 490–496.

38. "Pierre Coullet and Charles Tresser, who also used the technique of renormalization": Coullet, P., and Tresser, C., Itérations d'endomorphismes et groupe de renormalization. *Comptes Rendues de Académie des Sciences, Paris A*, 287, 1978, pp. 577–580.

Chapter 3

40. "The law that entropy increases": Eddington, A. E., *The Nature of the Physical World*. Macmillan, New York, 1928, p. 74.

40. "complex systems sense, store, and deploy more information": Cohen, I., Informational landscapes in art, science, and evolution. *Bulletin of Mathematical Biology*, 68, 2006, p. 1218.

40. "evolution can perform its tricks": Beinhocker, E. D., *The Origin of Wealth: Evolution, Complexity, and the Radical Remaking of Economics*. Cambridge, MA: Harvard Business School Press, 2006, p. 12.

41. "Although they differ widely in their physical attributes": Gell-Mann, M., *The Quark and the Jaguar*. New York: W. H. Freeman, 1994, p. 21.

43. "Why the second law should": Rothman, T., The evolution of entropy. In *Science à la Mode*. Princeton, NJ: Princeton University Press, 1989, p. 82.

45. "the hot system": Maxwell, quoted in Leff, H. S. and Rex, A. F., *Maxwell's Demon: Entropy, Information, Computing*. Princeton University Press. Second edition 2003, Institute of Physics Pub., 1990, p. 5.

45. "In a famous paper": Szilard, L., On the decrease of entropy in a thermodynamic system by the intervention of intelligent beings. *Zeitschrift fuer Physik*, 53, 1929, pp. 840–856.

46. "the mathematician Charles Bennett showed": Bennett's arguments are subtle; the details can be found in Bennett, C. H., The thermodynamics of computation—a review. *International Journal of Theoretical Physics*, 21, 1982, pp. 905–940. Many of these ideas were independently discovered by the physicist Oliver Penrose (Leff, H. S. and Rex, A. F., *Maxwell's Demon: Entropy, Information, Computing*, Taylor & Francis, 1990; second edition Institute of Physics Pub., 2003).

47. "the demon remains controversial to this day": E.g., see Maddox, J., Slamming the door. *Nature*, 417, 2007, p. 903.

49. "repellent to many other scientists": Evidently Boltzmann was himself a feisty critic of the work of others. As William Everdell reports, Boltzmann wrote a paper entitled "On a thesis of Schopenhauer," but later wrote that he had wanted to call it "Proof that Schopenhauer Was a Degenerate, Unthinking, Unknowing, Nonsense-Scribbling Philosopher, Whose Understanding Consisted Solely of Empty Verbal Trash." Everdell, W. R., *The First Moderns: Profiles in the Origins of Twentieth-Century Thought*. Chicago, IL: University of Chicago Press, 1998, p. 370.

50. "Boltzmann defined the *entropy* of a macrostate": This version of Boltzmann's entropy assumes all microstates that correspond to a given macrostate are equally probable. Boltzmann also gave a more general formula that defines entropy for non-equiprobable microstates.

51. "The actual equation": In the equation for Boltzmann's entropy, $S = k \log W$, S is entropy, W is the number of possible microstates corresponding to a given macrostate, and k is "Boltzmann's constant," a number used to put entropy into standard units.

52. "In his 1948 paper 'A Mathematical Theory of Communication'": Shannon, C., A mathematical theory of communication. *The Bell System Technical Journal*, 27, 1948, pp. 379–423, 623–656.

55. "efforts to marry communication theory": Pierce, J. R., *An Introduction to Information Theory: Symbols, Signals, and Noise*. New York: Dover, 1980, p. 24. (First edition, 1961.)

Chapter 4

56. "Quo facto": Leibniz, G. (1890). In C. Gerhardt (Ed.), *Die Philosophischen Schriften von Gottfried Wilheml Liebniz*, Volume VII. Berlin: Olms. Translation from Russell, B., *A History of Western Philosophy*, Touchstone, 1967, p. 592. (First edition, 1901.)

56. "computation in cells and tissues": E.g., Paton, R., Bolouri, H., Holcombe, M., Parish, J. H., and Tateson. R., editors. *Computation in Cells and Tissues: Perspectives and Tools of Thought*, Berlin: Springer-Verlag, 2004.

56. "immune system computation": Cohen, I. R., Immune system computation and the immunological homunculus. In O. Nierstrasz et al. (Editors), *MoDELS 2006, Lecture Notes in Computer Science* 4199. Springer-Verlag, 2006, pp. 499–512.

56. "the nature and limits of distributed computation in markets": lecture by David Pennock entitled "Information and complexity in securities markets," Institute for Computational and Mathematical Engineering, Stanford University, November 2005.

56. "emergent computation in plants": Peak, D., West, J. D., Messinger, S. M., and Mott, K. A., Evidence for complex, collective dynamics and emergent, distributed computation in plants. *Proceedings of the National Academy of Sciences, USA*, 101(4), 2004, pp. 918–922.

59. "there is no such thing as an unsolvable problem": Quoted in Hodges, A., *Alan Turing: The Enigma*, New York: Simon & Schuster, 1983, p. 92.

60. "Gödel's proof is complicated": For excellent expositions of the proof, see Nagel, E. and Newman, J. R., *Gödel's Proof*. New York: New York University, 1958; and Hofstadter, D. R., *Gödel, Escher, Bach: an Eternal Golden Braid*. New York: Basic Books, 1979.

60. "This was an amazing new turn": Hodges, A., *Alan Turing: The Enigma*. New York: Simon & Schuster, 1983, p. 92.

60. "Turing killed off the third": Another mathematician, Alonzo Church, also proved that there are undecidable statements in mathematics, but Turing's results ended up being more influential.

60. "his answer, again, was 'no' ": Turing, A. M., On computable numbers, with an application to the *Entscheidungsproblem*. *Proceedings of the London Mathematical Society*, 2(42), 1936, pp. 230–265.

69. "According to his biographer Hao Wang": Wang, H., *Reflections on Kurt Gödel*. Cambridge, MA: MIT Press, 1987.

Chapter 5

71. "All great truths begin as blasphemies": Shaw, G. B., *Annajanska, the Bolshevik Empress*. London: Kessinger Publishing, 1919/2004, p. 15.

71. "Subject to decay are all componded things": Quoted in Bowker, J. (editor), *The Cambridge Illustrated History of Religions*. Cambridge, UK: Cambridge University Press, 2002, p. 76.

71. "The earth shall wax old like a garment": Isaiah 51:6. *The Holy Bible, King James Version*.

71. "O! How shall summer's honey breath hold out": From Shakespeare, Sonnet 65.

72. "If I were to give an award for the single best idea anyone has ever had": Dennett, D. R., *Darwin's Dangerous Idea*. New York: Simon & Schuster, 1995, p. 21.

73. "Organic life beneath the shoreless waves": Darwin, E., *The Temple of Nature; or, The Origin of Society: A Poem with Philosophical Notes*. London: J. Johnson, 1803.

74. "Lamarck ... had few clear facts": Quoted in Grinnell, G. J., The rise and fall of Darwin's second theory. *Journal of the History of Biology*, 18 (1), 1985, p. 53.

74. "if [the] instinctual life of animals permits of any explanation at all": Freud, S., *Moses and Monotheism*. New York: Vintage Books, 1939, pp. 128–129. Quoted in Cochrane, E., Viva Lamarck: A brief history of the inheritance of acquired characteristics. *Aeon* 2:2, 1997, pp. 5–39.

75. "You care for nothing but shooting, dogs, and rat-catching": Darwin, C. and Barlow, N. D., *The Autobiography of Charles Darwin*. Reissue edition, New York: W. W. Norton, 1958/1993, p. 28.

76. "Darwin also read Adam Smith's free-market manifesto": Darwin writes in a letter to W. D. Fox (January 1829), "My studies consist of Adam Smith & Locke." *The Correspondence of Charles Darwin*, Volume 1 (F. Burkhardt and S. Smith, editors). Cambridge, U.K.: Cambridge University Press, 1985, p. 72.

76. "the type of beak was adapted": For further reading on this, see Weiner, J., *The Beak of the Finch: A Story of Evolution in Our Time*. New York: Knopf, 1994.

77. "I am almost convinced": Quoted in Bowlby, J., *Charles Darwin: A New Life*. New York W. W. Norton, 1992, p. 254.

77. "Plato ... says": Barrett, P. (editor), *Charles Darwin's Notebooks, 1836-1844: Geology, Transmutation of Species, Metaphysical Enquiries*. Ithaca, NY: Cornell University Press, 1987, p. 551.

77. "[A]ll my originality": Darwin's letter to Charles Lyell, June 18, 1858. Listed in *The Darwin Correspondence Online Database*, [http://www.darwinproject.ac.uk], Letter 2285.

77. "base and paltry": Darwin's letter to Charles Lyell, June 25, 1858. Ibid, Letter 2294.

78. "I freely acknowledge that Mr. Matthew": Quoted in Darwin, C., *The Autobiography of Charles Darwin*. Lanham, MD: Barnes & Noble Publishing, Endnote 21, 2005, p. 382. Originally published 1887.

78. "How extremely stupid not to have thought of that!": Quoted in Provine, W. B., *The Origins of Theoretical Population Genetics*. Chicago: University of Chicago Press, 1971, p. 4.

82. "This unification of Darwinism and Mendelism": The term "Modern Synthesis" came from Julian Huxley's influential book *Evolution: The Modern Synthesis*, New York, London: Harper, 1942. The Modern Synthesis has also been called the *Neo-Darwinian Synthesis*, the *Modern Evolutionary Synthesis*, and, for those particularly in the know, simply the *Synthesis*.

83. "Nobody could ever again look at the evolutionary process without very consciously standing on the edifice of the Synthesis": Tattersall, I., *Becoming Human: Evolution and Human Uniqueness*. New York: Harvest Books, 1999, p. 83.

86. "Motoo Kimura proposed a theory of 'neutral evolution'": For a discussion of Kimura's theory see Dietrich, M. R., The origins of the neutral theory of molecular evolution. *Journal of the History of Biology*, 27 (1), 1994, pp. 21–59.

86. "Manfred Eigen and Peter Schuster observed": For a good exposition of Eigen and Schuster's work, see Eigen, M., *Steps Towards Life*. Oxford: Oxford University Press, 1992.

86–87. "the synthetic theory ... is effectively dead": Gould, S. J., Is a new and general theory of evolution emerging? *Paleobiology*, 6, 1980, p. 120.

87. "The view of evolution due to the Modern Synthesis 'is one of the greatest myths'": Eldredge, N. and Tattersall, I., *The Myths of Human Evolution*. New York: Columbia University Press, 1982, p. 43.

87. "I am of the opinion that nothing is seriously wrong with the achievements": Mayr, E., An overview of current evolutionary biology. In Warren, L. and Koprowski, H. (editors), *New Perspectives on Evolution*. New York: Wiley-Liss, 1991, p. 12.

87. "The theory of evolution by cumulative natural selection": Dawkins, R., *The Extended Phenotype* (Reprint edition). Oxford University Press, 1989, p. 317. Originally published 1982.

Chapter 6

92. "the cell's flash cards": Hofstadter, D. R. The Genetic Code: Arbitrary? In *Metamagical Themas*. New York: Basic Books, 1985, p. 681.

Chapter 7

95. "what the term *gene* refers to": See Pearson, H., "What is a gene?" *Nature*, vol. 441, 2006, pp. 399–401.

95. "The physicist Seth Lloyd published a paper in 2001": Lloyd, S., Measures of complexity: A non-exhaustive list. *IEEE Control Systems Magazine*, August 2001.

98. "This is called the *algorithmic information content*": A detailed reference to Kolmogorov, Chaitin, and Solmonoff's ideas is Li, M. and Vitanyi, P., *An Introduction to Kolmogorov Complexity and Its Applications*, 2nd Edition. New York: Springer-Verlag, 1997.

98. "Murray Gell-Mann proposed a related measure": Gell-Mann, M. What is complexity? *Complexity*, 1 (1), 1995, pp. 16–19.

100. "the subjectivity of its definition remains a problem": See, e.g., McAllister, J. W., Effective complexity as a measure of information content. *Philosophy of Science* 70, 2003, pp. 302–307.

100. "Logically deep objects": Bennett, C. H., How to define complexity in physics, and why. In W. H. Zurek (editor), *Complexity, Entropy, and the Physics of Information*, Reading, MA: Addison-Wesley, 1990, p. 142.

100. "It is an appealing idea": Lloyd, S., The calculus of intricacy. *The Sciences*, 30, 1990, p. 42.

101. "Seth Lloyd and Heinz Pagels proposed": Lloyd, S. and Pagels, H., Complexity as thermodynamic depth. *Annals of Physics*, 188, 1988, pp. 186–213.

101. "the most plausible scientifically determined" and "the total amount of thermodynamic and informational resources": Lloyd, S., The calculus of intricacy. *The Sciences*, 30, 1990, p. 42.

101. "As pointed out by some critics": Crutchfield, J. P. and Shalizi, C. R., Thermodynamic depth of causal states: When paddling around in Occam's pool shallowness is a virtue. *Physical Review E*, 59 (1), 1999, pp. 275–283.

102. "Stephen Wolfram, for example, has proposed": Wolfram, S., Universality and complexity in cellular automata. *Physica D*, 10, 1984, pp. 1–35.

102. "However, as Charles Bennett and others have argued": e.g., see Bennett, C. H., Dissipation, information, computational complexity and the definition of organization. In D. Pines (editors), *Emerging Syntheses in Science*. Redwood City, CA: Addison-Wesley, 1985, pp. 215–233.

102. "*statistical complexity*": Crutchfield, J. P. and Young, K., Inferring statistical complexity, *Physics Review Letters* 63, 1989, pp. 105–108.

103. "the atomic structure of complicated crystals": Varn, D. P., Canright, G. S., and Crutchfield, J. P., Discovering planar disorder in close-packed structures from X-ray diffraction: Beyond the fault model. *Physical Review B*, 66, 2002, pp. 174110-1–174110-4.

103. "the firing patterns of neurons": Haslinger, R., Klinkner, K. L., and Shalizi, C. R., The computational structure of spike trains. Unpublished manuscript, 2007.

103. "the universe is fractal-like": Mandelbrot, B. B., *The Fractal Geometry of Nature*. New York: W. H. Freeman, 1977.

103. "in general a fractal is a geometric shape": Strogatz, S., *Nonlinear Dynamics and Chaos*. Reading, MA: Addison-Wesley, 1994.

107. "fractal dimension": A great introduction to fractals and the concept of fractal dimension is Mandelbrot's book *The Fractal Geometry of Nature*. New York: W. H. Freeman, 1977.

108. "I'll do a calculation out of your sight": For the Koch curve, $3^{dimension} = 4$. To solve for dimension, take the logarithm (using any base) of both sides:

$$\log(3^{dimension}) = dimension \times \log(3) = \log(4). \text{ Thus } dimension$$
$$= \log(4)/\log(3) \approx 1.26.$$

108. "the cascade of detail": Bovill, C., *Fractal Geometry in Architecture and Design*. Birkhäuser Boston, 1996, p. 4.

109. "The Architecture of Complexity": Simon, H. A., The architecture of complexity. *Proceedings of the American Philosophical Society*, 106 (96), 1962, pp. 467–482.

109. "the complex system being composed of subsystems": Ibid, p. 468.

111. "Daniel McShea ... has proposed a hierarchy scale": McShea, D. W., The hierarchical structure of organisms: A scale and documentation of a trend in the maximum. *Paleobiology*, 27 (2), 2001, pp. 405–423.

Part II

113. "Nature proceeds little by little": Quoted in Grene, M. and Depew, D., *The Philosophy of Biology: An Episodic History*. Cambridge University Press, 2004, p. 14.

113. "[W]e all know intuitively what life is": Lovelock, J. E, *The Ages of Gaia*. New York: W. W. Norton, 1988, p. 16.

Chapter 8

115. "Self-Reproducing Computer Programs": Parts of this chapter were adapted from Mitchell, M., Life and evolution in computers. *History and Philosophy of the Life Sciences*, 23, 2001, pp. 361–383.

115. "... there has been some fascinating research": See, for example, Luisi, P. L., *The Emergence of Life: From Chemical Origins to Synthetic Biology*. Cambridge: Cambridge University Press, 2006; or Fry, I., *The Emergence of Life on Earth: A Historical and Scientific Overview*. Piscataway, NJ: Rutgers University Press, 2000.

116. "the field of *artificial life*": For more information about the field of artificial life, see, for example, Langton, C. G., *Artificial Life: An Overview*. Cambridge, MA: MIT Press, 1997; or Adami, C., *Introduction to Artificial Life*. New York: Springer, 1998.

123. "The complete work was eventually published": Von Neumann, J., *Theory of Self-Reproducing Automata* (edited and completed by A. W. Burks). Urbana: University of Illinois Press, 1966. For descriptions of von Neumann's self-replicating automaton, see Burks, A. W., Von Neumann's self-reproducing automata. In A. W. Burks (editor), *Essays on Cellular Automata*. Urbana: University of Illinois Press, 1970; or Mitchell, M., Computation in cellular automata: A selected review. In T. Gramss et al. (editors), *Nonstandard Computation*, 1998, pp. 95–140. Weinheim, Germany: Wiley-VCH. For an account of self-replication in DNA and how it relates to mathematical logic and self-copying computer programs, see Hofstadter, D. R., *Gödel, Escher, Bach: an Eternal Golden Braid*. New York: Basic Books, 1979, pp. 495–548.

123. "reproductive potentialities of the machines of the future' ": Quoted in Heims, S. J., *John Von Neumann and Norbert Wiener: From Mathematics to the Technologies of Life and Death*. Cambridge, MA: MIT Press, 1980, pp. 212–213.

123. "their respective nonfiction": Kurzweil, R., *The Age of Spiritual Machines: When Computers Exceed Human Intelligence*. New York: Viking, 1999; and Moravec, H., *Robot: Mere Machine to Transcendent Mind*. New York: Oxford University Press, 1999.

124. "now famous article in *Wired*": Joy, B., Why the future doesn't need us. *Wired*, April 2000.

124. "some simple self-reproducing robots": Zykov, V. Mytilinaios, E., Adams, B., and Lipson, H., Self-reproducing machines. *Nature*, 435, 2005, pp. 163–164.

124. "When his mother once stared rather aimlessly": Macrae, N., *John von Neumann*. New York: Pantheon, 1992, p. 52.

124. "the greatest paper on mathematical economics": Quoted in Macrae, N., *John von Neumann*. New York: Pantheon, 1992, p. 23.

125. "the most important document ever written on computing and computers": Goldstine, H. H., *The Computer, from Pascal to von Neumann*. Princeton, NJ: Princeton University Press, first edition, 1972, p. 191.

125. "Five of Hungary's six Nobel Prize winners": Macrae, N., *John von Neumann*. New York: Pantheon, 1992, p. 32.

126. "The [IAS] School of Mathematics": Quoted in Macrae, N., *John von Neumann*. New York: Pantheon, 1992, p. 324.

126. "to have no experimental science": Quoted in Regis, E., *Who Got Einstein's Office? Eccentricity and Genius at the Institute for Advanced Study*. Menlo Park, CA: Addison-Wesley, 1987, p. 114.

126. "The snobs took revenge": Regis, E., *Who Got Einstein's Office? Eccentricity and Genius at the Institute for Advanced Study*. Menlo Park, CA: Addison-Wesley, 1987, p. 114.

Chapter 9

127. "*evolutionary computation*": For a history of early work on evolutionary computation, see Fogel, D. B., *Evolutionary Computation: The Fossil Record*. New York: Wiley-IEEE Press, 1998.

128. "That's where genetic algorithms came from": John Holland, quoted in Williams, S. Unnatural selection. *Technology Review*, February 2005.

130. "automating parts of aircraft design," Hammond, W. E. *Design Methodologies for Space Transportation Systems*, 2001, Reston, VA: American Institute of Aeronautics and Astronautics, Inc., p. 548.

130. "analyzing satellite images": See, e.g., Harvey, N. R., Theiler, J., Brumby, S. P., Perkins, S. Szymanski, J. J., Bloch, J. J., Porter, R. B., Galassi, M., and Young, A. C. Comparison of GENIE and conventional supervised classifiers for mulitspectral image feature extraction. *IEEE Transactions on Geoscience and Remote Sensing*, 40, 2002, pp. 393–404.

130. "automating assembly line scheduling." Begley, S. Software au naturel. *Newsweek*, May 8, 1995.

130. "computer chip design": Ibid.

130. "realistic computer-animated horses": See Morton, O., Attack of the stuntbots. *Wired*, 12.01, 2004.

130. "realistic computer-animated stunt doubles": "Virtual Stuntmen Debut in Hollywood Epic *Troy*," news release, NaturalMotion Ltd. [http://www.naturalmotion.com/files/nm_troy.pdf].

130. "discovery of new drugs": See, e.g., Felton, M. J., Survival of the fittest in drug design. *Modern Drug Discovery*, 3(9), 2000, pp. 49–50.

130. "detecting fraudulent trades": Bolton, R. J. and Hand, D. J., Statistical fraud detection: A review. *Statistical Science*, 17(3), 2002, pp. 235–255.

130. "analysis of credit card data": Holtham, C., Fear and opportunity. *Information Age*, July 11, 2007.

130. "forecasting financial markets": See, e.g., Williams, F., Artificial intelligence has a small but loyal following. *Pensions and Investments*, May 14, 2001.

130. "portfolio optimization": Coale, K., Darwin in a box. *Wired*, June 14, 1997.

130. "artwork created by an interactive genetic algorithm": see [http://www.karlsims.com].

130. "I will take you through a simple extended example": This example is inspired by a project at the MIT Artificial Intelligence Lab, in which a robot named "Herbert" wandered around the halls and offices collecting empty soda cans and taking them to the recycling bin. See Connell, J. H., *Minimalist Mobile Robotics: A Colony-Style Architecture for an Artificial Creature*. San Diego: Academic Press, 1990.

132. "This means that there are 243 different possible situations": There are five different sites each with three possible types of contents, thus there are $3 \times 3 \times 3 \times 3 \times 3 = 243$ different possible situations.

142. "Evolutionary algorithms are a great tool": Jason Lohn, quoted in Williams, S., Unnatural selection. *Technology Review*, February 2005.

Part III

143. "The proper domain of computer science": Quoted in Lewin, R., *Complexity: Life at the Edge of Chaos*. New York: Macmillan, 1992, p. 48.

Chapter 10

145. "a recent article in *Science* magazine": Shouse, B., Getting the behavior of social insects to compute. *Science*, 295(5564), 2002, 2357.

145. "'Is the brain a computer?'": Churchland, P. S., Koch, C., and Sejnowski, T. J., What is computational neuroscience? In E. L. Schwartz (editor), *Computational Neuroscience*. Cambridge, MA: MIT Press, 1994, pp. 46–55.

148. "The answer is ... 2^{512}": As will be described later in the chapter, to define a rule, you must specify the update state for the center lightbulb given all possible configurations of its local neighborhood. Since a local neighborhood consists of eight neighbors plus the center bulb itself, and each bulb can be either on or off, the number of possible configurations of a local neighborhood is $2^9 = 512$. For each configuration, one can assign either "on" or "off" as the update state, so the number of possible assignments to all 512 configurations is $2^{512} \approx 1.3 \times 10^{154}$.

149. "The Game of Life": Much of what is described here can be found in the following sources: Berlekamp, E., Conway, J. H., and Guy, R., *Winning Ways for Your Mathematical Plays*, Volume 2. San Diego: Academic Press, 1982; Poundstone, W., *The Recursive Universe*. William Morrow, 1984; and many of the thousands of Web sites devoted to the Game of Life.

150. "John Conway also sketched a proof": Berlekamp, E., Conway, J. H., and Guy, R., *Winning Ways for Your Mathematical Plays*, volume 2. San Diego: Academic Press, 1982.

150. "later refined by others": e.g., see Rendell, P., Turing universality of the game of Life. In A. Adamatzky (editor), *Collision-Based Computing*, pp. 513–539. London: Springer-Verlag, 2001.

154. "a review on how to convert base 2 numbers to decimal": Recall that for decimal (base 10) number, say, 235, each "place" in the number corresponds to a power of 10: $235 = 2 \times 10^2 + 3 \times 10^1 + 5 \times 10^0$ (where $10^0 = 1$). In base 2, each place corresponds to a power of 2. For example, 235 in base 2 is 11101011:

$$11101011 = 1 \times 2^7 + 1 \times 2^6 + 1 \times 2^5 + 0 \times 2^4 + 1 \times 2^3 + 0 \times 2^2 + 1 \times 2^1 + 1 \times 2^0 = 235.$$

155. "The Rule 30 automaton is the most surprising thing I've ever seen in science": Quoted in Malone, M. S., God, Stephen Wolfram, and everything else. *Forbes ASAP*, November 27, 2000. {http://members.forbes.com/asap/2000/1127/162.html}

155. "In fact, Wolfram was so impressed by rule 30": "Random Sequence Generators" U.S. Patent 4691291, September 1, 1987.

156. "class 4 involves a mixture": Wolfram, S., *A New Kind of Science*. Champaign, IL, Wolfram Media, 2002, p. 235.

156. "Matthew Cook ... finally proved that rule 110 was indeed universal": Cook, M., Universality in elementary cellular automata. *Complex Systems* 15(1), 2004, 1–40.

156. "*A New Kind of Science*": Wolfram, S., *A New Kind of Science*. Champaign; IL: Wolfram Media, 2002, p. 235.

158. "you would be able to build such a computer to solve the halting problem": See Moore, C., Recursion theory on the reals and continuous-time computation. *Theoretical Computer Science*, 162, 1996, pp. 23–44.

158. "definite ultimate model for the universe": Wolfram, S., *A New Kind of Science*. Champaign, IL: Wolfram Media, 2002, p. 466.

158. "I'm guessing it's really short": Stephen Wolfram, quoted in Levy, S., The man who cracked the code to everything .... *Wired*, Issue 10.06, June 2002.

158–159. "Konrad Zuse and Edward Fredkin had both theorized": See Zuse, K., *Rechnender Raum* Braunschweig: Friedrich Vieweg & Sohn, 1969 (English translation: *Calculating Space*. MIT Technical Translation AZT-70-164-GEMIT, Massachusetts Institute of Technology (Project MAC), Cambridge, MA, 02139, February 1970); and Wright, R., Did the universe just happen? *Atlantic Monthly*, April 1988, pp. 29–44.

Chapter 11

160. "Computing with Particles": A detailed description of our work on cellular automata and particles can be found in Crutchfield, J. P., Mitchell, M., and Das, R., Evolutionary design of collective computation in cellular automata. In J. P. Crutchfield and P. K. Schuster (editors), *Evolutionary Dynamics—Exploring the Interplay of Selection, Neutrality, Accident, and Function*. New York: Oxford University Press, 2003, pp. 361–411.

160. "an article by the physicist Norman Packard": Packard, N. H., Adaptation toward the edge of chaos. In J. A. S. Kelso, A. J. Mandell, M. F. Shlesinger, eds., *Dynamic Patterns in Complex Systems*. Singapore: World Scientific, 1988, pp. 293–301.

160. "majority classification": The majority classification task is also known in the cellular automata literature as "density classification."

166. "Jim Crutchfield had earlier invented": See, e.g., Crutchfield, J. P., and Hanson, J. E., Turbulent pattern bases for cellular automata. *Physica D* 69, 1993, pp. 279–301.

168. "Twenty Problems in the Theory of Cellular Automata": Wolfram, S., Twenty problems in the theory of cellular automata. *Physica Scripta*, T9, 1985, pp. 170–183.

168. "Botanist Keith Mott, physicist David Peak": See Peak, D., West, J. D., Messinger, S. M., and Mott, K. A., Evidence for complex, collective dynamics and emergent, distributed computation in plants. Proceedings of the National Academy of Sciences, USA, 101 (4), 2004, pp. 918–922.

Chapter 12

169. "Information Processing in Living Systems": Parts of this chapter were adapted from Mitchell, M., Complex systems: Network thinking. *Artificial Intelligence*, 170 (18), 2006, pp. 1194–1212.

170. "we need to answer the following": My questions are related to the three levels of description of information processing proposed by David Marr, described in his book *Vision*: Marr, D. *Vision*. San Francisco, Freeman, 1982. Questions similar to mine were formulated by Ron McClamrock; see McClamrock, R., Marr's three levels: A re-evaluation. *Minds and Machines*, 1 (2), 1991, pp. 185–196.

172. "The Immune System": Two excellent, readable overviews of the immune system are Sompayrac, L. M., *How the Immune System Works*, 2nd edition, Blackwell Publishing, 1991; and Hofmeyr, S. A., An interpretive introduction to the immune system. In L. A. Segel and I. R. Cohen (editors), *Design Principles for the Immune System and Other Distributed Autonomous Systems*. New York: Oxford University Press, 2001.

176. "A third mechanism has been hypothesized": For more details, see Lesley, R. Xu, Y., Kalled, S. L., Hess, D. M., Schwab, S. R., Shu, H.-B., and Cyster, J. G., Reduced competitiveness of autoantigen-engaged B cells due to increased dependence on BAFF. *Immunity*, 20 (4), 2004, pp. 441–453.

177. "foraging for food works roughly as follows": For more detailed descriptions of ant foraging, see Bonabeau, E., Dorigo, M., and Theraulaz, G., *Swarm Intelligence: From Natural to Artificial Systems*. New York: Oxford University Press, 1999.

177. "The ecologist Deborah Gordon has studied task allocation": See, e.g., Gordon, D. M., Task allocation in ant colonies. In L. A. Segel and I. R. Cohen. (editors), *Design Principles*

for the Immune System and Other Distributed Autonomous Systems. New York: Oxford University Press., 2001.

178. "construction of bridges or shelters": e.g., see Lioni, A., Sauwens, C., Theraulaz, G., and Deneubourg, J.-L., Chain formation in *OEcophylla longinoda*. *Journal of Insect Behavior*, 14 (5), 2001, pp. 679–696.

181. "All three systems described above use randomness": The role of randomness in complex adaptive systems is also explored in Millonas, M. M., The importance of being noisy. *Bulletin of the Santa Fe Institute*, Summer, 1994.

181. "Eventually, the ants will have established a detailed map": Ziff, E. and Rosenfield, I., Evolving evolution. *The New York Review of Books*, 53, 8, May 11, 2006.

182. "'parallel terraced scan'": Hofstadter D., *Fluid Concepts and Creative Analogies*. New York: Basic Books, 1995, p. 92.

183. "Maintaining a correct balance": This is discussed in Holland, J. H., *Adaptation in Natural and Artificial Systems*. Cambridge, MA: MIT Press, 1992 (first edition, 1975); and Hofstadter, D. R. and Mitchell, M., The Copycat project: A model of mental fluidity and analogy-making. In K. Holyoak and J. Barnden (editors), *Advances in Connectionist and Neural Computation Theory, Volume 2: Analogical Connections*, 1994, pp. 31–112.

184. "*who* or *what* actually perceives the meaning": Some of the many books and articles addressing these issues from a philosophical standpoint are the following: Hofstadter, D. R., *Gödel, Escher, Bach: an Eternal Golden Braid*. New York: Basic Books, 1979; Dennett, D. R., *Consciousness Explained*. Boston: Little, Brown, 1991; Bickhard, M. H., The biological foundations of cognitive science. In *Mind 4: Proceedings of the 4th Annual Meeting of the Cognitive Science Society of Ireland*. Dublin, Ireland: J. Benjamins, 1999; Floridi, L., Open problems in the philosophy of information. *Metaphilosophy*, 35 (4), 2004, pp. 554–582; and Hofstadter, D., *I am a Strange Loop*. New York: Basic Books, 2007.

184. "artificial immune systems": See, e.g., Hofmeyr, S. A. and Forrest, S., Architecture for an artificial immune system. *Evolutionary Computation*, 8 (4), 2000, pp. 443–473.

184. "ant colony optimization algorithms": See, e.g., Dorigo, M. and Stützle, T., *Ant Colony Optimization*, MIT Press, 2004.

Chapter 13

186. "How to Make Analogies": Parts of this chapter were adapted from Mitchell, M., *Analogy-Making as Perception*, MIT Press, 1993; and Mitchell, M., Analogy-making as a complex adaptive system. In L. Segel and I. Cohen (editors), *Design Principles for the Immune System and Other Distributed Autonomous Systems*. New York: Oxford University Press, 2001.

187. "About a year ago, the *Sacramento Bee*": Lohr, S., This boring headline is written for Google. *New York Times*, April 9, 2006.

187. "At conferences you are hearing the phrase 'human-level AI' ": Eric Horvitz, quoted in Markoff, J., Brainy robots start stepping into daily life. *New York Times*, July 18, 2006.

187. "Easy things are hard": Minsky, M., *The Society of Mind*. New York: Simon & Schuster, 1987, p. 29.

188. "All perception of truth is the detection of an analogy": Thoreau, H. D. (with L. D. Walls, editor). *Material Faith: Thoreau on Science*. New York: Mariner Books, 1999, p. 28.

189. "a relatively new book written by a Computer Science professor": Hofstadter, D. R., *Gödel, Escher, Bach: an Eternal Golden Braid*. New York: Basic Books, 1979.

208. "an even more impressive array of successor programs": For descriptions of several of these programs, including Copycat, see Hofstadter D., *Fluid Concepts and Creative Analogies*. New York: Basic Books, 1995.

208. "the barrier of meaning": Rota, G.-C., In memoriam of Stan Ulam—The barrier of meaning. *Physica D*, 2 (1–3), 1986, pp. 1–3.

Chapter 14

210. "necessary and sufficient components of all models in physics": Garber, D., Descartes, mechanics, and the mechanical philosophy. *Midwest Studies in Philosophy* 26 (1), 2002, pp. 185–204.

210. "he pictured the Earth like a sponge": Kubrin, D., Newton and the cyclical cosmos: Providence and the mechanical philosophy. *Journal of the History of Ideas*, 28 (3), 1967.

211. "intuition pumps": Dennett, D. R., *Elbow Room: The Varieties of Free Will Worth Wanting.* Cambridge, MA: MIT Press, 1984, p. 12.

213. "Merrill Flood and Melvin Drescher, invented the Prisoner's Dilemma": For an entertaining and enlightening discussion of the Prisoner's Dilemma, and, more generally, game theory and its history and applications, see Poundstone, W., *Prisoner's Dilemma.* New York: Doubleday, 1992.

214. "the pursuit of self-interest for each": Axelrod, R., *The Evolution of Cooperation.* New York: Basic Books, 1984, p. 7.

214. "the tragedy of the commons": Hardin, G., The tragedy of the commons. *Science*, 162, 1968, pp. 1243–1248.

215. "Under what conditions will cooperation emerge": Axelrod, R., *The Evolution of Cooperation.* New York: Basic Books, 1984, p. 3.

215. "Thomas Hobbes, who concluded that cooperation could develop": Hobbes' arguments about centralized governments can be found in Hobbes, T., *Leviathan.* First published in 1651; 1991 edition edited by R. Tuck. Cambridge University Press, 1991.

215. "Albert Einstein similarly proposed": Einstein's thoughts about world government and many other issues can be found in a collection of his writings, Einstein, A., *Out of My Later Years.* First published in 1950; revised edition published in 2005 by Castle Books.

218. "Axelrod experimented with adding *norms*": Axelrod, R., An evolutionary approach to norms. *American Political Science Review*, 80 (4), 1986, pp. 1095–1111.

219. "Meta-norms can promote and sustain cooperation": Axelrod, R., An evolutionary approach to norms. *American Political Science Review*, 80 (4), 1986, pp. 1095–1111.

219. "Nowak performed computer simulations": Nowak, M. A. and May, R. M., Evolutionary games and spatial chaos. *Nature*, 359 (6398), 1992, pp. 826–829.

219. "We believe that deterministically generated spatial structure": Ibid.

220. "chaotically changing": Ibid.

220. "That territoriality favours cooperation": Sigmund, K., On prisoners and cells, *Nature*, 359 (6398), 1992, p. 774.

221. "John Holland has likened such models to 'flight simulators'": Holland, J. H., *Emergence: From Chaos to Order.* Perseus Books, 1998, p. 243.

221. "proposals for improving peer-to-peer networks": e.g., Hales, D. and Arteconi, S., SLACER: A Self-Organizing Protocol for Coordination in Peer-to-Peer Networks. *IEEE Intelligent Systems*, 21 (2), 2006, pp. 29–35.

221. "preventing fraud in electronic commerce": e.g., see Kollock, P., The production of trust in online markets. In E. J. Lawler, M. Macy, S. Thyne, and H. A. Walker (editors), *Advances in Group Processes, Vol. 16.* Greenwich, CT: JAI Press, 1999.

221. "... work by Martin Nowak": Nowak, M. A., Five rules for the evolution of cooperation. *Science*, 314 (5805), 2006, pp. 1560–1563.

222. "New Energy Finance ... recently put out a report": Liebreich, M., How to Save the Planet: Be Nice, Retaliatory, Forgiving, & Clear. White Paper, New Energy Finance, Ltd., 2007. [http://www.newenergyfinance.com/docs/Press/NEF_WP_Carbon-Game-Theory_05.pdf]

222. "impact on policy-making research regarding government response to terrorism, arms control, and environmental governance policies": For example, see Cupitt, R. T., Target rogue

behavior, not rogue states. *The Nonproliferation Review*, 3, 1996, pp. 46–54; Cupitt, R. T. and Grillot, S. R., COCOM is dead, long live COCOM: Persistence and change in multilateral security institutions. *British Journal of Political Science* 27, 7, pp. 361–89; and Friedheim, R. L., Ocean governance at the millennium: Where we have been, where we should go: Cooperation and discord in the world economy. *Ocean and Coastal Management*, 42 (9), 1999, pp. 747–765.

222. "areas ranging from the maintenance of biodiversity to the effectiveness of bacteria in producing new antibiotics": E.g., Nowak, M. A. and Sigmund, K., Biodiversity: Bacterial game dynamics. *Nature*, 418, 2002, pp. 138–139; Wiener, P., Antibiotic production in a spatially structured environment. *Ecology Letters*, 3(2), 2000, pp. 122–130.

222. "All models are wrong": Box, G.E.P. and Draper, N. R., *Empirical Model Building and Response Surfaces*. New York: Wiley 1997, p. 424.

223. "Replication is one of the hallmarks": Axelrod R., Advancing the art of simulation in the social sciences. In Conte, R., Hegselmann, R., Terna, P. (editors), *Simulating Social Phenomena*. (Lecture Notes in Economics and Mathematical Systems 456). Berlin: Springer-Verlag, 1997.

223. "Bernardo Huberman and Natalie Glance re-implemented": Huberman, B. A. and Glance, N. S., Evolutionary games and computer simulations. *Proceedings of the National Academy of Science, USA*, 90, 1993, pp. 7716–7718.

223. "A similar result was obtained independently by Arijit Mukherji, Vijay Rajan, and James Slagle": Mukherji, A., Rajan, V., and Slagle, J. R., Robustness of cooperation. *Nature*, 379, 1996, pp. 125–126.

223. "Nowak, May, and their collaborator Sebastian Bonhoeffer replied": Nowak, M. A., Bonhoeffer, S., and May, R. M., Spatial games and the maintenance of cooperation. *Proceedings of the National Academy of Sciences, USA*, 91, 1994, pp. 4877–4881; Nowak, M. A., Bonhoeffer, S., and May, R. M., Reply to Mukherji et al. *Nature*, 379, 1996, p. 126.

223. "Jose Manuel Galan and Luis Izquierdo published results": Galan, J. M. and Izquierdo, L. R., Appearances can be deceiving: Lessons learned re-implementing Axelrod's 'Evolutionary Approaches to Norms.' *Journal of Artificial Societies and Social Simulation*, 8 (3), 2005, [http://jasss.soc.surrey.ac.uk/8/3/2.html].

224. "The art of model building": Anderson, Nobel Prize acceptance speech, 1977.

Part IV

225. "In Ersilia": From Calvino, I. *Invisible Cities*. New York: Harcourt Brace Jovanovich, 1974, p. 76. (Translated by W. Weaver.)

Chapter 15

227. "The Science of Networks": Parts of this chapter were adapted from Mitchell, M., Complex systems: Network thinking. *Artificial Intelligence*, 170 (18), 2006, pp. 1194–1212.

228. "Milgram wrote of one example": From Milgram, S., The small-world problem. *Psychology Today* 1, 1967, pp. 61–67.

228. "Later work by psychologist Judith Kleinfeld": see Kleinfeld, Could it be a big world after all? Society, 39, 2002.

228. an "urban myth": Kleinfeld, J. S., Six degrees: Urban myth? *Psychology Today*, 74, March/April 2002.

228. "When people experience an unexpected social connection": Kleinfeld, J. S., Could it be a big world after all? The "six degrees of separation" myth. *Society*, 39, 2002.

229. "the 'new science of networks'": E.g., Barabási, A.-L., *Linked: The New Science of Networks*. Cambridge, MA: Perseus, 2002.

230. "the publication of two important papers": Watts, D. J. and Strogatz, S. H., Collective dynamics of 'small world' networks. *Nature* 393, 1998, pp. 440–442; Barabási,

A.-L. and Albert, R., Emergence of scaling in random networks, *Science*, 286, 1999, pp. 509–512.

231. "No one descends with such fury": Watts, D. J., *Six Degrees: The Science of a Connected Age*. New York: W. W. Norton & Co, 2003, p. 32.

233. "network thinking is poised": Barabási, A.-L. *Linked: The New Science of Networks*. Cambridge, MA: Perseus, 2002, p. 222.

236. "123 incoming links": All the in-link counts in this chapter were obtained from [http://www.microsoft-watch.org/cgi-bin/ranking.htm]. The count includes only in-links from outside the Web page's domain.

236. "mathematically define the concept of 'small-world network'": See Watts, D. J. and Strogatz, S. H., Collective dynamics of 'small world' networks. *Nature*, 393, 1998, pp. 440–442.

237. "The average path length of the regular network of figure 15.4 turns out to be 15": This value was calculated using the formula $l = N/2k$. Here l is the average path length, N is the number of nodes, and k is the degree of each node (here 2). See Albert, R. and Barabási, A.-L., Statistical mechanics of complex networks. *Reviews of Modern Physics*, 74, 2002, pp. 48–97.

238. "the average path-length has shrunk to about 9": This value was estimated from the results given in Newman, M. E. J., Moore, C., and Watts, D. J., Mean-field solution of the small-world network model. *Physical Review Letters*, 84, 1999, pp. 3201–3204.

238. "only a few random links can generate a very large effect": Watts, D. J., *Six Degrees: The Science of a Connected Age*. New York: W. W. Norton, 2003, p. 89.

238. "small-world property": The formal definition of the small-world property is that, even though relatively few long-distance connections are present, the shortest path length (number of link hops) between two nodes scales logarithmically or slower with network size n for fixed average degree.

238. "Kevin Bacon game": See, e.g., [http://en.wikipedia.org/wiki/Six_Degrees_of_Kevin_Bacon].

238. "neuroscientists had already mapped out every neuron and neural connection": For more information, see Achacoso, T. B. and Yamamoto, W. S., *AY's Neuroanatomy of C. Elegans for Computation*. Boca Raton, FL: CRC Press, 1991.

239. "do not actually have the kinds of degree distributions": The Watts-Strogatz model produces networks with exponential degree distributions, rather than the much more commonly observed power-law degree distributions in real-world networks. For details, see Albert, R. and Barabási, A.-L., (2002). Statistical mechanics of complex networks. *Reviews of Modern Physics*, 74, 2002, pp. 48–97.

240. "report about Washington State apple prices": [http://www.americanfruitgrower.com/e_notes/page.php?page=news].

240. "information about the Great Huon Apple Race of Tasmania": [http://www.huonfranklincottage.com.au/events.htm].

241. "this rule actually fits the data": The in-link degree distribution of the Web is fit reasonably well with the power law $k^{-2.3}$ and cutoff $k_{min} = 3684$ (see Clauset, A., Shalizi, C. R., and Newman, M. E. J., Power-law distributions in empirical data. Preprint, 2007 [http://arxiv.org/abs/0706.1062].) The expression k^{-2} given in this chapter is an approximation used for simplifying the discussion; plots of the $k^{-2.3}$ distribution look very similar to those given in the chapter.

Chapter 16

248. "neuroscientists have mapped the connectivity structure": e.g., see Bassett, D. S. and Bullmore, D., Small-world brain networks. *The Neuroscientist*, 12, 2006, pp. 512–523; and

Stam, C. J. and Reijneveld, J. C., Graph theoretical analysis of complex networks in the brain. *Nonlinear Biomedical Physics*, 1(1), 2007, p. 3.

248. "Genetic Regulatory Networks": For more details on the application of network ideas to genetic regulation, see Barabási, A.-L. and Oltvai, Z. N., Network biology: Understanding the cell's functional organization. *Nature Reviews: Genetics*, 5, 2004, pp. 101–113.

249. "Albert-László Barabási and colleagues looked in detail at the structure of metabolic networks": Jeong, H., Tombor, B., Albert, R., Oltvai, Z. N., and Barabási, A.-L., The large-scale organization of metabolic networks. *Nature*, 407, 2000, pp. 651–654. Examples of other work on the structure of metabolic networks are Fell, D. A. and Wagner, A., The small world of metabolism. *Nature Biotechnology*, 18, 2000, pp. 1121–1122; and Burgard, A. P., Nikolaev, E. V., Schilling, C. H., and Maranas, C. D., Flux coupling analysis of genome-scale metabolic network reconstructions. *Genome Research*, 14, 2004, pp. 301–312.

250. "Although later studies debunked the theory": See, e.g., Robbins, K. E., Lemey, P., Pybus, O. G., Jaffe, H. W., Youngpairoj, A. S., Brown, T. M., Salemi, M., Vandamme, A. M., and Kalish, M. L., U.S. human immunodeficiency virus type 1 epidemic: Date of origin, population history, and characterization of early strains. *Journal of Virology*, 77 (11), 2003, pp. 6359–6366.

250. "Recently, a group consisting of": Liljeros, F., Edling, C. R., Nunes Amaral, L. A., Stanely, H. E., and Aberg, Y., The web of human sexual contacts. *Nature*, 411, 2001, pp. 907–908.

250. "similar results have been found in studies of other sexual networks": e.g., Schneeberger, A., Mercer, C. H., Gregson, S. A., Ferguson, N. M., Nyamukapa, C. A., Anderson, R. M., Johnson, A. M., and Garnett, G. P., Scale-free networks and sexually transmitted diseases: A description of observed patterns of sexual contacts in Britain and Zimbabwe. *Sexually Transmitted Diseases*, 31 (6), 2004, pp. 380–387.

250. "A very clever yet simple method was proposed": Cohen, R., ben-Avraham, D., and Havlin, S., Efficient immunization strategies for computer networks and populations. *Physics Review Letters*, 91 (24), 2003, p. 247901.

250. "one should target anti-virus methods": Newman, M. E. J., Forrest, S., and Balthrop, J., Email networks and the spread of computer viruses. *Physical Review E*, 66, 2002, p. 035101.

252. "the ecology research community has recently seen a lot of debate": See, e.g., Montoya, J. M. and Solé, R. V., Small world patterns in food webs. *Journal of Theoretical Biology*, 214 (3), 2002, pp. 405–412; Dunne, J. A., Williams, R. J., and Martinez, N. D., Food-web structure and network theory: The role of connectance and size. *Proceedings of the National Academy of Science, USA*, 99 (20), 2002, pp. 12917–12922; and Dunne, J. A., The network structure of food webs. In M. Pascual and J. A. Dunne (editors), *Ecological Networks: Linking Structure to Dynamics in Food Webs*. New York: Oxford University Press, 2006, pp. 27–86.

252. "Where Do Scale-Free Networks Come From?" Parts of this section were adapted from Mitchell, M., Complex systems: Network thinking. *Artificial Intelligence*, 170 (18), 2006, pp. 1194–1212.

252. "Albert-László Barabási and Réka Albert proposed": Barabási, A.-L. and Albert, R., Emergence of scaling in random networks, *Science*, 286, 1999, pp. 509–512.

253. "this process and its power-law outcome had been discovered independently": Yule, G. U., A mathematical theory of evolution, based on the conclusions of Dr. J. C. Willis. *Philosophical Transactions of the Royal Society of London*, Ser. B 213, 1924, pp. 21–87; Simon, H. A., On a class of skew distribution functions." *Biometrika* 42 (3-4), 1955, p. 425; and Price, D. J., Networks of scientific papers. *Science* 149, 1965, pp. 510–515.

253. "the growth of so-called scientific citation networks": e.g., see Redner, S., How popular is your paper? An empirical study of the citation distribution. *European Physical Journal B*, 4(2), 1998, pp. 131–134.

253. "what the writer Malcolm Gladwell called *tipping points*": Gladwell, M., *The Tipping Point: How Little Things Can Make a Big Difference.* Boston: Little, Brown, 2000.

254. "A number of networks previously identified to be 'scale-free' ": Clauset, A., Shalizi, C. R., and Newman, M. E. J., Power-law distributions in empirical data. Preprint, 2007, [http://arxiv.org/abs/0706.1062].

254. "Current assessments of the commonality of power-laws": Keller, E. F., Revisiting 'scale-free' networks. *BioEssays*, 27, 2005, pp. 1060–1068.

254. "Our tendency to hallucinate": Shalizi, C., Networks and Netwars, 2005. Essay at [http://www.cscs.umich.edu/~crshalizi/weblog/347.html].

254. "there turn out to be nine and sixty ways": Shalizi, C., Power Law Distributions, 1/f noise, Long-Memory Time Series, 2007. Essay at [http://cscs.umich.edu/~crshalizi/notebooks/power-laws.html].

254. "a new hypothesized mechanism that resulted in power law distributions": For surveys of some such mechanisms, see Mitzenmacher, M., A brief history of generative models for power law and lognormal distributions. *Internet Mathematics*, 1(2), 2003, pp. 226–251; and Newman, M. E. J., Power laws, Pareto distributions and Zipf's law. *Contemporary Physics*, 46, 2005, pp. 323–351.

256. "The reported cause of the shutdown": The cascading failure and its causes are described in detail in the U.S.-Canada Power System Outage Task Force's Final Report on the August 14, 2003 Blackout in the United States and Canada: Causes and Recommendations [https://reports.energy.gov/].

256. "The computer system of the US Customs and Border protection agency": see Schlossberg, D. "LAX Computer Crash Strands International Passengers." *ConsumerAffairs.com*, August 13, 2007, [http://www.consumeraffairs.com/news04/2007/08/lax_computers.html]; and Schwartz, J., "Who Needs Hackers?" *New York Times*, September 12, 2007.

256. "Long-Term Capital Management": see, e.g., Government Accounting Office, *Long-Term Capital Management: Regulators Need to Focus Greater Attention on Systemic Risk.* Report to Congressional Request, 1999, [http://www.gao.gov/cgi-bin/getrpt?GGD-00-3]; and Coy, P., Woolley, S., Spiro, L. N., and Glasgall, W., Failed wizards of Wall Street. *Business Week*, September 21, 1998.

257. "The threat is complexity itself": Andreas Antonopoulos, quoted in Schwartz, J., "Who Needs Hackers?" *New York Times*, September 12, 2007.

257. "Self-Organized Criticality": for an introduction to SOC, see Bak, P., *How Nature Works: The Science of Self-Organized Criticality.* New York: Springer, 1996.

257. "Highly Optimized Tolerance": For an introduction to HOT, see Carlson, J. M. and Doyle, J., Complexity and robustness. *Proceedings of the National Academy of Science, USA* 99, 2002, pp. 2538–2545.

257. "Next to the mysteries of dynamics on a network": Watts, D. J., *Six Degrees: The Science of a Connected Age.* New York: W. W. Norton, 2003, p. 161.

Chapter 17

260. "the surface area scales as the volume raised to the two-thirds power": Let V denote volume, S denote surface area, and r denote radius. V is proportional to r^3, so cube-root(V) is proportional to the radius. Surface area is proportional to radius^2, and thus to cube-root(V)^2, which is $V^{\frac{2}{3}}$.

261. "If you plot a power law on a double logarithmic plot, it will look like a straight line, and the slope of that line will be equal to the power law's exponent": In the example here, the power law is

$$\text{metabolic rate} \propto \text{body mass}^{3/4}.$$

Taking the logarithm of both sides, we get

$$\log(\textit{metabolic rate}) \propto 3/4 \log(\textit{body mass}).$$

This is the equation of a straight line with slope 3/4, if we plot log(*metabolic rate*) against log (*body mass*), which is actually what is plotted in figure 16.2

264. "Enquist later described the group's math results as 'pyrotechnics'": Grant, B., The powers that be. *The Scientist*, 21 (3), 2007.

265. "You really have to think in terms of two separate scales": G. B. West, quoted in Mackenzie, D., Biophysics: New clues to why size equals destiny. *Science*, 284 (5420), 1999, pp. 1607–1609.

266. "The mathematical details of the model": A technical, but not too difficult to understand, description of the Metabolic Scaling model is given in West, G. B. and Brown, J. H., Life's universal scaling laws. *Physics Today*, 57 (9), 2004, pp. 36–43.

266. "Although living things occupy a three-dimensional space": West, G. B., Brown, J. H., and Enquist, B. J., The fourth dimension of life: Fractal geometry and allometric scaling of organisms. *Science*, 284, pp. 1677–1679.

267. "the potential to unify all of biology": Grant, B., The powers that be. *The Scientist*, 21 (3), 2007.

267. "as potentially important to biology as Newton's contributions are to physics": Niklas, K. J., Size matters! *Trends in Ecology and Evolution* 16 (8), 2001, p. 468.

267. "We see the prospects for the emergence of a general theory of metabolism": West, G. B. and Brown, J. H., The origin of allometric scaling laws in biology from genomes to ecosystems: Towards a quantitative unifying theory of biological structure and organization. *Journal of Experimental Biology* 208, 2005, pp. 1575–1592.

268. "Still others argue that Kleiber was wrong all along": A review of various critiques of metabolic scaling theory is given in Agutter P. S and Wheatley, D. N., Metabolic scaling: Consensus or Controversy? *Theoretical Biology and Medical Modeling*, 18, 2004, pp. 283–289.

268. "The more detail that one knows about the particular physiology involved": H. Horn, quoted in Whitfield, J., All creatures great and small. *Nature*, 413, 2001, pp. 342–344.

268. "It's nice when things are simple": H. Müller-Landau, quoted in Grant, B., The powers that be. *The Scientist*, 21 (3), 2007.

268. "There have been arguments that the mathematics in metabolic scaling theory is incorrect": e.g., Kozlowski, J. and Konarzweski, M., Is West, Brown and Enquist's model of allometric scaling mathematically correct and biologically relevant? *Functional Ecology*, 18, 2004, pp. 283–289.

269. "The authors of metabolic scaling theory have strongly stood by their work": E.g., see West, G. B., Brown, J. H., and Enquist, B. J., Yes, West, Brown and Enquist's model of allometric scaling is both mathematically correct and biologically relevant. (Reply to Kozlowski and Konarzweski, 2004.) *Functional Ecology*, 19, 2005, pp. 735–738; and Borrell, B., Metabolic theory spat heats up. *The Scientist* (News), November 8, 2007. [http://www.the-scientist.com/news/display/53846/].

269. "Part of me doesn't want to be cowered": G. West, quoted in Grant, B., The powers that be. *The Scientist*, 21 (3), 2007.

269. "I suspect that West, Enquist et al. will continue repeating their central arguments": H. Müller-Landau, quoted in Borrell, B., Metabolic theory spat heats up. *The Scientist* (News), November 8, 2007. [http://www.the-scientist.com/news/display/53846/].

269. "more normal than 'normal' ": "The presence of [power-law] distributions": Willinger, W., Alderson, D., Doyle, J. C., and Li, L., More 'normal' than normal: Scaling

distributions and complex systems. In R. G. Ingalls et al., *Proceedings of the 2004 Winter Simulation Conference*, pp. 130–141. Piscataway, NJ: IEEE Press, 2004.

271. "This relation is now called *Zipf's law*": Zipf's original publication on this work is a book: Zipf, G. K., *Selected Studies of the Principle of Relative Frequency in Language*. Cambridge, MA: Harvard University Press, 1932.

271. "Benoit Mandelbrot ... had a somewhat different explanation": Mandelbrot. B., An informational theory of the statistical structure of languages. In W. Jackson (editor), *Communicaiton Theory*, Woburn, MA: Butterworth, 1953, pp. 486–502.

272. "Herbert Simon proposed yet another explanation": Simon, H. A., On a class of skew distribution functions." *Biometrika* 42 (3–4), 1955, p. 425.

272. "Evidently Mandelbrot and Simon had a rather heated argument": Mitzenmacher, M., A brief history of generative models for power law and lognormal distributions. *Internet Mathematics*, 1 (2), 2003, pp. 226–251.

272. "the psychologist George Miller showed": Miller, G. A., Some effects of intermittent silence. *The American Journal of Psychology*, 70, 1957, pp. 311–314.

Chapter 18

275. " 'mobile genetic elements' ": A technical article on the proposed role of mobile genetic elements on brain diversity is Muotri, A. R., Chu, V. T., Marchetto, M. C. N., Deng, W., Moran, J. V. and Gage, F. H., Somatic mosaicism in neuronal precursor cells mediated by L1 retrotransposition. *Nature*, 435, 2005, pp. 903–910.

275. "Recently a group of science philosophers and biologists performed a survey": Reported in Pearson, H., What is a gene? *Nature*, 441, 2006, pp. 399–401.

275. "The more expert scientists become in molecular genetics": Ibid.

275. "the absence of necessary methylation": See, e.g., Dean, W., Santos, F., Stojkovic, M., Zakhartchenko, V., Walter, J., Wolf, E., and Reik, W., Conservation of methylation reprogramming in mammalian development: Aberrant reprogramming in cloned embryos. *Proceedings of the National Academy of Science, USA*, 98 (24), 2001, pp. 13734–13738.

276. "a large proportion of the DNA that is transcribed into RNA is *not* subsequently translated into proteins": See, e.g., Mattick, J. S., RNA regulation: A new genetics? *Nature Reviews: Genetics*, 5, 2004, pp. 316–323; Grosshans, H. and Filipowicz, W., The expanding world of small RNAs. *Nature*, 451, 2008, pp. 414–416.

276. "The significance of non-coding RNA": For a discussion of some of the current research and controversies in this area, see Hüttenhofer, A., Scattner, P., and Polacek, N., Non-coding RNAs: Hope or Hype? *Trends in Genetics*, 21 (5), 2005, pp. 289–297.

277. "The presumption that genes operate independently": Caruso, D., A challenge to gene theory, a tougher look at biotech. *New York Times*, July 1, 2007.

277. "encode a specific functional product": U.S. patent law, quoted in ibid.

277. "Evidence of a networked genome shatters the scientific basis": Ibid.

277. "more than 90% of our DNA is shared with mice": Published estimates of DNA overlap between humans and different species differ. However, it seems that "over 90% shared" is a fairly safe statement.

278. "a 'black box' that somehow transformed genetic information": Carroll, S. B., *Endless Forms Most Beautiful: The New Science of Evo Devo and the Making of the Animal Kingdom*. New York: W. W. Norton, 2005, p. 7.

280. "The irony ... is that what was dismissed as junk": In the video Gene Regulation; *Science*, 319, no 5871, 2008.

280. "To verify that the BMP4 gene itself could indeed trigger the growth": From Yoon, C. K., From a few genes, life's myriad shapes. *New York Times*, June 26, 2007.

281. "In a strange but revealing experiment": This work is described in Travis, J., Eye-opening gene. *Science News Online*, May 10, 1997.

281. "This conclusion is still quite controversial among evolutionary biologists": For discussion of this controversy, see, e.g., Erwin, D. H., The developmental origins of animal bodyplans. In S. Xiao and A. J. Kaufman (editors), *Neoproterozoic Geobiology and Paleobiology*, New York: Springer, 2006, pp. 159–197.

282. "a world-class intellectual riffer": Horgan, J., From complexity to perplexity. *Scientific American*, 272, June 1995, pp. 74–79.

284. "the genomic networks that control development": Kauffman, S. A., *At Home in the Universe*. New York: Oxford University Press, 1995, p. 26.

285. "life exists at the edge of chaos": Ibid, p. 26.

286. "Most biologists, heritors of the Darwinian tradition": Ibid, p. 25.

286. "a 'candidate fourth law of thermodynamics' ": Kauffman, S. J., *Investigations*. New York: Oxford University Press, 2002, p. 3.

286. "Kauffman's book: The Origins of Order": Kauffman, S. A., *The Origins of Order*. New York: Oxford University Press, 1993.

286. "a fundamental reinterpretation of the place of selection": Burian, R. M. and Richardson, R. C., Form and order in evolutionary biology: Stuart Kauffman's transformation of theoretical biology. In *PSA: Proceedings of the Biennial Meeting of the Philosophy of Science Association*, Vol. 2: Symposia and Invited Papers, 1990, pp. 267–287.

286. "His approach opens up new vistas": Ibid.

286. "the first serious attempt to model a complete biology": Bak, P., *How Nature Works: The Science of Self-Organized Criticality*. New York: Springer, 1996.

286. "dangerously seductive": Dover, G. A., On the edge. *Nature*, 365, 1993, pp. 704–706.

286. "There are times when the bracing walk through hyperspace": Ibid.

287. "noise has a significant effect on the behavior of RBNs": E.g., see Goodrich, C. S. and Matache, M. T., The stabilizing effect of noise on the dynamics of a Boolean network. *Physica A*, 379(1), 2007, pp. 334–356.

288. "It has essentially become a matter of social responsibility": Hoelzer, G. A., Smith, E., and Pepper, J. W., On the logical relationship between natural selection and self-organization. *Journal of Evolutionary Biology*, 19 (6), 2007, pp. 1785–1794.

288. "Evolutionary biologist Dan McShea has given me a useful way to think about these various issues": D. W. McShea, personal communication.

288. "Evolutionary biology is in a state of intellectual chaos": D. W. McShea, personal communication.

Part V

289. "I will put Chaos into fourteen lines": In Millay, E. St. Vincent, *Mine the Harvest: A Collection of New Poems*. New York: Harper, 1949, p. 130.

Chapter 19

291. "John Horgan published an article": Horgan, J., From complexity to perplexity. *Scientific American*, 272, June 1995, pp. 74–79.

292. "*The End of Science*": Horgan, J., *The End of Science: Facing the Limits of Knowledge in the Twilight of the Scientific Age*. Reading, MA: Addison-Wesley, 1996.

293. "[T]he hope that physics could be complete": Crutchfield, J. P., Farmer, J. D., Packard, N. H., and Shaw, R. S., Chaos. *Scientific American*, 255, December 1986.

293. "Gravitation is not responsible": This quotation is very commonly attributed to Einstein. However, it is apparently a (more elegant) rephrasing of what he actually said: "Falling in love is not at all the most stupid thing that people do—but gravitation cannot be held responsible for it." Quoted in Dukas, H. and Hoffmann B. (editors), *Albert Einstein, The Human Side: New Glimpses from His Archives*. Princeton, NJ: Princeton University Press, 1979, p. 56.

293. "Recently, ideas about complexity": Gordon, D. M., Control without hierarchy. *Nature*, 446 (7132), 2007, p. 143.

296. "a new discipline of *cybernetics*": Fascinating histories of the field of cybernetics can be found in Aspray, W., *John von Neumann and the Origins of Modern Computing*. Cambridge, MA: MIT Press, 1990; and Heims, S. *The Cybernetics Group*. Cambridge, MIT Press, 1991.

296. "the entire field of control and communication": Wiener, N. *Cybernetics*. Cambridge, MA: MIT Press, 1948, p. 11.

297. "H. Ross Ashby's 'Design for a Brain' ": Ashby, R. H., *Design for a Brain*. New York: Wiley, 1954.

297. "Warren McCulloch and Walter Pitts' model of neurons": McCulloch, W. and Pitts, W., A logical calculus of ideas immanent in nervous activity. *Bulletin of Mathematical Biophysics* 5, 1942, pp. 115–133.

297. "Margaret Mead and Gregory Bateson's application of cybernetic ideas": See, e.g., Bateson, G., *Mind and Nature: A Necessary Unity*. Cresskill, NJ: Hampton Press, 1979.

297. "Norbert Wiener's books *Cybernetics* and *The Human Use of Human Beings*": Wiener, N. *Cybernetics: Or the Control and Communication in the Animal and the Machine*. Cambridge, MA: MIT Press, 1948; Wiener, N. *The Human Use of Human Beings*. Boston: Houghton Mifflin, 1950.

297. "The two most important historical events": Gregory Bateson, quoted in Heims, S., *The Cybernetics Group*. Cambridge: MIT Press, 1991, p. 96.

297. "vacuous in the extreme": Max Delbrück, quoted in Heims, S., *The Cybernetics Group*. Cambridge: MIT Press, 1991, p. 95.

297. "bull sessions with a very elite group": Leonard Savage, quoted in Heims, S., *The Cybernetics Group*, 1991, p. 96.

297. "in the end Wiener's hope": Aspray, W., *John von Neumann and the Origins of Modern Computing*. Cambridge: MIT Press, 1990, pp. 209–210.

297. "General System Theory": see Von Bertalanffy, L., *General System Theory: Foundations, Development, Applications*, New York: G. Braziller, 1969; or Rapoport, A. *General System Theory: Essential Concepts and Applications*, Cambridge, MA: Abacus Press, 1986.

297. "the formulation and deduction of those principles . . .": Von Bertanlanffy, L., An outline of general system theory. *The British Journal for the Philosophy of Science*, 1(92), 1950, pp. 134–165.

298. "biologists Humberto Maturana and Francisco Varela attempted . . .": See, e.g., Maturana, H. R. and Varela, F. J., *Autopoiesis and Cognition: The Realization of the Living*. Boston, MA: D. Reidel Publishing Co., 1980.

298. "Hermann Haken's *Synergetics* and Ilya Prigogine's theories of *dissipative structures* and *nonequilibrium systems* . . . ". See Haken, H., *The Science of Structure: Synergetics*, New York: Van Nostrand Reinhold, 1984; and Prigogine, I. *From Being to Becoming: Time and Complexity in the Physical Sciences*, San Francisco: W. H. Freeman, 1980.

298. "vocabulary of complexity," "A number of concepts that deal with mechanisms . . .": Nicolis, G. and Prigogine, I., *Exploring Complexity*, New York: W. H. Freeman and Co., 1989, p. x.

301. "I think we may be missing the conceptual equivalent of calculus . . .": Strogatz, S., *Sync: How Order Emerges from Chaos in the Universe, Nature, and Daily Life*. New York: Hyperion, 2004, p. 287.

302. "He was hampered by the chaos of language": Gleick, J., *Isaac Newton*, New York: Pantheon Books, 2003, pp. 58–59.

303. "The physicist Per Bak introduced the notion of *self-organized criticality*": See Bak, P., *How Nature Works: The Science of Self-Organized Criticality*. New York: Copernicus, 1996.

303. "The physicist Jim Crutchfield has proposed a theory of *computational mechanics*": See, e.g., Crutchfield, J. P., The calculi of emergence. *Physica D*, 75, 1994, 11–54.

303. "the simplicity on the other side of complexity": The famous quotation "I do not give a fig for the simplicity on this side of complexity, but I would give my life for the simplicity on the other side of complexity" is usually attributed to Oliver Wendell Holmes, but I could not find the source of this in his writings. I have also seen this quotation attributed to the poet Gerald Manley Hopkins.

303. "One doesn't discover new lands": Gide, A., *The Counterfeiters*. Translated by D. Bussy. New York: Vintage, 1973, p. 353. Original: *Journal des Faux-Monnayeurs*. Paris: Librairie Gallimard, 1927.

NOTES | 325

BIBLIOGRAPHY

Achacoso, T. B. and Yamamoto, W. S. *AY's Neuroanatomy of C. Elegans for Computation*. Boca Raton, FL: CRC Press, 1991.

Adami, C. *Introduction to Artificial Life*, Springer, 1998.

Agutter P. S and Wheatley D. N. Metabolic scaling: Consensus or Controversy? *Theoretical Biology and Medical Modeling*, 18, 2004, pp. 283–289.

Albert, R. and Barabási, A-L. Statistical mechanics of complex networks. *Reviews of Modern Physics*, 74, pp. 48–97, 2002.

Ashby, H. R. *Design for a Brain*. New York: Wiley, 1954.

Aspray, W. *John von Neumann and the Origins of Modern Computing*. Cambridge, MA: MIT Press, 1990.

Aubin, D. and Dalmedico, A. D. Writing the history of dynamical systems and chaos: *Longue Durée* and revolution, disciplines, and cultures. *Historia Mathematica* 29, 2002, pp. 273–339.

Axelrod, R. *The Evolution of Cooperation*. New York: Basic Books, 1984.

Axelrod, R. An evolutionary approach to norms. *American Political Science Review*, 80 (4), 1986, pp. 1095–1111.

Axelrod R. Advancing the art of simulation in the social sciences. In Conte, R., Hegselmann, R., Terna, P. (editors). *Simulating Social Phenomena*. (Lecture Notes in Economics and Mathematical Systems 456). Berlin: Springer-Verlag, 1997.

Bak, P. *How Nature Works: The Science of Self-Organized Criticality*. New York: Springer, 1996.

Barabási, A.-L. *Linked: The New Science of Networks*. Cambridge, MA: Perseus, 2002.

Barabási, A.-L. and Albert, R. Emergence of scaling in random networks. *Science*, 286, 1999, pp. 509–512.

Barabási, A.-L. and Oltvai, Z. N. Network biology: Understanding the cell's functional organization. *Nature Reviews: Genetics*, 5, 2004, pp. 101–113.

Barrett, P. (editor). *Charles Darwin's Notebooks, 1836-1844: Geology, Transmutaiton of Species, Metaphysical Enquiries*. Ithaca, NY: Cornell University Press, 1987.

Bassett, D. S. and Bullmore, D. Small-world brain networks. *The Neuroscientist*, 12, 2006, pp. 512–523.

Bateson, G. H. *Mind and Nature: A Necessary Unity*. Skokie, IL: Hampton Press, 1979.

Beinhocker, E. D. *The Origin of Wealth: Evolution, Complexity, and the Radical Remaking of Economics*. Cambridge, MA: Harvard Business School Press, 2006.

Bennett, C. H. The thermodynamics of computation—a review. *International Journal of Theoretical Physics*, 21, 1982, pp. 905–940.

Bennett, C. H. Dissipation, information, computational complexity and the definition of organization. In Pines, D. (editor), *Emerging Syntheses in Science*. Redwood City, CA: Addison-Wesley, 1985, pp. 215–233.

Bennett, C. H. How to define complexity in physics, and why. In W. H. Zurek (editor), *Complexity, Entropy, and the Physics of Information*. Reading, MA: Addison-Wesley, 1990, pp. 137–148.

Berlekamp, E., Conway, J. H., and Guy, R. *Winning Ways for Your Mathematical Plays*, volume 2. San Diego, CA: Academic Press, 1982.

Bickhard, M. H. The biological foundations of cognitive science. In *Mind 4: Proceedings of the 4th Annual Meeting of the Cognitive Science Society of Ireland*. Dublin, Ireland: J. Benjamins, 1999.

Bolton, R. J. and Hand, D. J. Statistical fraud detection: A review. *Statistical Science*, 17(3), 2002, pp. 235–255.

Bonabeau, E. Control mechanisms for distributed autonomous systems: Insights from social insects. In L. A. Segel and I. R. Cohen (editors), *Design Principles for the Immune System and Other Distributed Autonomous Systems*. New York: Oxford University Press, 2001.

Bonabeau, E., Dorigo, M., and Theraulaz, G. *Swarm Intelligence: From Natural to Artificial Systems*. New York: Oxford University Press, 1999.

Borrell, B. Metabolic theory spat heats up. *The Scientist* (News), November 8, 2007. [http://www.the-scientist.com/news/display/53846/].

Bovill, C. *Fractal Geometry in Architecture and Design*. Boston: Birkhäuser. 1996.

Bowker, J. (editor). *The Cambridge Illustrated History of Religions*. Cambridge, UK: Cambridge University Press, 2002.

Bowlby, J. *Charles Darwin: A New Life*. New York: Norton, 1992.

Box, G. E. P. and Draper, N. R. *Empirical Model Building and Response Surfaces*. New York: Wiley, 1997.

Burgard, A. P., Nikolaev, E. V., Schilling, C. H., and Maranas, C. D. Flux coupling analysis of genome-scale metabolic network reconstructions. *Genome Research*, 14, 2004, pp. 301–312.

Burian, R. M. and Richardson, R. C. Form and order in evolutionary biology: Stuart Kauffman's transformation of theoretical biology. In *PSA: Proceedings of the Biennial Meeting of the Philosophy of Science Association*, Vol. 2: Symposia and Invited Papers, 1990, 267–287.

Burkhardt, F. and Smith, S. (editors). *The Correspondence of Charles Darwin*, Volume 1. Cambridge, UK: Cambridge University Press, 1985.

Burks, A. W. Von Neumann's self-reproducing automata. In A. W. Burks (editor), *Essays on Cellular Automata*. Urbana: University of Illinois Press, 1970.

Calvino, I. *Invisible Cities*. New York: Harcourt Brace Jovanovich, 1974. (Translated by W. Weaver.)

Carlson, J. M. and Doyle, J. Complexity and robustness. *Proceedings of the National Academy of Science, USA* 99, 2002, pp. 2538–2545.

Carroll, S. B. *Endless Forms Most Beautiful: The New Science of Evo Devo and the Making of the Animal Kingdom*. New York: Norton, 2005.

Caruso, D. A challenge to gene theory, a tougher look at biotech. *New York Times*, July 1, 2007.

Churchland, P. S., Koch, C., and Sejnowski, T. J. What is computational neuroscience? In E. L. Schwartz (editor), *Computational Neuroscience*. Cambridge, MA: MIT Press, 1994, pp. 46–55.

Clauset, A., Shalizi, C. R., and Newman, M. E. J. Power-law distributions in empirical data. Preprint, 2007, [http://arxiv.org/abs/0706.1062].

Coale, K. Darwin in a box. *Wired*, June 14, 1997.

Cochrane, E. Viva Lamarck: A brief history of the inheritance of acquired characteristics. *Aeon* 2(2), 1997, 5–39.

Cohen, I. Informational landscapes in art, science, and evolution. *Bulletin of Mathematical Biology*, 68, 2006, pp. 1213–1229.

Cohen, I. Immune system computation and the immunological homunculus. In O. Niestrasz, J. Whittle, D. Harel, and G. Reggio (editors), *MoDELS 2006, Lecture Notes in Computer Science*, 4199, 2006, pp. 499–512. Berlin: Springer-Verlag.

Cohen, R., ben-Avraham, D., and Havlin, S. Efficient immunization strategies for computer networks and populations. *Physics Review Letters*, 91(24), 2003, p. 247901.

Connell, J. H. *Minimalist Mobile Robotics: A Colony-Style Architecture for an Artificial Creature*. San Diego, CA: Academic Press, 1990.

Cook, M. Universality in elementary cellular automata. *Complex Systems* 15(1), 2004, pp. 1–40.

Coullet, P. and Tresser, C. Itérations d'endomorphismes et groupe de renormalization. *Comptes Rendues de Académie des Sciences, Paris A*, 287, 1978, pp. 577–580.

Coy, P., Woolley, S., Spiro, L. N., and Glasgall, W. Failed wizards of Wall Street. *Business Week*, September 21, 1998.

Crutchfield, J. P. The calculi of emergence. *Physica D*, 75, 1994, 11–54.

Crutchfield, J. P., Farmer, J. D., Packard, N. H., and Shaw, R. S. Chaos. *Scientific American*, 255, December 1986.

Crutchfield, J. P., and Hanson, J. E. Turbulent pattern bases for cellular automata. *Physica D* 69, 1993, pp. 279–301.

Crutchfield, J. P., Mitchell, M., and Das, R. Evolutionary design of collective computation in cellular automata. In J. P. Crutchfield and P. K. Schuster (editors), *Evolutionary Dynamics—Exploring the Interplay of Selection, Neutrality, Accident, and Function*. New York: Oxford University Press, 2003, pp. 361–411.

Crutchfield, J. P. and Shalizi, C. R. Thermodynamic depth of causal states: When paddling around in Occam's pool shallowness is a virtue. *Physical Review E*, 59 (1), 1999, pp. 275–283.

Crutchfield, J. P. and Young, K. Inferring statistical complexity. *Physical Review Letters* 63, 1989, pp. 105–108.

Cupitt, R. T. Target rogue behavior, not rogue states. *The Nonproliferation Review*, 3, 1996, pp. 46–54.

Cupitt, R. T. and Grillot, S. R. COCOM is dead, long live COCOM: Persistence and change in multilateral security institutions. *British Journal of Political Science* 27, 1997, pp. 361–389.

Darwin, C. *The Autobiography of Charles Darwin*. Lanham, MD: Barnes & Noble Publishing, 2005. Originally published 1887.

Darwin, C. and Barlow, N. D. *The Autobiography of Charles Darwin* (Reissue edition). New York: Norton, 1993. [Originally published 1958.]

Darwin, E. *The Temple of Nature; or, The Origin of Society: A Poem with Philosophical Notes*. London: J. Johnson, 1803.

Dawkins, R. *The Extended Phenotype* (Reprint edition). New York: Oxford University Press, 1989. (Originally published 1982.)

Dean, W., Santos, F., Stojkovic, M., Zakhartchenko, V., Walter, J., Wolf, E., and Reik, W. Conservation of methylation reprogramming in mammalian development: Aberrant reprogramming in cloned embryos. *Proceedings of the National Academy of Science, USA*, 98(24), 2001, pp. 13734–13738.

Dennett, D. R. *Elbow Room: The Varieties of Free Will Worth Wanting*. Cambridge, MA: MIT Press, 1984.

Dennett, D. R. *Consciousness Explained*. Boston: Little, Brown & Co, 1991.

Dennett, D. R. *Darwin's Dangerous Idea*. New York: Simon & Schuster, 1995.

Descartes, R. *A Discourse on the Method*. Translated by Ian Maclean. London: Oxford World's Classics, Oxford University Press, 2006 [1637].

Dietrich, M. R. The origins of the neutral theory of molecular evolution. *Journal of the History of Biology*, 27(1), 1994, pp. 21–59.

Dorigo, M. and Stützle, T. *Ant Colony Optimization*, MIT Press, 2004.

Dover, G. A. On the edge. *Nature*, 365, 1993, pp. 704–706.

Dukas, H. and Hoffmann B. (editors). *Albert Einstein, The Human Side : New Glimpses from His Archives*. Princeton University Press, 1979.

Dunne, J. A. The network structure of food webs. In M. Pascual and J. A. Dunne (editors). *Ecological Networks: Linking Structure to Dynamics in Food Webs*. New York: Oxford University Press, 2006, pp. 27–86.

Dunne, J. A. Williams, R. J. and Martinez, N. D. Food-web structure and network theory: The role of connectance and size. *Proceedings of the National Academy of Science, USA*, 99(20), 2002, pp. 12917–12922.

Eddington, A. E. *The Nature of the Physical World*. New York: Macmillan, 1928.

Eigen, M. How does information originate? Principles of biological self-organization. In S. A. Rice (editor), *For Ilya Prigogine*. New York: Wiley 1978, pp. 211–262.

Eigen, M. *Steps Towards Life*. Oxford: Oxford University Press, 1992.

Einstein, A. *Out of My Later Years* (revised edition). Castle Books, 2005 [originally published 1950, New York: Philosophical Books.]

Eldredge, N. and Tattersall, I. *The Myths of Human Evolution*. New York: Columbia University Press, 1982.

Erwin, D. H. The developmental origins of animal bodyplans. In S. Xiao and A. J. Kaufman (editors), *Neoproterozoic Geobiology and Paleobiology*. New York: Springer, 2006, pp. 159–197.

Everdell, W. R. *The First Moderns: Profiles in the Origins of Twentieth-Century Thought*. Chicago, IL: University of Chicago Press, 1998.

Feigenbaum, M. J. Universal behavior in nonlinear systems. *Los Alamos Science*, 1(1), 1980, pp. 4–27.

Fell, D. A. and Wagner, A. The small world of metabolism. *Nature Biotechnology*, 18, 2000, pp. 1121–1122.

Felton, M. J. Survival of the fittest in drug design. *Modern Drug Discovery*, 3(9), 2000, pp. 49–50.

Floridi, L., Open problems in the philosophy of information. *Metaphilosophy*, 35(4), 2004, pp. 554–582.

Fogel, D. B. *Evolutionary Computation: The Fossil Record*. New York: Wiley-IEEE Press, 1998.

Forrest, S. *Emergent Computation*. Cambridge, MA: MIT Press, 1991.

Franks, N. R. Army ants: A collective intelligence. *American Scientist*, 77(2), 1989, pp. 138–145.

Freud, S. *Moses and Monotheism*. New York: Vintage Books, 1939.

Friedheim, R. L. Ocean governance at the millennium: Where we have been, where we should go: Cooperation and discord in the world economy. *Ocean and Coastal Management*, 42(9), pp. 747–765, 1999.

Fry, I. *The Emergence of Life on Earth: A Historical and Scientific Overview*. Piscataway, NJ: Rutgers University Press, 2000.

Galan, J. M. and Izquierdo, L. R. Appearances can be deceiving: Lessons learned re-implementing Axelrod's 'Evolutionary Approaches to Norms.' *Journal of Artificial Societies and Social Simulation*, 8(3), 2005, [http://jasss.soc.surrey.ac.uk/8/3/2.html].

Garber, D. Descartes, mechanics, and the mechanical philosophy. *Midwest Studies in Philosophy* 26 (1), 2002, pp. 185–204.

Gell-Mann, M. *The Quark and the Jaguar*. New York: Freeman, 1994.

Gell-Mann, M. What is complexity? *Complexity*, 1(1), 1995, pp. 16–19.

Gide, A. *The Counterfeiters*. Translated by D. Bussy. New York: Vintage, 1973, p. 353. Original: *Journal des Faux-Monnayeurs*, Librairie Gallimard, Paris, 1927.

Gladwell, M. *The Tipping Point: How Little Things Can Make a Big Difference*. Boston: Little, Brown, 2000.

Gleick, J. *Chaos: Making a New Science.* New York: Viking, 1987.
Gleick, J. *Isaac Newton,* New York: Pantheon Books, 2003.
Goldstine, H. H. *The Computer, from Pascal to von Neumann.* Princeton, NJ: Princeton University Press, 1993. First edition, 1972.
Goodrich, C. S. and Matache, M. T. The stabilizing effect of noise on the dynamics of a Boolean network, *Physica A*, 379(1), 2007, pp. 334–356.
Gordon, D. M. Task allocation in ant colonies. In L. A. Segel and I. R. Cohen (editors), *Design Principles for the Immune System and Other Distributed Autonomous Systems.* New York: Oxford University Press, 2001.
Gordon, D. M. Control without hierarchy. *Nature*, 446(7132), 2007, p. 143.
Gould, S. J. Is a new and general theory of evolution emerging? *Paleobiology*, 6, 1980, pp. 119–130.
Gould, S. J. Sociobiology and the theory of natural selection. In G.W. Barlow and J. Silverberg (editors), *Sociobiology: Beyond Nature/Nurture?,* pp. 257–269. Boulder, CO: Westview Press Inc., 1980.
Government Accounting Office. *Long-Term Capital Management: Regulators Need to Focus Greater Attention on Systemic Risk.* Report to Congressional Request, 1999 [http://www.gao.gov/cgi-bin/getrpt?GGD-00-3].
Grant, B. The powers that be. *The Scientist*, 21(3), 2007.
Grene, M. and Depew, D., *The Philosophy of Biology: An Episodic History.* Cambridge, U.K.: Cambridge University Press, 2004.
Grinnell, G. J. The rise and fall of Darwin's second theory. *Journal of the History of Biology*, 18(1), 1985, pp. 51–70.
Grosshans, H. and Filipowicz, W. The expanding world of small RNAs. *Nature*, 451, 2008, pp. 414–416.
Hales, D. and Arteconi, S. SLACER: A Self-Organizing Protocol for Coordination in Peer-to-Peer Networks. *IEEE Intelligent Systems*, 21(2), 2006, pp. 29–35.
Hardin, G. The tragedy of the commons. *Science*, 162, 1968, pp. 1243–1248.
Heims, S. *The Cybernetics Group.* Cambridge, MA: MIT Press, 1991.
Heims, S. J. *John von Neumann and Norbert Wiener: From Mathematics to the Technologies of Life and Death.* Cambridge: MIT Press, 1980.
Hobbes, T. *Leviathan.* Cambridge, U.K.: Cambridge University Press, (1651/1991).
Hodges, A. *Alan Turing: The Enigma.* New York: Simon & Schuster, 1983.
Hoelzer, G. A. Smith, E., and Pepper, J. W., On the logical relationship between natural selection and self-organization. *Journal of Evolutionary Biology*, 19(6), 2007, pp. 1785–1794.
Hofmeyr, S. A. An interpretive introduction to the immune system. In L. A. Segel and I. R. Cohen (editors), *Design Principles for the Immune System and Other Distributed Autonomous Systems.* New York: Oxford University Press, 2001.
Hofmeyr, S. A. and Forrest, S. Architecture for an artificial immune system. *Evolutionary Computation*, 8(4), 2000, pp. 443–473.
Hofstadter, D. R. *Gödel, Escher, Bach: an Eternal Golden Braid.* New York: Basic Books, 1979.
Hofstadter, D. R. Mathematical chaos and strange attractors. Chapter 16 in *Metamagical Themas.* New York: Basic Books, 1985.
Hofstadter, D. R. The Genetic Code: Arbitrary? Chapter 27 in *Metamagical Themas.* Basic Books, 1985.
Hofstadter D. *Fluid Concepts and Creative Analogies.* New York: Basic Books, 1995.
Hofstadter, D. *I am a Strange Loop.* New York: Basic Books, 2007.
Hofstadter, D. R. and Mitchell, M. The Copycat project: A model of mental fluidity and analogy-making. In K. Holyoak and J. Barnden (editors), *Advances in Connectionist and Neural Computation Theory, Volume 2: Analogical Connections*, 1994, pp. 31–112.

Holland, J. H. *Adaptation in Natural and Artificial Systems.* Cambridge, MA: MIT Press, 1992. (First edition, 1975.)

Holland, J. H. *Emergence: From Chaos to Order.* Perseus Books, 1998.

Hölldobler, B. and Wilson, E. O. *The Ants.* Cambridge, MA: Belknap Press, 1990.

Holtham, C. Fear and opportunity. *Information Age,* July 11, 2007. [http://www.information-age.com/article/2006/february/fear_and_opportunity].

Horgan, J. From complexity to perplexity. *Scientific American,* 272, June 1995, pp. 74–79.

Horgan, J. *The End of Science: Facing the Limits of Knowledge in the Twilight of the Scientific Age.* Reading, MA: Addison-Wesley, 1996.

Huberman, B. A. and Glance, N. S. Evolutionary games and computer simulations. *Proceedings of the National Academy of Science, USA,* 90, 1993, pp. 7716–7718.

Hüttenhofer, A., Scattner, P., and Polacek, N. Non-coding RNAs: Hope or Hype? *Trends in Genetics,* 21 (5), 2005, pp. 289–297.

Huxley, J. *Evolution: The Modern Synthesis.* New York, London: Harper & Brothers, 1942.

Jeong, H., Tombor, B., Albert, R., Oltvai, Z. N., and Barbási, A.-L. The large-scale organization of metabolic networks. *Nature,* 407, 2000, pp. 651–654.

Joy, B. Why the future doesn't need us. *Wired,* April 2000.

Kadanoff, Leo P. Chaos: A view of complexity in the physical sciences. In *From Order to Chaos: Essays: Critical, Chaotic, and Otherwise.* Singapore: World Scientific, 1993.

Kauffman, S. A. *The Origins of Order.* New York: Oxford University Press, 1993.

Kauffman, S. A. *At Home in the Universe.* New York: Oxford University Press, 1995.

Kauffman, S. A. *Investigations.* New York: Oxford University Press, 2002.

Keller, E. F. Revisiting 'scale-free' networks. *BioEssays,* 27, 2005, pp. 1060–1068.

Kleinfeld, J. S. Could it be a big world after all? The "six degrees of separation" myth. *Society,* 39, 2002.

Kleinfeld, J. S. Six degrees: Urban myth? *Psychology Today,* 74, March/April 2002.

Kollock, P. The production of trust in online markets. In E. J. Lawler, M. Macy, S. Thyne, and H. A. Walker (editors), *Advances in Group Processes,* 16. Greenwich, CT: JAI Press, 1999.

Kozlowski, J. and Konarzweski, M. Is West, Brown and Enquist's model of allometric scaling mathematically correct and biologically relevant? *Functional Ecology,* 18, 2004, pp. 283–289.

Kubrin, D. Newton and the cyclical cosmos: Providence and the mechanical philosophy. *Journal of the History of Ideas,* 28(3), 1967.

Kurzweil, R. *The Age of Spiritual Machines: When Computers Exceed Human Intelligence.* New York: Viking, 1999.

Langton, C. G. *Artificial Life: An Overview.* Cambridge, MA: MIT Press, 1997.

Laplace, P. S. *Essai Philosophique Sur Les Probabilites.* Paris: Courcier, 1814.

Lee, B. and Ajjarapu, V. Period-doubling route to chaos in an electrical power system. *IEE Proceedings,* Part C, 140, 1993, pp. 490–496.

Leff, H. S. and Rex, A. F. *Maxwell's Demon: Entropy, Information, Computing.* Princeton, NJ: Princeton University Press. Second edition 2003, Institute of Physics Pub., 1990.

Leibniz, G. In C. Gerhardt (Ed.), *Die Philosophischen Schriften von Gottfried Wilhelm Leibniz.* Volume vii. Berlin: Olms, 1890.

Lesley, R. Xu, Y., Kalled, S. L., Hess, D. M., Schwab, S. R., Shu, H.-B., and Cyster, J. G. Reduced competitiveness of autoantigen-engaged B cells due to increased dependence on BAFF. *Immunity,* 20(4), 2004, pp. 441–453.

Levy, S. The man who cracked the code to everything. *Wired,* Issue 10.06, June 2002.

Lewin, R. *Complexity: Life at the Edge of Chaos.* New York: Macmillan, 1992.

Li, M. and Vitanyi, P. *An Introduction to Kolmogorov Complexity and Its Applications.* 2nd Edition. New York: Springer-Verlag, 1997.

Li, T. Y. and Yorke, J. A. Period three implies chaos. *American Mathematical Monthly* 82, 1975, p. 985.

Liebreich, M. How to Save the Planet: Be Nice, Retaliatory, Forgiving, & Clear. White Paper, New Energy Finance, Ltd., 2007 [http://www.newenergyfinance.com/docs/Press/NEF_WP_Carbon-Game-Theory_05.pdf].

Liljeros, F., Edling, C. R., Nunes Amaral, L. A., Stanely, H. E., and Aberg, Y. The web of human sexual contacts. *Nature*, 441, 2001, pp. 907–908.

Lioni, A., Sauwens, C., Theraulaz, G., and Deneubourg, J.-L. Chain formation in O*Ecophylla longinoda*. *Journal of Insect Behavior*, 14(5), 2001, pp. 679–696.

Liu, H. A brief history of the concept of chaos, 1999 [http://members.tripod.com/~huajie/Paper/chaos.htm].

Lloyd, S. The calculus of intricacy. *The Sciences*, 30, 1990, pp. 38–44.

Lloyd, S. Measures of complexity: A non-exhaustive list. *IEEE Control Systems Magazine*, August 2001.

Lloyd, S. and Pagels, H. Complexity as thermodynamic depth. *Annals of Physics*, 188, 1988, pp. 186–213.

Locke, J. *An Essay concerning Human Understanding*. Edited by P. H. Nidditch. Oxford: Clarendon Press, 1690/1975.

Lohr, S. This boring headline is written for Google. *New York Times*, April 9, 2006.

Lorenz, E. N. Deterministic nonperiodic flow. *Journal of Atmospheric Science*, 357, 1963, pp. 130–141.

Lovelock, J. E. *The Ages of Gaia*. New York: Norton, 1988.

Luisi, P. L. *The Emergence of Life: from Chemical Origins to Synthetic Biology*. Cambridge, U.K.: Cambridge University Press, 2006.

Mackenzie, D. Biophysics: New clues to why size equals destiny. *Science*, 284(5420), 1999, pp. 1607–1609.

Macrae, N. *John von Neumann*. New York: Pantheon Books, 1992.

Maddox, J. Slamming the door. *Nature*, 417, 2007, p. 903.

Malone, M. S. God, Stephen Wolfram, and everything else. *Forbes ASAP*, November 27, 2000. [http://members.forbes.com/asap/2000/1127/162.html].

Mandelbrot. B. An informational theory of the statistical structure of languages. In W. Jackson (editor), *Communication Theory*, Woburn, MA: Butterworth, 1953, pp. 486–502.

Mandelbrot, B. B. *The Fractal Geometry of Nature*. New York: W. H. Freeman, 1977.

Markoff, J. Brainy robots start stepping into daily life, *New York Times*, July 18, 2006.

Marr, D. *Vision*. San Francisco: Freeman, 1982.

Mattick, J. S. RNA regulation: A new genetics? *Nature Reviews: Genetics*, 5, 2004, pp. 316–323.

Maturana, H. R. and Varela, F. J. *Autopoiesis and Cognition : The Realization of the Living*, Boston: D. Reidel Publishing Co., 1980.

Maxwell, J. C. *Theory of Heat*. London: Longmans, Green and Co, 1871.

May, R. M. Simple mathematical models with very complicated dynamics. *Nature*, 261, 459–467, 1976.

Mayr, E. An overview of current evolutionary biology. In *New Perspectives on Evolution*, 1991, pp. 1–14.

McAllister, J. W. Effective complexity as a measure of information content. *Philosophy of Science* 70, 2003, pp. 302–307.

McClamrock, R. Marr's three leves: A re-evaluation. *Minds and Machines*, 1(2), 1991, pp. 185–196.

McCulloch, W. and Pitts, W. A logical calculus of ideas immanent in nervous activity. *Bulletin of Mathematical Biophysics* 5, 1942, pp. 115–133.

McShea, D. W. The hierarchical structure of organisms: A scale and documentation of a trend in the maximum. *Paleobiology*, 27(2), 2001, pp. 405–423.

Metropolis, N., Stein, M. L., and Stein, P. R. On finite limit sets for transformations on the unit interval. *Journal of Combinatorial Theory*, 15(A), 1973, pp. 25–44.

Milgram, S. The small-world problem. *Psychology Today* 1, 1967, pp. 61–67.

Millay, E. St. Vincent. *Mine the Harvest: A Collection of New Poems.* New York: Harper, 1949.

Miller, G. A. Some effects of intermittent silence. *The American Journal of Psychology*, 70, 1957, pp. 311–314.

Millonas, M. M. The importance of being noisy. *Bulletin of the Santa Fe Institute*, Summer, 1994.

Minsky, M. *The Society of Mind*, Simon & Schuster, 1987.

Mitchell, M. Computation in cellular automata: A selected review. In T. Gramss et al. (editors), *Nonstandard Computation*. Weinheim, Germany: Wiley-VCH, 1998, pp. 95–140.

Mitchell, M. *Analogy-Making as Perception*, MIT Press, 1993.

Mitchell, M. Analogy-making as a complex adaptive system. In L. Segel and I. Cohen (editors), *Design Principles for the Immune System and Other Distributed Autonomous Systems*. New York: Oxford University Press, 2001.

Mitchell, M. Life and evolution in computers. *History and Philosophy of the Life Sciences*, 23, 2001, pp. 361–383.

Mitchell, M. Complex systems: Network thinking. *Artificial Intelligence*, 170(18), 2006, pp. 1194–1212.

Mitchell, M. Crutchfield, J. P., and Das, R. Evolving cellular automata to perform computations: A review of recent work. In *Proceedings of the First International Conference on Evolutionary Computation and its Applications (EvCA '96)*. Moscow: Russian Academy of Sciences, 1996, pp. 42–55.

Mitzenmacher, M. A brief history of generative models for power law and lognormal distributions. *Internet Mathematics*, 1(2), 2003, pp. 226–251.

Montoya, J. M. and Solé, R. V. Small world patterns in food webs. *Journal of Theoretical Biology*, 214(3), 2002, pp. 405–412.

Moore, C. Recursion theory on the reals and continuous-time computation. *Theoretical Computer Science*, 162, 1996, pp. 23–44.

Moravec, H. *Robot: Mere Machine to Transcendent Mind*. New York: Oxford University Press, 1999.

Morton, O. Attack of the stuntbots. *Wired*, 12.01.2004.

Mukherji, A., Rajan, V., and Slagle, J. R. Robustness of cooperation. *Nature*, 379, 1996, pp. 125–126.

Muotri, A. R., Chu, V. T., Marchetto, M. C. N., Deng, W., Moran, J. V. and Gage, F. H. Somatic mosaicism in neuronal precursor cells mediated by L1 retrotransposition. *Nature*, 435, 2005, pp. 903–910.

Nagel, E. and Newman, J. R. *Gödel's Proof*. New York: New York University Press, 1958.

Newman, M.E.J. Power laws, Pareto distributions and Zipf's law. *Contemporary Physics*, 46, 2005, pp. 323–351.

Newman, M.E.J., Moore, C., and Watts, D. J. Mean-field solution of the small-world network model. *Physical Review Letters*, 84, 1999, pp. 3201–3204.

Newman, M.E.J., Forrest, S., and Balthrop, J. Email networks and the spread of computer viruses. *Physical Review E*, 66, 2002, p. 035101.

Nicolis, G. and Progogine, I. *Exploring Complexity*. New York: W.H. Freeman, 1989.

Niklas, K. J. Size matters! *Trends in Ecology and Evolution* 16(8), 2001, p. 468.

Nowak, M. A. Five rules for the evolution of cooperation. *Science*, 314(5805), 2006, pp. 1560–1563.

Nowak, M. A., Bonhoeffer, S., and May, R. M. Spatial games and the maintenance of cooperation. *Proceedings of the National Academy of Sciences, USA*, 91, 1994, pp. 4877–4881.

Nowak, M. A., Bonhoeffer, S., and May, R. M. Reply to Mukherji et al. *Nature*, 379, 1996, p. 126.

Nowak, M. A. and May, R. M. Evolutionary games and spatial chaos. *Nature*, 359(6398), 1992, pp. 826–829.

Nowak, M. A. and Sigmund, K. Biodiversity: Bacterial game dynamics. *Nature*, 418, 2002, pp. 138–139.

Packard, N. H. Adaptation toward the edge of chaos. In J.A.S. Kelso, A. J. Mandell, M. F. Shlesinger (editors), *Dynamic Patterns in Complex Systems*, pp. 293–301. Singapore: World Scientific, 1988.

Pagels, H. *The Dreams of Reason*. New York: Simon & Schuster, 1988.

Paton, R., Bolouri, H., Holcombe, M., Parish, J. H., and Tateson, R. (editors). *Computation in Cells and Tissues: Perspectives and Tools of Thought*. Berlin: Springer-Verlag, 2004.

Peak, D., West, J. D., Messinger, S. M., and Mott, K. A. Evidence for complex, collective dynamics and emergent, distributed computation in plants. *Proceedings of the National Academy of Sciences, USA*, 101(4), 2004, pp. 918–922.

Pearson, H. What is a gene? *Nature*, 441, 2006, pp. 399–401.

Pierce, J. R. *An Introduction to Information Theory: Symbols, Signals, and Noise*. New York: Dover, 1980. (First edition, 1961.)

Pines, D. (editor). *Emerging Syntheses in Science*. Reading, MA: Addison-Wesley, 1988.

Poincaré, H. *Science and Method*. Translated by Francis Maitland. London: Nelson and Sons, 1914.

Poundstone, W. *The Recursive Universe*. William Morrow, 1984.

Poundstone, W. *Prisoner's Dilemma*. New York: Doubleday, 1992.

Price, D. J. Networks of scientific papers. *Science* 149, 1965, pp. 510–515.

Provine, W. B. *The Origins of Theoretical Population Genetics*. University of Chicago Press, 1971.

Rapoport, A. *General System Theory: Essential Concepts and Applications*, Cambridge, MA: Abacus Press, 1986.

Redner, S. How popular is your paper? An empirical study of the citation distribution. *European Physical Journal B*, 4(2), 1998, pp. 131–134.

Regis, E. *Who Got Einstein's Office? Eccentricity and genius at the Institute for Advanced Study*. Menlo Park, CA: Addison-Wesley, 1987.

Rendell, P. Turing universality of the game of Life. In A. Adamatzky (editor), *Collision-Based Computing*. London: Springer-Verlag, 2001, pp. 513–539.

Robbins, K. E., Lemey, P., Pybus, O. G., Jaffe, H. W., Youngpairoj, A. S., Brown, T. M., Salemi, M., Vandamme, A. M. and Kalish, M. L., U.S. human immunodeficiency virus type 1 epidemic: Date of origin, population history, and characterization of early strains. *Journal of Virology*, 77(11), 2003, pp. 6359–6366.

Rota, G-C, In memoriam of Stan Ulam—The barrier of meaning. *Physica D*, 2, 1986, pp. 1–3.

Rothman, T. The evolution of entropy. Chapter 4 in *Science à la Mode*. Princeton, NJ: Princeton University Press, 1989.

Russell, B. *A History of Western Philosophy*, Touchstone, 1967 (First edition, 1901.)

Schlossberg, D. LAX Computer Crash Strands International Passengers. *ConsumerAffairs.com*, August 13, 2007. [http://www.consumeraffairs.com/news04/2007/08/lax_computers.html].

Schneeberger, A. Mercer, C. H., Gregson, S. A., Ferguson, N. M., Nyamukapa, C. A., Anderson, R. M., Johnson, A. M., and Garnett, G. P. Scale-free networks and sexually transmitted diseases: A description of observed patterns of sexual contacts in Britain and Zimbabwe. *Sexually Transmitted Diseases*, 31(6), 2004, pp. 380–387.

Schwartz, J. Who needs hackers? *New York Times*, September 12, 2007.

Selvam, A. M. The dynamics of deterministic chaos in numerical weather prediction models. *Proceedings of the American Meteorological Society, 8th Conference on Numerical Weather Prediction*, Baltimore, MD, 1988.

Shalizi, C. Networks and Netwars, 2005. Essay at [http://www.cscs.umich.edu/~crshalizi/weblog/347.html].

Shalizi, C. Power Law Distributions, 1/f noise, Long-Memory Time Series, 2007. Essay at [http://cscs.umich.edu/~crshalizi/notebooks/power-laws.html].

Shannon, C. A mathematical theory of communication. *The Bell System Technical Journal*, 27, 1948, pp. 379–423, 623–656.

Shaw, G. B. *Annajanska, the Bolshevik Empress*. Whitefish, MT: Kessinger Publishing, 2004. (Originally published 1919.)

Shouse, B. Getting the behavior of social insects to compute. *Science*, 295(5564), 2002, p. 2357.

Sigmund, K. On prisoners and cells. *Nature*, 359(6398), 1992, p. 774.

Simon, H. A. On a class of skew distribution functions. *Biometrika* 42(3–4), 1955, p. 425.

Simon, H. A. The architecture of complexity. *Proceedings of the American Philosophical Society*, 106(6), 1962, pp. 467–482.

Sompayrac, L. M. *How the Immune System Works*, 2nd edition, Blackwell Publishing, 1991.

Stam, C. J. and Reijneveld, J. C. Graph theoretical analysis of complex networks in the brain. *Nonlinear Biomedical Physics*, 1(1), 2007, p. 3.

Stoppard, T. *Arcadia*. New York: Faber & Faber, 1993.

Strogatz, S. *Nonlinear Dynamics and Chaos*. Reading, MA: Addison-Wesley, 1994.

Strogatz, S. *Sync: How Order Emerges from Chaos in the Universe, Nature, and Daily Life*. New York: Hyperion, 2004, p. 287.

Szilard, L. On the decrease of entropy in a thermodynamic system by the intervention of intelligent beings. *Zeitschrift fuer Physik*, 53, 1929, pp. 840–856.

Tattersall, I. *Becoming Human: Evolution and Human Uniqueness*. New York: Harvest Books, 1999.

Travis, J. Eye-opening gene. *Science News Online*, May 10, 1997.

Turing, A. M. On computable numbers, with an application to the *Entscheidungsproblem*. *Proceedings of the London Mathematical Society*, 2(42), 1936, pp. 230–265.

Ulam, S. M. and von Neumann, J. On combination of stochastic and deterministic processes (abstract). *Bulletin of the American Mathematical Society*, 53, 1947, 1120.

Varn, D. P., Canright, G. S., and Crutchfield, J. P. Discovering planar disorder in close-packed structures from X-ray diffraction: Beyond the fault model. *Physical Review B*, 66, 2002, pp. 174110-1–174110-4.

Verhulst, P.-F. Recherches mathematiques sur la loi d'accroissement de la population. *Nouv. mem. de l'Academie Royale des Sci. et Belles-Lettres de Bruxelles* 18, 1845, pp. 1–41.

Von Bertalanffy, L. An outline of general system theory. *The British Journal for the Philosophy of Science*, 1(92), 1950, 134–165.

Von Bertalanffy, L. *General System Theory: Foundations, Development, Applications*, New York: G. Braziller, 1969.

Von Neumann, J. *Theory of Self-Reproducing Automata* (edited and completed by A. W. Burks). Urbana: University of Illinois Press, 1966.

Wagner, N. R. The logistic equation in random number generation. *Proceedings of the Thirtieth Annual Allerton Conference on Communications, Control, and Computing*, University of Illinois at Urbana-Champaign, 1993, pp. 922–931.

Wang, H. *Reflections on Kurt Gödel*. Cambridge, MA: MIT Press, 1987.

Watts, D. J. *Six Degrees: The Science of a Connected Age*. New York: Norton, 2003.

Watts, D. J. and Strogatz, S. H. Collective dynamics of 'small world' networks. *Nature*, 393, 1998, pp. 440–442.

Weiner, J. *The Beak of the Finch: A Story of Evolution in Our Time.* New York: Knopf, 1994.

Wiener, N. *The Human Use of Human Beings.* Boston: Houghton Mifflin, 1950.

Wiener, N. *Cybernetics: Or the Control and Communication in the Animal and the Machine.* Cambridge, MA: MIT Press, 1948.

Wiener, P. Antibiotic production in a spatially structured environment. *Ecology Letters*, 3(2), 2000, pp. 122–130.

West, G. B., Brown, J. H., and Enquist, B. J. The fourth dimension of life: Fractal geometry and allometric scaling of organisms. *Science*, 284, 1999, pp. 1677–1679.

West, G. B. and Brown, J. H. Life's universal scaling laws. *Physics Today*, 57(9), 2004, p. 36.

West, G. B. and Brown, J. H. The origin of allometric scaling laws in biology from genomes to ecosystems: Towards a quantitative unifying theory of biological structure and organization. *Journal of Experimental Biology* 208, 2005, pp. 1575–1592.

West, G. B., Brown, J. H., and Enquist, B. J. Yes, West, Brown and Enquist's model of allometric scaling is both mathematically correct and biologically relevant. (Reply to Kozlowski and Konarzweski, 2004.) *Functional Ecology*, 19, 2005, pp. 735–738.

Westfall, R. S. *Never at Rest: A Biography of Isaac Newton.* Cambridge, U.K.: Cambridge University Press, 1983.

Whitfield, J. All creatures great and small. *Nature*, 413, 2001, pp. 342–344.

Williams, F. Artificial intelligence has a small but loyal following. *Pensions and Investments*, May 14, 2001.

Williams, S. Unnatural selection. *Technology Review*, February 2005.

Willinger, W., Alderson, D., Doyle, J. C., and Li, L. More 'normal' than normal: Scaling distributions and complex systems. In R. G. Ingalls et al., *Proceedings of the 2004 Winter Simulation Conference*, pp. 130–141. Piscataway, NJ: IEEE Press, 2004.

Wolfram, S. Universality and complexity in cellular automata. *Physica D*, 10, 1984, pp. 1–35.

Wolfram, S. Twenty problems in the theory of cellular automata. *Physica Scripta*, T9, 1985, pp. 170–183.

Wolfram, S. *A New Kind of Science.* Champaign, IL: Wolfram Media, 2002.

Wright, R. Did the universe just happen? *Atlantic Monthly*, April 1988, pp. 29–44.

Yoon, C. K. From a few genes, life's myriad shapes. *New York Times*, June 26, 2007.

Yule, G. U. A mathematical theory of evolution, based on the conclusions of Dr. J. C. Willis. *Philosophical Transactions of the Royal Society of London*, Ser. B, 213, 1924, pp. 21–87.

Ziff, E. and Rosenfield, I. Evolving evolution. *The New York Review of Books*, 53(8), May 11, 2006.

Zipf, G. K. *Selected Studies of the Principle of Relative Frequency in Language.* Cambridge, MA: Harvard University Press, 1932.

Zuse, K. *Rechnender Raum.* Braunschweig: Friedrich Vieweg & Sohn, 1969. English translation: *Calculating Space.* Cambridge, MA: MIT Technical Translation AZT-70-164-GEMIT, Massachusetts Institute of Technology (Project MAC), 02139, February 1970.

Zykov, V. Mytilinaios, E., Adams, B., and Lipson, H. Self-reproducing machines. *Nature*, 435, 2005, pp. 163–164.

INDEX

adaptation, 13, 128
  balance between unfocused and focused exploration in, 183–184
  challenges to centrality of, 86
  Darwin's observations of, 76–77
  as expanded beyond biological realm, 300
  Holland's general principles for, 128, 184
  Lamarckian, 73
  in Modern Synthesis, 83, 86
  as requisite of life, 116
  role in Darwin's theory, 78–79
  role of information in, 146, 170
  See also evolution; natural selection
Albert, Réka, 230, 252, 294
algebraic topology, 21
algorithm, 129, 145
  ant-colony optimization, 184
  genetic (*see* genetic algorithms)
  PageRank, 258
  pseudo-random number generation, 133, 155, 306
  for Turing machine, 63
algorithmic information content, 98–99
allele, 80–82
alternative splicing, 275
amino acids, 89–92, 140, 275
analogy
  between ant colonies and brains, 5
  as central to intelligence, 188, 208
  conceptual slippage in, 188, 191–193, 196–197, 202, 206
  definition of, 187
  in definition of Shannon entropy, 54
  between DNA and self-copying program, 122
  between effective complexity and scientific theory formation, 99
  examples of, 187–188
  letter-string microworld for, 190–193
  as modeled by the Copycat program, 193–208
Anderson, Phillip, 234
ant colonies, 3–5, 145, 176–178, 180–184, 195, 212
  information processing (or computation) in, 176–178, 179–185, 195–196
ant colony optimization algorithms, 184
antibodies, 8–9, 172, 174–175, 195
anticodons, 91–92
antigens, 173–175, 180–183, 195
Antonopoulos, Andreas, 257
Aristotle, 16–17, 113
arrow of time, 43
artificial immune systems, 184
artificial intelligence (AI), x, 55, 185, 187, 190, 208, 227, 298
artificial life, 115–116, 292, 298
Ashby, W. Ross, 296–297
Aspray, William, 297
attractors, 30, 32, 34–35, 38, 103
  in random Boolean networks, 285, 287

autonomy
  as requisite for life, 116
autopoiesis, 298
Axelrod, Robert, 214–219, 222–224

Bak, Per, 303
Barabási, Albert-László, 230, 232–233, 249, 252–254, 294
basal metabolic rate. *See* metabolic rate
base pairs. *See* bases (genetic)
bases (genetic), 90–93, 96, 278. *See also* nucleotides
Bateson, Gregory, 296–297
B cells, 9, 172–176, 195
Beagle, H.M.S., 75–76
Beinhocker, Eric, 40
bell-curve distribution, 243–244, 269
Bennett, Charles, 46–47, 100–102
bifurcation, 34–36, 38, 285, 298
bifurcation diagram, 34–36, 103
biological constraints, 85–86, 281, 287
biologically inspired computation, 184–185, 207. *See also* genetic algorithms
bit of information, 45, 54
Boltzmann, Ludwig, 47–51, 307
Boltzmann entropy, 50–51, 307
Bonhoeffor, Sebastian, 223
Boole, George, 283
Boolean function, 283
Boolean networks. *See* random Boolean networks
Box, George, 222
brain, 5–7, 125, 168
  as a computer, 56, 69, 145, 158, 168
  as a network, 229, 238–239, 247–248
Brillouin, Leon, 46
Brown, James, 262–267, 294, 300
Buddha, 71
Buffon, Louis Leclerk de, 72
Burks, Alice, 57
Burks, Arthur, 57, 123, 145

CA. *See* cellular automata
calculus, 18, 301–302
  of complexity, 301–303
Calvino, Italo, 225
Carnot, Sadi, 302
Carroll, Sean, 278
carrying capacity, 25, 27

cascading failure, 255–258
C. elegans, 158, 238, 247
cellular automata
  architecture of 146–148
  classes of behavior in, 155–156
  computation in, 157–158, 161, 164–168, 171–172, 303
  elementary, 152–153 (*see also* rule 110 cellular automaton; rule 30 cellular automaton)
  as evolved by genetic algorithms, 161–164
  as idealized models of complex systems, 148–149, 211
  information processing in, 157–158, 161, 164–168, 171–172, 303
  as models for the universe, 158–159
  numbering of, 153–154
  particles in, 166–168, 171–172
  as pseudo-random number generators, 155
  rules, 147–149
  space-time diagrams of, 153–155, 162, 164–165, 167
  as substrate for self-reproducing automata, 149
  as universal computers, 149–151, 156
central processing unit (CPU), 145–146, 160–161
chaos, 20–22, 28, 31–39, 211, 273, 284, 293, 300
  edge of, 284–285
  in the logistic map, 31–33
  onset of, 35–36
  period-doubling route to, 34–35
  in random Boolean Networks, 284–285
  revolutionary ideas from, 38
  universal properties of, 34–38, 294
characteristic scale (of a distribution), 243–244
chromosomes, 88–89, 96, 274–275
citric acid cycle, 179
classical mechanics, 19, 48
Clausius, Rudolph, 47, 51
clockwork universe, 19, 33
clustering (in networks), 235–236, 238–240, 245, 252, 255
coarse graining, 101, 183
codons, 90–92
coevolution of Web and search engines, 10

338 | INDEX

Cohen, Irun, 40
colonial organisms, 110
complex adaptive systems
    distinction from complex systems, 13
    See also complexity
complexity (or complex systems)
    as algorithmic information content, 98–99
    "calculus" of, 301–303
    central question of sciences of, 13
    common properties of, 294–295
    as computational capacity, 102
    definitions of, 13, 94–111
    as degree of hierarchy, 109–111
    effective, 98–100
    in elementary cellular automata, 155
    as entropy, 96–98
    as fractal dimension, 102–109
    future of, 301–303
    Horgan's article on, 291–292
    Latin root, 4
    as logical depth, 100–101
    measurement of, 13, 94–111
    problems with term, 95, 299, 301
    roots of sciences of, 295–298
    science of versus sciences of, 14, 95
    significance of in science, 300
    as size, 96
    source of biological, 233, 248–249, 273–288
    statistical, 102–103
    as thermodynamic depth, 101–102
    as a threat, 257
    unified theories of, 293, 299
    universal computation as upper limit on, 157
    universal principles for, 299
    vocabulary for, 293, 298, 301–303
complex systems. *See* complexity
computable problem (or process), 157
computation
    biologically inspired, 184–185, 207 (*see also* genetic algorithms)
    in the brain, 168
    in cellular automata, 157–158, 161, 164–168, 171–172, 303
    courses on theory of, 67
    defined as Turing machine (*see* Turing machines)
    definite procedures as, 63–64, 146
    definitions of, 57
    evolutionary (*see* genetic algorithms)
    limits to, 68
    linked to life and evolution, 115
    in logical depth, 100
    in natural systems, *xi*, 56–57, 145–146, 156–158, 169–170, 172, 179–185
    non-von-Neumann-style, 149, 151, 171
    reversible, 46–47
    in stomata networks, 168
    in traditional computers, 170–171
    universal (*see* universal computation)
    von-Neumann-style, 146, 169–171, 209
    See also information processing
computational capacity, 102
computational mechanics, 303
computer models
    caveats for, 222–224, 291
    of genetic regulatory networks, 282–284
    period-doubling route to chaos in, 37
    prospects of, 158, 220–222
    replication of, 223–224
    of weather, 22, 37
    See also models
computing. *See* computation
conceptual slippage, 188, 191–193, 196–197, 202, 206
consciousness, 4, 6, 184, 189
convergent evolution, 280
Conway, John, 149–151
Cook, Matthew, 156
Copernicus, 17
Copycat program, 193
    analogies with biological systems, 208
    codelets, 197–198
    as example of idea model, 211
    example run of, 198–206
    frequencies of answers in, 206–207
    parallel terraced scan in, 197–198
    perception of meaning by, 208
    perceptual structures in, 198–206
    Slipnet, 196
    temperature, 198
    Workspace, 197
Coullet, Pierre, 28
Crick, Francis, 89, 93, 274
Crutchfield, James, 94, 102–103, 164, 166, 293, 303
cybernetics, 125, 296–299
cytokines, 176, 179–180, 208

Darwin, Charles, 72–79, 81, 97, 124, 128, 273, 280
Darwin, Erasmus, 73
Darwinian evolution (or Darwinism)
  in computers (*see* genetic algorithms)
  defense of, 288
  in the immune system, 9
  major ideas of, 78
  origin of the theory of, 72–79
  principles of, under Modern Synthesis, 83
  of strategies in the Prisoner's dilemma, 216
  synthesis with Mendel's theory, 81–84
  See also natural selection
Das, Rajarshi, 160
Dawkins, Richard, 87
de Buffon, Louis Leclerk, 72
decision problem. *See Entscheidungsproblem*
definite procedures, 58–61, 63–68
  defined as Turing machines, 63–64
degree distributions, 235–236, 239–240, 251–255
  of scale-free networks, 245, 248
  of the Web, 240–244, 265
Delbrück, Max, 297
Dennett, Daniel, 72, 211
deoxyribonucleic acid. *See* DNA
Descartes, René, *ix*, 210
dimension
  spatial, 107–108
  fractal, 107–109, 264–265
diploid organism, 88
DNA, 79, 89
  complete sequence of, 276
  engineering, 274
  information content of, 98–99
  junk, 96, 99, 278, 280
  mechanics of, 90–93
  methylation of, 276
  new ideas about, 274–276
  patents on, 277
  rate of change in, 267
  self-replication of, 122
  shuffling of in lymphocytes, 174
  similarity of (in different species), 277–278
  switches in, 278–280
  transcription of, 90–93, 249
  translation of, 91–93
  of yeast, 96
  See also genes
dominant allele, 80–82
double logarithmic plot, 261, 320
Dow Jones Industrial Average, 11
Draper, Norman, 222
Drescher, Melvin, 213
Dugas, Gaetan, 250
dynamical systems theory, 15–16, 38–39, 285, 298, 303
  logistic map as illustration of ideas in, 27–33
  roots of, 16–19
  See also chaos
dynamics. *See* dynamical systems theory
Dyson, Freeman, 126

economic networks, 230
economies, 4, 9–10, 222
Eddington, Arthur, 40
edge of chaos, 284–285
effective complexity, 98–100, 102
effective measure complexity, 102
Eigen, Manfred, 86
Einstein, Albert, 69, 72, 124, 210, 215, 293, 295
Eisenstein, Robert, 94
Eldredge, Niles, 84–85, 87
emergence, *xii*, 13, 286, 293, 301, 303
  in cellular automata, 155
  of complexity, 4, 155, 286
  of cooperation, 215–220
  general theories of, 303
  of parallel terraced scan, 195–196
  predicting, 301
  of randomness, 38
  in statistical mechanics, 48
  of thinking and consciousness, 189
  vagueness of definition of, *xii*, 293–294, 301
  of Web's degree distribution, 252
emergent behavior. *See* emergence
energy
  definition of, 41–42
  heat as, 47
  in metabolism, 178–179, 258, 265
  as a primitive component of reality, 169, 293
  relation to entropy, 42, 47

in thermodynamics, 42–43
for work done by natural selection, 79
Enigma, 69
Enquist, Brian, 262–267, 294, 300
entropy
　Boltzmann, 50–51, 307
　complexity as, 96–98
　decrease of in evolution, 79
　definition of, 42
　link with information, 45–47
　in literature, 71
　in Maxwell's demon paradox, 43–45
　Shannon, 51–54
　in thermodynamics, 42–43, 47–48
Entscheidungsproblem (decision problem), 58–60
　Turing's solution to, 65–69
epigenetics, 276–277
Erdös, Paul, 125
Evo-Devo, 277–281
evolution
　challenges to Modern Synthesis principles of, 84–87
　in computers (*see* genetic algorithms)
　of cooperation (*see* evolution of cooperation)
　increase in complexity under, 109–110
　Kauffman's theories of self-organization in, 281–287
　major ideas of Darwinian, 78–79
　Modern Synthesis principles of, 83
　modification of genetic switches as major force in, 279–280
　by natural selection (*see* natural selection)
　neutral, 86
　optimization of biological fuel-transport networks by, 265–266
　origins of Darwin's theory of, 75–79
　pre-Darwinian notions of, 72–75
　principles of, under Modern Synthesis, 83
　as requisite for life, 116
evolutionary computation. *See* genetic algorithms
evolutionary developmental biology. *See* Evo-Devo
evolution of cooperation, 212–220
　effect of norms and metanorms on, 218–219, 223–224
　effect of spatial structure on, 219–220, 223

　as example of common principles in complex systems, 294
　general conditions for, 217–221
　exploration and exploitation, balance between, 184, 195, 294
　expression (of genes), 92, 278

Farmer, Doyne, 94, 293
Feigenbaum, Mitchell, 28, 35–38
Feigenbaum-Coullet-Tressor theory, 38
Feigenbaum's constant, 35–38
finches, beaks of, 76, 289
fine-grained exploration 182–183
Fisher, Ronald, 82–83, 128
Flood, Merrill, 213
food webs, 251–252
Forrest, Stephanie, 94
"fourth law" of thermodynamics, 286
fractal, 103–106
　dimension, 107–109, 264–265
　networks, 266–267, 294–295
　relation to power laws, 264–265, 268–269
　space-filling, 266
fractal dimension, 107–109, 264–265
fraction of carrying capacity, 27
Franks, Nigel, 3–4
Fredkin, Edward, 159
Freud, Sigmund, 74

GA. *See* genetic algorithm
Gabor, Denis, 46, 125
Galan, Jose Manuel, 223
Galápagos Islands, 76, 280
Galileo, 17–19
Galton, Francis, 82
Game of Life, 149–151, 156
　simulating a universal computer in, 150–151
Gaussian (or normal) distribution, 243–244, 269
Gehring, Walter, 281
Gell-Mann, Murray, 41, 98–99, 151
general relativity, 210
general system theory, 297–298
genes
　alternative splicing in, 275
　definition of, 89–90
　difficulty with definition of, 95, 274–277
　expression of, 92, 278

genes (*continued*)
  for controlling beak size and shape in birds, 280
  for development of eyes, 280–281
  jumping, 275
  master, 278–281
  nonlinearity of, 276–277
  random Boolean networks as models of, 282–287
  regulation of (*see* genetic regulatory networks)
  RNA editing of, 275
  status of patents on, 277
  switches for, 278–280
  transcription of, 90–91
  translation of, 91–92
genetic algorithms
  applications of, 129–130, 142
  balancing exploration and exploitation in, 184
  evolving cellular automata with, 160, 162–164
  evolving Prisoner's dilemma strategies with, 217–218
  as example of idea model, 211
  origin of, 128
  recipe for, 128–129
  Robby the Robot as illustration of, 130–142
genetic code, 89–90, 93
genetic drift, 82–83
genetic engineering, 277
genetic regulatory networks, 229, 248–249, 275–281
  genetic switches in, 278–280
  as modeled by random Boolean networks, 282–287
  noise in, 249
genetics
  basics of, 88–93
  implications of Evo-Devo on, 277–281
  new ideas about, 274–277
  population, 82
  See also genes; genetic code; genetic regulatory networks; genetic switches
genetic switches, 278–280
Gershenson, Carlos, 299
Gide, André, 303
Gladwell, Malcolm, 253
Glance, Natalie, 223

Gleick, James, 302
glycolysis, 179, 249
Gödel, Escher Bach: an Eternal Golden Braid (Douglas Hofstadter), ix, 5, 121, 189
Gödel, Kurt, 59–60, 68–70
Gödel's theorem, 59–60
Google, 236, 239–240. 244–245
Gordon, Deborah, 177, 293–295
Gould, Stephen Jay, 84–87, 278
Grand Unified Theories (GUTs), 292–293
Grassberger, Peter, 102
gravity, universal law of, 19, 209–210, 269

Haken, Hermann, 298
Haldane, J.B.S., 82
Halting problem, 66–67
  solvability by nondigital computers, 158
  Turing's solution to, 67–68, 121
Hardin, Garrett, 214
Heisenberg, Werner, 20
heredity
  chromosomes as carriers of, 89
  Mendel's results on, 79–81
  See also inheritance
hierarchy (as a measure of complexity), 109–111
Highly Optimized Tolerance (HOT), 257, 269
Hilbert, David, 57–60, 68
Hilbert's problems 57–59
historical contingency, 85–86
H.M.S. *Beagle*, 75–76
Hobbes, Thomas, 215, 221
Hoelzer, Guy, 287
Hofstadter, Douglas, ix, xi, 5, 92–93, 121, 189–193, 195, 208.
Holland, John, 127–128, 184, 221, 294
Holley, Robert, 93
Horgan, John, 291–292, 294
Horvitz, Eric, 187
Hraber, Peter, 160
Huberman, Bernardo, 223
Hübler, Alfred, 94
hubs (network), 236, 240, 245, 248, 250, 252
Human Genome Project, 96, 276
Hungarian phenomenon, 125
Huxley, Thomas, 78, 128

idea models, 38–39, 211–212, 214, 220–222

immune system, 6, 8–9, 172–176
   analogy with Copycat, 207–208
   artificial, 184
   information processing (or computation) in, 56, 172–176, 179–185, 195–196
in-degree, 240–241
   distribution in Web, 245
   in random Boolean networks, 283–284
infinite loop, 66
infinite loop detector, 66
information
   acquisition of meaning in complex systems, 184, 208
   bit of, 45
   as central topic in cybernetics, 296
   in cellular automata, 171–172
   in cellular automaton particles, 165–168
   in complex systems, 40–41, 146, 157–158, 179–185
   content, 52–54, 96–97, 99, 102 (*see also* algorithmic information content)
   dual use of in self-replication, 121–122
   flow in networks, 236, 239, 248, 255–257
   ontological status of, 169
   role in explaining Zipf's law, 271
   Shannon, 51–55, 57, 169
   in solution to Maxwell's demon paradox, 45–47
   statistical representations of, 179–180, 300
   in traditional computers, 170–180
information processing
   as giving rise to *meaning* and *purpose*, 184, 296
   in ant colonies, 176–178, 179–185
   in biological metabolism, 178–185
   in the brain, 248
   in cellular automata, 157–158, 161, 164–168, 171–172, 303
   common principles of, 295
   in the immune system, 172–176,
   in genetic regulatory networks, 276
   as a level of description in biology, 208
   in markets, 10
   in natural systems, xi, 56–57, 145–146, 156–158, 169–170, 172, 179–185
   in traditional computers, 170–171
   See also computation; information
information theory, 45, 47, 52, 55

in definitions of complexity, 96–100
in explaining Zipf's law, 271
inheritance
   blending, 81
   ideas about medium of, 80
   Lamarckian, 73–74
   Mendelian, 81–82, 89
   modern views of, 274–277
inheritance of acquired characteristics, 73–74
   Mendel's disconfirmation of, 79
in-links, 240–244
   in random Boolean networks, 283
insect colonies. *See* ant colonies
Institute for Advanced Study (IAS), 69, 124–126, 151–152
instruction pointer, 118
intuition pumps, 211
invisible hand, 10, 72, 76
irreversibility, 43
Izquierdo, Luis, 223

Joy, Bill, 124
jumping genes, 275
junk DNA, 96, 99, 278, 280

Kadanoff, Leo, 37
Kauffman, Stuart, 249, 281–282, 284–287
Keller, Evelyn Fox, 254
Kemeny, John, 125
Kepler, Johannes, 17–19
Kepler's laws, 19
Kevin Bacon game, 238
Kimura, Motoo, 86
kinematics, 19
Kleiber, Max, 260
Kleiber's law, 260–261
   in critiques of metabolic scaling theory, 268
   as explained by metabolic scaling theory, 264
Kleinfeld, Judith, 228
Koch curve, 103–108, 264–265
   as example of idea model, 211
   fractal dimension of, 107–108, 264
Korana, Har Gobind, 93
Kurzweil, Ray 123

Lamarck, Jean-Baptist, 72–74
Lamarckian inheritance, 73–74
   Mendel's disconfirmation of, 79

Landauer, Rolf, 47
Langton, Chris, 143
Laplace, Pierre Simon, 19–20, 33, 68
Lax, Peter, 125
Leclerk de Buffon, 72
Leibniz, Gottfried, 56, 59, 61, 68
Lendaris, George, 227
life
 artificial, 115–116, 292, 298
 autopoiesis as key feature for, 298
 as counterexample to second law of thermodynamics, 71
 expanded notions of, 300
 Game of (*see* Game of Life)
 requisites for, 116
 tape of, 86
linearity, 22–25
 versus nonlinearity of genes, 276–277
linear system. *See* linearity
Lipson, Hod, 124
Lloyd, Seth, 95–96, 100–101
Locke, John, 3
logical depth, 100–101
Logic of Computers group, 127
logistic map, 27–33
 bifurcation diagram for, 34
 as example of idea model, 211
logistic model, 25–27
 as example of idea model, 211
log-log plot, 261
Lohn, Jason, 142
Long Term Capital Management, 256–257
Lorenz, Edward, 22
Lovelock, James, 113
Lyell, Charles, 76–78
lymphocytes, 8–9, 172–176, 180–183. *See also* B cells; T cells

MacRae, Norman, 125
macrophage, 9
macrostate, 49–51, 54, 101, 307
Macy foundation meetings, 295–297
majority classification task, 160–161
 cellular automaton evolved for, 162–164, 171
Malthus, Thomas, 76
Mandelbrot, Benoit, 103, 271–272
master genes, 278–281
Mathematica, 154, 158
Matthew, Patrick, 78

Maturana, Humberto, 298
Maxwell, James Clerk, 20, 43–47
Maxwell's demon, 43–47, 169
 as example of idea model, 211
Maxwell's equations, 43, 210
May, Robert, 28, 33, 219–220, 223
Mayr, Ernst, 87
McCulloch, Warren, 296–297
McShea, Daniel, 110, 288
Mead, Margaret, 296–297
meaning (in complex systems), 171, 184, 208
mechanics, classical, 19, 48
meiosis, 88–89
Mendel, Gregor, 79–81
 ideas considered as opposed to Darwin's, 81–82
Mendelian inheritance, 79–81, 89, 276
messenger RNA, 90–93, 122, 275
metabolic pathways, 178–179, 249
 feedback in, 181–182
metabolic networks, 110, 229, 249–250, 254
metabolic rate, 258–262, 265–267
 scaling of (*see* metabolic scaling theory)
metabolic scaling theory, 264–266
 controversy about, 267–269
 as example of common principles in complex systems, 294–295
 scope of, 266–267
metabolism, 79, 110, 116, 178–184, 249,
 information processing (or computation) in, 178–185
 rate of, 258–262, 265–267
 as requisite for life, 116
 scaling of (*see* metabolic scaling theory)
metanorms model, 219, 222–224
Metropolis, Nicholas, 28, 35–36
Michelson, Albert, *ix*
microstate, 49–51, 54, 307
microworld, 191
 letter-string, 191–193
Milgram, Stanley, 227–229
Millay, Edna St. Vincent, 289
Miller, George, 272
Miller, John, 94
Minsky, Marvin, 187
MIT (Massachusetts Institute of Technology) Artificial Intelligence Lab, 190
mitosis, 88–8, 92

mobile genetic elements, 275
models, 209–210
  computer (*see* computer models)
  idea (*see* idea models)
  mathematical, 25
Modern Synthesis, 81–84
  challenges to, 84–87
molecular revolution, 274
Moravec, Hans, 123
Morgan, Thomas Hunt, 89
Mott, Keith, 168
mRNA, 90–93, 122, 275
Mukherji, Arijit, 223
mutations
  in DNA, 89, 93
  in genetic algorithms, 129
  in the immune system, 9, 174–175, 181
  via "jumping genes," 275
  knockout, 140
  role in Evo-Devo, 280
  role in Modern Synthesis, 83
mutation theory, 81
Myrberg, P. J., 35

natural selection
  challenges to primacy of, 85–87, 285–288, 300
  in Darwinian evolution, 72, 77–79
  in immune system, 9, 175
  in Modern Synthesis, 83
  relation to *meaning*, 184
  versus random genetic drift, 82–83
near-decomposability, 109–110
negative selection, 176
networks
  clustering in, 235–236, 238–240, 245, 252, 255
  definition of, 234
  degree distribution of, 235
  examples of, 229–230, 234–236, 247–251
  hubs in, 236, 240, 245, 248, 250, 252
  information spreading in, 255–258
  path-length in, 237–239, 245, 257, 318
  regular, 236–239
  resilience in, 245–246
  See also genetic regulatory networks; metabolic networks; random Boolean networks; scale-free networks; scientific citation networks; social networks; small-world networks
neurons, 6–7, 15, 189
  information processing with, 161, 168
  McCulloch and Pitts model of, 297
  as network nodes, 229, 238, 247–248
neutral evolution (theory of), 86
New Energy Finance, 222
New Kind of Science, A (Stephen Wolfram), 156–159
Newman, Max, 60
Newton, Isaac, *ix*, 17–19
  invention of calculus, 302
  lack of definition of *force*, 95
  law of gravity, 209–210, 269
  laws, 19
Newton's laws, 19
New York Stock Exchange, 11
Nicolis, Grégoire, 298
Nirenberg, Marshall, 93
noncoding regions, 96. *See also* junk DNA; genetic switches
noncoding RNA, 276, 279
noncomputable problem (or process), 157–158. *See also* uncomputability
nonlinearity, 22–27, 300
  of genes, 276–277
non-von-Neumann-style architecture, 149, 151, 171
normal (or Gaussian) distribution, 243–244, 269
norms, social, 218–219
norms model, 218–219, 222
  Galan and Izquierdo's reimplementation of 223–224
Nowak, Martin, 219–223
nucleotides, 90–93, 96, 122, 275

Occam's Razor, 99–100
onset of chaos, 35–36
Origins of Order, The (Stuart Kauffman), 285–286
out-links, 240

Packard, Norman, 160–161, 293
Pagels, Heinz, 1, 101
PageRank algorithm, 240, 244
parallel terraced scan, 182–183, 195–197
particles (in cellular automata), 166–168, 171–172

path length (network), 237–239, 245, 257, 318
pathogens, 8, 172–176, 180, 182, 195
   effect on gene transcription and regulation, 249
   representation of population of, 180
pathways, metabolic. *See* metabolic pathways
payoff matrix, 214–215
Peak, David, 168
Pepper, John, 287
period doubling route to chaos, 34–35
perpetual motion machine, 43
phenotype, 82, 90, 276
   of strategy evolved by genetic algorithm, 136
pheromones, 177–181, 183–184, 195
Pierce, John, 55
Pitts, Walter, 296–297
Plato, 77
Poincaré, Henri, 21–22
population genetics, 82
power law
   definition of, 245
   degree distribution of the Web as a, 240–245
   on double logarithmic (log-log) plot, 261
   in metabolic networks, 249
   in metabolic scaling, 260–266
   origin of, 252
   quarter-power scaling laws as, 262
   relationship to fractals, 264–265
   skepticism about, 253–255
   Zipf's law as, 270–272
   See also scale-free distribution
pre-Darwinian notions of evolution, 72–75
preferential attachment, 252–254, 257
   as example of common principles in complex systems, 294
Prigogine, Ilya, 298, 303
Principle of Computational Equivalence, 156–158
   as example of common principles in complex systems, 294
Prisoner's dilemma, 213–218
   adding social norms to, 218–219
   adding spatial structure to, 219–220
   payoff matrix for, 214–215
proof by contradiction, 66
pseudo-random number generators, 33, 98, 133, 155, 306

punctuated equilibria (theory of), 85–86, 278
purpose (in complex systems), 184, 296, 301

quantum mechanics, 20, 33, 48, 95
   renormalization in, 36
   role of observer in, 46
quarter-power scaling laws, 262, 267
   skepticism about universality of, 268
quasi-species, 86

Rajan, Vijay, 223
random access memory (RAM), 145–146, 161
random Boolean networks (RBNs), 282–287
   attractors in, 285
   effect of noise on, 287
   global state of, 285
   as models of genetic regulatory networks, 284, 287
   regimes of behavior in, 284
random genetic drift, 82–83
randomness
   in ant colonies, 177
   in biological metabolism, 178
   complexity as mixture of order and, 98, 102, 156
   in Copycat, 198, 207–208
   from deterministic chaos, 33, 38, 300
   as essential in adaptive information processing, 181–184, 195–196, 295, 300
   evolutionary role of, 77–78, 83
   in historical contingency, 85
   in immune system, 174
random number generators, 33, 98, 133, 155, 306
Rapoport, Anatol, 217, 297
receptors (on lymphocytes), 8, 173–176, 181, 183
recessive allele, 80–82
recombination, 81, 83, 89, 101
   in genetic algorithms, 129
reductionism, *ix–x*
   linearity, nonlinearity, and, 23
regular network, 236–239
   average path length of, 318
regulatory T cells, 176
renormalization, 36, 38
reversibility, 43
reversible computing, 46–47

ribonucleic acid. *See* RNA
ribosomes, 91–93, 122, 274
RNA, 86, 89–93, 122, 274–276
　in genetic regulation, 278–279
　noncoding, 276, 279
RNA editing, 275
RNA polymerase, 90
Robby the robot, 130–142
Rosenfield, Israel, 181
Rothman, Tony, 43
Rubner, Max, 258, 260, 266, 268
rule 110 cellular automaton, 153–157
rule 30 cellular automaton, 154–156
rules, cellular automata, 147–149

Santa Fe Institute, *x, xi*, 94, 156, 160, 164, 254, 264, 282, 291
　Complex Systems Summer School, 94, 300
Savage, Leonard, 297
scale-free distribution, 240–245
　versus bell-curve distribution, 243–245
　See also power-law distribution
scale-free networks, 239–240
　degree distribution of, 240–245 (*see also* power-law distribution)
　examples of, 247–252
　origin of, 252–254, 257
　resilience in, 245–246
　skepticism about, 253–255
scaling, 258–264. *See also* metabolic scaling theory
Schuster, Peter, 86
Scientific American, 291–292
scientific citation networks, 253
search engines, 10, 239–240
second law of thermodynamics, 40, 42–43
　Boltzmann's interpretation of, 50
　and evolution, 71
　Maxwell's demon and, 43–47
selection. *See* natural selection
self-awareness, 184, 189
self-copying computer program, 119–121
　deeper meaning of, 121–122
　difference between self-replication of DNA and, 122
　difference between von Neumann's self-reproducing automaton and, 122–123
self-organization, 13. *See also* emergence
　in definition of complex system, 13, 40
　examples of, 10, 40
　in Kauffman's theories, 285–286
　vagueness of definition of, *xii*, 293–294, 301
Self-Organized Criticality (SOC), 257, 269, 303
self-replication in DNA, 122
self-reproducing automata (also self-replicating automaton), 118, 122–124, 212, 297
　as example of common principle for complex systems, 294
　as example of idea model, 211
　in extending notion of *life*, 300
　self-copying program as illustration of, 118 (*see also* self-copying computer program)
　as universal computers, 156
　von Neumann's design for, 122–124, 127
self-reproduction
　in computers (*see* self-reproducing automata)
　logic of, 149, 211
　as requisite for life, 116
self-similarity
　in fractals, 103–106, 265–266
　as quantified by fractal dimension, 108
　relation to hierarchy, 109
　in scale-free (or power law) distributions, 242–243, 245, 265
sensitive dependence on initial conditions, 20
　as defining chaotic systems, 20–22, 34, 38
　in logistic map, 31–33
　in random Boolean networks, 284–285
sexual recombination, 81, 83, 89, 101
　in genetic algorithms, 129
Shakespeare, William, 71, 270–271
Shalizi, Cosma, 254
Shannon, Claude, 51–54, 296
Shannon entropy. *See* Shannon information
Shannon information (or entropy), 51–55, 57, 169
　in explanation of Zipf's law, 271

Shannon information (*continued*)
  as measure of complexity, 96–98
  relation to statistical complexity, 102
Shaw, George Bernard, 71
Shaw, Robert, 293
Sigmund, Karl, 220
Simon, Herbert, 109–110, 272
sine map, 36
six degrees of separation, 228
Slagle, James, 223
slippage, conceptual, 188, 191–193, 196–197, 202, 206
small-world networks, 236–239
  examples of, 247–252
  simplifying assumptions about, 254–255
  See also small-world property
small-world property, 238, 245, 248, 251
  formal definition of, 318
Smith, Adam, 10, 76
Smith, Eric, 94, 287
social networks, 227–230, 234–236, 238
social norms, 218–219, 220, 222
space-filling fractals, 266
spatial dimension, 107–108
statistical complexity, 102–103
statistical mechanics, 47–51
  influence on Stuart Kauffman's work, 286
Stein, Myron, 28, 35–36
Stein, Paul, 28, 35–36
stomata networks, 168
Stoppard, Tom, 15
strategy
  for majority classification task, 165
  for norms model, 218
  for Prisoner's dilemma, 216–217
  for Robby the Robot, 131–132
string theory, 210, 293
Strogatz, Steven, 230, 232, 236–239, 301
surface hypothesis, 260, 266, 268
survival instinct
  in Copycat, 208
  as requisite for life, 116, 127
Sutton, Walter, 89
switches, genetic, 278–280
synchronization, 248
Synergetics, 298
Synthesis, Modern, 81–84
  challenges to, 84–87
system science, 297–298
Szilard, Leo, 45–47, 125, 169

Tattersall, Ian, 83, 87
T cells, 9, 172, 174–175
  regulatory, 176
teleology, 296
Teller, Edward, 125
thermodynamic depth, 101
thermodynamics, 41–42, 48, 51, 258, 298, 302
  "fourth law" of, 286
  as inspiration for information theory, 55
  laws of, 42
  second law of (*see* second law of thermodynamics)
Thoreau, Henry David, 188
three-body problem, 21–22
tipping points, 253, 257
TIT FOR TAT strategy, 217–218
topology, algebraic, 21
tragedy of the commons, 214
trajectory, 31–32, 36
  of states in random Boolean network, 284
transcription (genetic), 90–93, 249
  in alternate splicing and RNA editing, 275
  in gene regulation, 278–279
  noise in, 249
transfer RNA, 91–93, 122
translation (genetic), 91–93
Tresser, Charles, 38
tRNA, 91–93, 122
Turing, Alan, 60–61, 63–65, 68–70, 209
  solution to the *Entscheidungsproblem*, 65–68
Turing machines, 60–63
  as definition of definite procedures, 63–64
  in definition of logical depth, 100
  encoding of, 64–65
  example of, 62
  as example of idea model, 211
  meaning of information in, 171
  simulation of in Game of Life, 150
  in solution to the *Entscheidungsproblem*, 65–68
  universal (*see* universal Turing machine)
Turing statement, 66
two-body problem, 21
two-person game, 214

Ulam, Stanislaw, *xi*, 28, 149
uncertainty principle, 20

uncomputability, 60, 158
  of the Halting problem, 65–68
  See also noncomputable problem (or process)
unified principles. *See* universal principles
unimodal map, 35, 36, 38
universal computation
  in cellular automata, 149–150, 156
  in defining complexity, 102, 156
  definition of, 149
  in nature, 157–158
  See also universal Turing machine
universal computer. *See* universal Turing machine
universal principles (of complex systems), 95, 292–295
  examples of proposals for, 294–295
  skepticism about, 293–295, 299
universal properties of chaotic systems, 34–38
universal Turing machine, 64–65,
  as blueprint for programmable computers, 65, 69
  cellular automata equivalent to, 149–150, 156
  in defining complexity, 102, 156

Varela, Francisco, 298
variable (in computer program), 119
von Neumann, John, 28, 117–118, 124–127, 146, 149, 156, 209, 211–212, 294, 296–297
  invention of cellular automata, 149
  self-reproducing automaton, 122–124, 156

von-Neumann-style architecture, 146, 169–171, 209

Wang, Hao, 69
Watson, James, 89, 93, 274
Watts, Duncan, 230–231, 236–239, 257
Web (World Wide), 10, 12, 186
  coevolution with search engines, 10
  degree distribution of, 240–245, 265, 318
  network structure of, 12, 229–230, 235–236, 240–245, 252–253, 265
  resilience of, 245
  search engines for, 239–240
West, Geoffrey, 263–267, 269, 294, 300
Wiener, Norbert, 209, 296–297
Wigner, Eugene, 125
Wilkins, Maurice, 93
Willinger, Walter, 269
Wolfram, Stephen, 102, 151–159, 168, 294, 303
work (as related to energy), 41–43, 51
  in evolution, 72, 79
  in Maxwell's demon, 43–47
World Wide Web. *See* Web (World Wide)

Yeast Genome Project, 96
Yoon, Carol Kaesuk, 280
Young, Karl, 102–103
Yukawa, Hideki, 188

Ziff, Edward, 181
Zipf, George Kingsley, 270
Zipf's law, 271
  explanations for, 271–272
Zuse, Konrad, 159

Lightning Source UK Ltd.
Milton Keynes UK
UKHW012031250121
377563UK00011B/331

9 780199 798100